Smart Crop Development

Vijay Gahlaut • Sachin Rustgi
Vandana Jaiswal
Editors

Smart Crop Development

Adapting Agriculture to Climate Change

Springer

Editors
Vijay Gahlaut
University Centre for Research and Development,
Chandigarh University
Mohali, Punjab, India

Sachin Rustgi
Department of Plant and Environmental Sciences, Clemson University
Pee Dee Research and Education Center
Florence, SC, USA

Vandana Jaiswal
Biotechnology Division
CSIR-Institute of Himalayan Bioresource Technology
Palampur, Himachal Pradesh, India

ISBN 978-981-95-7679-1 ISBN 978-981-95-7680-7 (eBook)
https://doi.org/10.1007/978-981-95-7680-7

This Springer imprint is published by the registered company Springer Nature Singapore Pte Ltd.
The registered company address is: 152 Beach Road, #21-01/04 Gateway East, Singapore 189721, Singapore

To the resilient farmers and visionary scientists who continue to nurture both knowledge and crops in the most challenging corners of our planet—your relentless pursuit of understanding the power of genetics and genomics in the face of climate change has deepened our scientific insight and strengthened the resilience of communities worldwide.

May this book stand as a tribute to your passion, perseverance, and the unbreakable human spirit that flourishes even under adversity. Your dedication to uncovering the genetic foundations of climate-resilient crops holds the promise of transforming agriculture and ensuring food security for generations to come.

With profound gratitude and respect, we dedicate this book, Smart Crop Development: Adapting Agriculture to Climate Change, *to you.*

With deepest respect,

Dr. Vijay Gahlaut
Dr. Sachin Rustgi
Dr. Vandana Jaiswal

Foreword 1

The agricultural landscape of the twenty-first century is greatly affected by climate variability and environmental stresses. Unpredictable weather and rising temperatures challenge agriculture. Meanwhile, growing demands on land and resources intensify the need to secure food supplies while maintaining environmental integrity. In response to the pressing and complex challenges posed by climate change, this book, *Smart Crop Development: Adapting Agriculture to Climate Change*, offers a timely and visionary perspective.

This book presents important advances in plant genomics, molecular breeding, integrative phenomics, and next-generation biotechnology. It offers a roadmap for developing climate-resilient crops and sustainable agriculture. The chapters cover genomic innovations for stress-resilient crops, the potential of crop wild relatives, and high-throughput phenotyping. New insights into plant proteomics and epigenetics are also explored. Through case studies on traditional staples, emerging pseudocereals, soybean, peanut, and microgreens, the book examines both the methods and motivations for resilience. It integrates nutritional quality, food safety, and entrepreneurship with molecular discoveries.

Addressing the climate challenge requires collaboration among scientists, breeders, farmers, policymakers, and entrepreneurs. This book supports that approach by highlighting open data sharing, regional phenotyping hubs, and participatory breeding. It also discusses advances in artificial intelligence, multi-omics, and digital tools. The book emphasizes that resilient agriculture depends on adaptive communities and equitable systems. It ensures innovations reach marginalizedfarmers and women, and that nutritional value is valued alongside yield.

Developed at the intersection of research, field practice, and global policy, these chapters are dedicated to those committed to sustainable and equitable food production in the face of climate uncertainty. We hope these insights inspire the next generation of researchers, practitioners, and leaders to advance smart, climate-adapted crops as the foundation of resilient and nutritious food systems.

Director, CSIR-Institute of Himalayan Bioresource Technology
Palampur, HP, India

Sudesh Kumar Yadav

Foreword 2

Climate change has emerged as one of the most formidable challenges confronting global agriculture, threatening food security, ecosystem stability, and rural livelihoods. Rising temperatures, erratic rainfall patterns, and increasing frequency of extreme weather events have intensified the urgency for scientific innovation in crop improvement. In this context, *Smart Crop Development: Adapting Agriculture to Climate Change* provides a timely and comprehensive synthesis of the latest advances in genomics, molecular breeding, phenomics, and omics-driven crop science aimed at building resilience in our food systems.

This volume brings together contributions from experts across diverse disciplines, reflecting the multidimensional nature of climate-smart agriculture. The chapters traverse the spectrum, from genomic innovations and the utilization of crop wild relatives, to integrative phenomics, proteomics, and epigenetics, culminating in the emerging roles of pseudocereals, millets, and microgreens as climate-resilient and nutritionally rich crops. The inclusion of frontier approaches such as molecular breeding, system-level proteomics, and AI-integrated multi-omics demonstrates the forward-looking vision of the editors and contributors.

By bridging fundamental science with applied strategies, this book not only documents technological progress but also highlights translational opportunities for breeders, biotechnologists, and policymakers. The emphasis on sustainability, nutritional quality, and inclusivity, particularly through the valorization of underutilized crops and microgreens, offers a holistic perspective on the future of agriculture under changing climatic regimes.

The insights presented here will serve as a valuable reference for researchers, students, and practitioners working at the intersection of crop genetics, biotechnology, and climate adaptation. It underscores that the path to global food and nutritional security lies in innovation, collaboration, and the intelligent application of science toward resilient agriculture.

I congratulate the editors and contributing authors for curating this scientifically rigorous and forward-thinking volume. Their collective effort significantly enriches the discourse on sustainable crop improvement and paves the way for future-ready agricultural systems capable of thriving in an era of climate uncertainty.

Kansas State University
Manhattan, KS, USA

P. V. Vara Prasad,

Preface

The challenges posed by climate change to global agriculture are unprecedented, and demand innovative, interdisciplinary solutions to ensure food security and sustainable crop production. This book, *Smart Crop Development: Adapting Agriculture to Climate Change*, brings together the latest research and innovations in areas such as genomics, breeding, phenomics, proteomics, epigenetics, and climate-resilient farming. It is designed to support researchers, plant breeders, and policymakers in developing practical solutions for these urgent global challenges.

Each chapter explores key aspects of building climate-smart crops from using advanced genetic tools and crop wild relatives to applying modern phenotyping technologies that connect plant traits to stress resilience. The book looks at how crops respond to major stresses like heat, drought, and salinity, covering a wide range of plants including cereals, pulses, oilseeds, millets, pseudocereals, and even emerging crops like microgreens. It also discusses how plants adapt through molecular processes such as protein responses and epigenetic changes, opening new paths for crop improvement and biotechnology. By combining different "omics" technologies with advanced data analysis and field-based approaches, this book illustrates the future of precision breeding in a changing climate. It also emphasizes sustainable and climate-smart farming practices that benefit farmers, especially those working with limited resources.

This collection represents the work of leading scientists dedicated to agricultural innovation. Our goal is to provide a comprehensive and inspiring resource for anyone working to create resilient crops capable of thriving under climate stress, ensuring a healthy and reliable food supply for future generations. We hope this book informs and inspires all who are committed to advancing climate-smart agriculture.

Mohali, Punjab, India — Vijay Gahlaut
Florence, SC, USA — Sachin Rustgi
Palampur, Himachal Pradesh, India — Vandana Jaiswal

Contents

About Editors and Contributors

About the Editors

Vijay Gahlaut, presently an Associate Professor at Chandigarh University's University Centre for Research & Development in Mohali, focuses his research on identifying novel genomic regions and genes governing abiotic stress regulation in crop plants. His expertise spans crop biotechnology, quantitative genetics, QTL/GWAS, and epigenetics. Employing diverse bioinformatic and genomic methods, he has successfully pinpointed crucial genomic regions, epialleles, and candidate genes in various crop species. This research significantly contributes to bolstering crop resilience against challenging environmental conditions, particularly water scarcity and heat stress. With over 90 published research papers in prestigious international journals and notable contributions as an Associate Editor and independent peer reviewer, his impactful achievements led to his induction as a Member of the National Academy of Sciences in India, acknowledging his profound influence in crop genetics and abiotic stress tolerance research.

Sachin Rustgi is Associate Professor of Molecular Breeding affiliated with the Department of Plant and Environmental Sciences at Clemson University. He also serves as a Faculty Scholar at the Clemson University School of Health Research and the Center for Human Genetics. Additionally, he holds the position of Adjunct Associate Professor at both the Department of Crop and Soil Sciences, Washington State University, Pullman, and the Department of Genetics and Plant Breeding, Ch. Charan Singh University, Meerut. His research is centered on critical areas such as developing dietary therapies for individuals with celiac disease or allergies to wheat and peanuts. His work also involves creating biomarkers for heat tolerance in crops like wheat, soybean, and peanut, as well as devising innovative strategies to combat major insect pests and fungal pathogens affecting primary crops in South Carolina. His impactful research has resulted in more than a hundred publications in esteemed international journals. In addition to his research contributions, he has authored or edited three books. Recognizing his academic excellence, he was honored with the Faculty Excellence in Research Award from the Washington State University College of Agricultural, Human, and Natural Resource Sciences. He has also received the See Fellow Award from the Society of Extension Education in Agra, India.

Vandana Jaiswal, a Scientist at CSIR-IHBT in Palampur, India, is dedicated to unraveling the genetics of economically vital traits and enhancing Himalayan crops through molecular methods. Focusing on saffron, a pivotal Himalayan crop with inherent challenges like triploidy and seed sterility, she pioneers genetic improvements utilizing molecular techniques. Her groundbreaking work extends to more than 40 research papers and book chapters, demonstrating her profound contribution to the field. In addition, she serves as an Associate Editor and independent peer reviewer for several journals. Acknowledging her excellence, she has been honored with prestigious awards, including the CSIR-Young Scientist Award, DST-INSPIRE Faculty Award, Early Career Research Award, and Women Scientist Awards from various Indian government agencies. Her innovative research underscores her commitment to advancing agricultural genetics and crop improvement in the Himalayan region.

Contributors

Aastha Biotechnology, Division, CSIR-Institute of Himalayan Bioresource Technology, Palampur, Himachal Pradesh, India

Ansh Department of Biotechnology, Sharda School of Biosciences and Technology, Sharda University, Greater Noida, Uttar Pradesh, India

Arushi Arora Gurudev Singh Khush Institute of Genetics, Plant Breeding and Biotechnology, Punjab Agricultural University, Ludhiana, Punjab, India

Deepak Bhamare School of Agricultural Biotechnology, Punjab Agricultural University, Ludhiana, Punjab, India

Varun Bhardwaj Department of Botany, University of Delhi, New Delhi, India

Snesha Umesh Bhomkar School of Biological Sciences and Biotechnology, Goa University, Goa, India

Pratikshya Borah Department of Botany, The Assam Royal Global University, Guwahati, India

Anita Choudhary Biotechnology, Division, CSIR-Institute of Himalayan Bioresource Technology, Palampur, Himachal Pradesh, India

Monika Chouhan Department of Biotechnology, CSIR-Institute of Himalayan Bioresource Technology (IHBT), Palampur, India

Academy of Scientific and Innovative Research (AcSIR), Ghaziabad, India

Vijay Gahlaut University Centre for Research and Development, Chandigarh University, Mohali, Punjab, India

Navin Chander Gahtyari ICAR-Vivekananda Parvatiya Krishi Anusandhan Sansthan (Vivekananda Institute of Hill Agriculture), Almora, Uttarakhand, India

Himanshi Gangwar Biotechnology Division, CSIR-Institute of Himalayan Bioresource Technology, Palampur, India

Prakash Ghasal ICAR-Vivekananda Parvatiya Krishi Anusandhan Sansthan (Vivekananda Institute of Hill Agriculture), Almora, Uttarakhand, India

Astha Gupta Department of Biotechnology, Sharda School of Biosciences and Technology, Sharda University, Greater Noida, Uttar Pradesh, India

Shrishty Gupta Department of Biotechnology, Sharda School of Biosciences and Technology, Sharda University, Greater Noida, Uttar Pradesh, India

Chetana Hasnu Department of Botany, The Assam Royal Global University, Guwahati, India

Department of Botany, Pandit Deendayal Upadhyaya Adarsha Mahavidyalaya, Tulungia, Bongaigaon, Assam, India

Nirmal Hedau ICAR-Vivekananda Parvatiya Krishi Anusandhan Sansthan (Vivekananda Institute of Hill Agriculture), Almora, Uttarakhand, India

Hrishikesh Ingole Department of Plant and Environmental Sciences, Clemson University, Pee Dee Research and Education Center, Florence, SC, USA

Kajal Jaiswal Department of Biotechnology, Sharda School of Biosciences and Technology, Sharda University, Greater Noida, Uttar Pradesh, India

Vandana Jaiswal Biotechnology, Division, CSIR-Institute of Himalayan Bioresource Technology, Palampur, Himachal Pradesh, India

Jyoti Prasad Kakati Department of Plant and Environmental Sciences, Clemson University, Pee Dee Research and Education Center, Florence, SC, USA

Lakshmi Kant ICAR-Vivekananda Parvatiya Krishi Anusandhan Sansthan (Vivekananda Institute of Hill Agriculture), Almora, Uttarakhand, India

Ayushi Kukreti Biotechnology, Division, CSIR-Institute of Himalayan Bioresource Technology, Palampur, Himachal Pradesh, India

Amit Kumar ICAR-Vivekananda Parvatiya Krishi Anusandhan Sansthan (Vivekananda Institute of Hill Agriculture), Almora, Uttarakhand, India

Amit Kumar Department of Genetics and Plant Breeding, Govind Ballabh Pant University of Agriculture and Technology, Pantnagar, Uttarakhand, India

Anand Kumar Faculty of Agriculture Sciences, GLA University, Mathura, Uttar Pradesh, India

Priya Kumari Biotechnology, Division, CSIR-Institute of Himalayan Bioresource Technology, Palampur, Himachal Pradesh, India

Manoj Kumar ICAR-Vivekananda Parvatiya Krishi Anusandhan Sansthan (Vivekananda Institute of Hill Agriculture), Almora, Uttarakhand, India

Paramdeep Kumar Academy of Scientific and Innovative Research (AcSIR), Ghaziabad, India

Praveen Kumar Department of Biotechnology, University Institute of Biotechnology, Chandigarh University, Mohali, Punjab, India

Rajiv Kumar Department of Biotechnology, CSIR-Institute of Himalayan Bioresource Technology (IHBT), Palampur, India

Academy of Scientific and Innovative Research (AcSIR), Ghaziabad, India

Ramesh Singh Pal ICAR-Vivekananda Parvatiya Krishi Anusandhan Sansthan (Vivekananda Institute of Hill Agriculture), Almora, Uttarakhand, India

Sachin Rustgi Department of Plant and Environmental Sciences, Clemson University, Pee Dee Research and Education Center, Florence, SC, USA

Dinesh Kumar Saini Department of Plant and Soil Science, Texas Tech University, Lubbock, TX, USA

Vishal Saini Department of Biotechnology, CSIR-Institute of Himalayan Bioresource Technology (IHBT), Palampur, India

Academy of Scientific and Innovative Research (AcSIR), Ghaziabad, India

Devender Sharma ICAR-Vivekananda Parvatiya Krishi Anusandhan Sansthan (Vivekananda Institute of Hill Agriculture), Almora, Uttarakhand, India

Jatin Sharma Department of Plant Breeding and Genetics, Punjab Agricultural University, Ludhiana, Punjab, India

Shubham Sharma Department of Plant Breeding and Genetics, Punjab Agricultural University, Ludhiana, Punjab, India

Pooja Kanwar Shekhawat Department of Genetics and Plant Breeding, Sardarkrushinagar Dantiwada Agricultural University, Dantiwada, Gujarat, India

Harshal Singh Department of Biotechnology, Sharda School of Biosciences and Technology, Sharda University, Greater Noida, Uttar Pradesh, India

Kumari Prabha Singh College of Basic Sciences and Humanities, School of Business Studies, Punjab Agricultural University, Punjab, India

Rachana Singh Department of Biotechnology, Sharda School of Biosciences and Technology, Sharda University, Greater Noida, Uttar Pradesh, India

Priyanka Upadhyay Department of Plant Breeding and Genetics, Dr. Rajendra Prasad Central Agricultural University, Pusa Bihar, India

Pragati Yadav Department of Botany, University of Delhi, New Delhi, India

Vikash Kumar Yadav School of Biological Sciences and Biotechnology, Goa University, Goa, India

Abbreviations

ABA	Abscisic Acid
ABRE	ABA-Responsive Element
ABTS	2,2'-Azino-Bis(3-Ethylbenzothiazoline-6-Sulfonic Acid)
AFLP	Amplified Fragment Length Polymorphism
AGO	Argonaute Proteins
AI	Artificial Intelligence
AOS	Allene Oxide Synthase
AP2ERF	APETALA2 Ethylene Response Factor
APX	Ascorbate Peroxidase
ASF1B	Anti-Silencing Function 1B
ATPase	Adenosine Triphosphatase
BAG	Bcl-2-Associated Athanogene Proteins
bHLH	Basic Helix-Loop-Helix domain
bZIP	Basic Leucine Zipper Motif
CAT	Catalase
CCS	Copper Chaperone for Superoxide Dismutase
CDPK	Calcium-Dependent Protein Kinase
CKX	Cytokinin Oxidase
ClpB	Caseinolytic Protease B
CMS	Cytoplasmic Male Sterility
CNV	Copy Number Variation
COR	Cold-Responsive Protein
CPK	Calcium Protein Kinase
CRISPR/Cas9	Clustered Regularly Interspaced Short Palindromic Repeats/ CRISPR-Associated Protein 9
CT	Canopy Temperature
CTD	Canopy Temperature Depression
CuZn-SOD	Copper-Zinc Superoxide Dismutase
CWRs	Crop Wild Relatives
CYPs	Cytochrome P450 Proteins
DCL	Dicer-Like Protein
DEG	Differentially Expressed Genes
DHAR	Dehydroascorbate Reductase
DIGE	Differential In-Gel Electrophoresis

DME	DNA Demethylase Enzyme
DML	Differentially Methylated Loci
DnaJs	DnaJ Chaperone Proteins
DNMTs	DNA Methyltransferases
DREB	Dehydration-Responsive Element Binding
DRM	Domains Rearranged Methyltransferase
dsRNA	Double-Stranded RNA
ELISA	Enzyme-Linked Immunosorbent Assay
ER	Endoplasmic Reticulum
ESPS	Early Soybean Production System
ESTs	Expressed Sequence Tags
FADs	Fatty Acid Desaturases
FAO	Food and Agriculture Organization
FGM	Fenugreek Microgreens
FKBP	FK506-Binding Proteins
FRET	Förster Resonance Energy Transfer
GA	Gibberellic Acid
GAE	Gallic Acid Equivalent
GBS	Genotyping-by-Sequencing
GBSS	Granule-Bound Starch Synthase
GC-MS	Gas Chromatography–Mass Spectrometry
GH3	Indole-3-Acetic Acid-Amido Synthetase
GID1B	Gibberellin Receptor
GLUT-2	Glucose Transporter 2
GNSS	Global Navigation Satellite System
GPOX	Guaiacol Peroxidase
GR	Glutathione Reductase
GS	Genomic Selection
GST	Glutathione S-Transferase
GWAS	Genome-Wide Association Studies
HAT	Histone Acetyltransferase
HDAC	Histone Deacetylase
HKT	High-Affinity Potassium Transporter
HMT	Histone Methyltransferase
HSCB	Heat Shock Cognate Binding Protein
HSFs	Heat Shock Factors
HSP	Heat Shock Protein
IAA	Indole-3-Acetic Acid
IBS	Irritable Bowel Syndrome
INM	Integrated Nutrient Management
IoT	Internet of Things
IRS-1	Insulin Receptor Substrate 1
ISSR	Inter-Simple Sequence Repeat
iTRAQ	Isobaric Tags for Relative and Absolute Quantification
JA	Jasmonic Acid

KEGG	Kyoto Encyclopedia of Genes and Genomes
LC-MS/MS	Liquid Chromatography-Mass Spectrometry/Mass Spectrometry
LDL	Low-Density Lipoprotein
LEA	Late Embryogenesis Abundant
lncRNA	Long Non-coding RNA
MAGIC	Multi-parent Advanced Generation Inter-Cross
MAPK	Mitogen Activated Protein Kinase
MAS	Marker-Assisted Selection
MDA	Malondialdehyde
MDHAR	Monodehydroascorbate Reductase
miRNA	Micro RNA
ML	Machine Learning.
MRM	Multiple Reaction Monitoring
MS	Mass Spectrometry
MYB	Myeloblastosis
Na/K	Sodium/Potassium
NAC	NAM, ATAF, and CUC transcription factor family
NAM	Nested Association Mapping
NCBI	National Center for Biotechnology Information
NDVI	Normalized Difference Vegetation Index
NFKβ	Nuclear Factor Kappa B
NGS	Next-Generation Sequencing
NHX	Na+/H+ Exchanger
NMR	Nuclear Magnetic Resonance
NO	Nitric Oxide
NPQ	Non-Photochemical Quenching
nsLTPs	Non-Specific Lipid Transfer Proteins
P5CS	Pyrroline-5-Carboxylate Synthetase
PAGE	Polyacrylamide Gel Electrophoresis
PAV	Presence-Absence Variation
pCO2	Partial Pressure of Carbon Dioxide
PCR	Polymerase Chain Reaction
PEP Carboxylase	Phosphoenolpyruvate Carboxylase
POD	Peroxidase
POL IV/V	RNA Polymerase IV/V
PP2C	Protein Phosphatase 2C
PPDB	Plant Proteome Database
PPFD	Photosynthetic Photon Flux Density
PPI	Protein–Protein Interaction
PPO	Polyphenol Oxidase
PSII	Photosystem II
PTM	Post-Translational Modification
QTL	Quantitative Trait Locus/Loci
RdDM	RNA-Directed DNA Methylation

RDR2	RNA-Dependent RNA Polymerase 2
RGB	Red-Green-Blue imaging
RIL	Recombinant Inbred Line
RNAi	RNA Interference
RNA-seq	RNA Sequencing
ROS	Reactive Oxygen Species
RTI	Respiratory Tract Infection
RuBP Carboxylase	Ribulose Bisphosphate Carboxylase
RWC	Relative Water Content
SFD	Seed Filling Duration
SGR	Seed Growth Rate
SILAC	Stable Isotope Labeling by Amino acids in Cell culture
siRNA	Small Interfering RNA
SNP	Single-Nucleotide Polymorphism
SnRK2	SNF1-Related Kinase 2
SOD	Superoxide Dismutase
SRM	Selected Reaction Monitoring
SSR	Simple Sequence Repeat
SUMO	Small Ubiquitin-like Modifier
SV	Structural Variant(s)
TAA	Total Antioxidant Activity
TAIR	The Arabidopsis Information Resource
TALEN	Transcription Activator-Like Effector Nuclease
TE	Transposable Element
TF	Transcription Factor
TMT	Tandem Mass Tag
TPFD	Total Photon Flux Density
UAV	Unmanned Aerial Vehicle
USDA	United States Department of Agriculture
UV	Ultraviolet
WGD	Whole Genome Duplication
WGS	Whole-Genome Sequencing
WOX	WUSCHEL-related homeobox
ZFN	Zinc-Finger Nuclease
ZIP	Zinc-Iron Permease

Genomic Innovations for Climate-Smart Crop Development

1

Snesha Umesh Bhomkar and Vikash Kumar Yadav

Abstract

Climate change threatens global agricultural sustainability. Accelerating climatic shifts are already reducing crop yields, food availability, and ecological stability across many regions worldwide. This chapter emphasizes the role of genomic science in advancing adaptive crop improvement, to help farmers confront these challenges and build resilience in agricultural practices. The focus is placed on the transformative impact of genome analysis techniques, which enable deeper exploration of genetic diversity and stress tolerance traits. Genome editing tools, especially CRISPR-Cas platforms, are discussed as powerful methods for precision breeding, allowing rapid and targeted modifications for stress tolerance and yield stability. Beyond sequence variation, the chapter briefly examines how various omics-based approaches broaden our understanding of plant adaptation, while bioinformatics and machine learning integrate these multi-omic datasets into predictive models for trait selection. Additionally, we address challenges such as high costs, phenotyping bottlenecks, regulatory complexities, and issues of equitable access, highlighting the importance of inclusive governance and open data sharing. By linking molecular insights with sustainable practices, genomic innovations provide a roadmap for climate-resilient agriculture that maintains productivity, while preserving biodiversity and supporting farmer livelihoods.

Keywords

Climate-smart crops · Genome editing · Multi-omic integration · Precision agriculture

S. U. Bhomkar · V. K. Yadav (✉)
School of Biological Sciences and Biotechnology, Goa University, Taleigao, Goa, India
e-mail: vikash@unigoa.ac.in

V. Gahlaut et al. (eds.), *Smart Crop Development*,
https://doi.org/10.1007/978-981-95-7680-7_1

1.1 Introduction

Climate change is generally defined as the long-term alteration of the Earth's climate system. It is primarily driven by anthropogenic activity like greenhouse gas emissions, in addition to natural factors. These changes lead to significant shifts in temperature, precipitation patterns, and frequency of extreme weather events (Bibi and Rahman 2023; Pachauri 2014). Agriculture, being one of the most climate-sensitive sectors, is directly impacted by these environmental shifts, making it particularly vulnerable. The climatic alterations have profound implications for agriculture. They influence crop growth, yield, and soil health. Additionally, it also impacts water availability and the global distribution of pests and diseases (Alkorta et al. 2017; Shuhadah et al. 2022; Yuan et al. 2024). Climate change imposes severe challenges on agriculture by intensifying abiotic stresses such as heat, drought, salinity, and rainfall, which together cause 30–50% global losses in agricultural productivity (Bibi and Rahman 2023). A report suggesting approximately a 2% reduction in maize yields under arid and high-temperature conditions has underscored the vulnerability of staple crops to climate-driven water stress (Lobell et al. 2011).

Rising temperatures accelerate soil degradation through loss of organic matter, increased bulk density, and compaction, thereby reducing fertility and inhibiting root growth and affecting the soil microbiota (Subedi et al. 2023). Increasing temperatures also affect plant physiology, leading to temperature-related growth limitations. Predictive models suggest that wheat yields could decline by up to 50%. This reduction is primarily attributed to accelerated leaf senescence caused by altered rainfall patterns and an average of two-degree increase in mean growing season temperatures (Asseng et al. 2011). Such projections highlight the sensitivity of wheat physiology to climate variability and the extreme implications for food production in a warming climate. Similarly, another simulation model indicated that rising temperatures increase the likelihood of reduced precipitation during the growing seasons of key staples such as winter wheat, rice, and maize (Arnell et al. 2019). Moreover, climate change broadens the range and severity of pests and pathogens, leading to significant crop losses and threatening global food security (Subedi et al. 2023). Changes in climatic conditions alter feeding patterns, dispersal rates, and population dynamics of various potential pests and pathogens, eventually leading to earlier infestations and crop damage (Dukes et al. 2009). It enhances the overwintering survival of insect pests such as the corn earworm and cotton bollworm, creating challenges for pest control and threatening maize yields, a crop of critical global importance (Pareek et al. 2017). This change further increases the risk of yield instability under changing climatic conditions. These escalating challenges underscore the urgent need for climate-smart crops capable of withstanding drought, heat, salinity, and pest pressures.

Equipping crops to survive in increasingly harsh climates invites for more innovative solutions. Genomics and molecular breeding provide a powerful pathway in creating climate-smart crops. These techniques have transformed how we approach crop improvement under climate stress. By unlocking the genetic blueprint of plants, genomics enables the discovery of key genes and alleles that regulate tolerance to both, biotic and abiotic stressors, which can then be incorporated into breeding programmes (Krishna et al. 2023; Varshney et al. 2020). Molecular breeding approaches such as marker-assisted backcrossing, genomic selection (GS), and haplotype-based breeding allow these traits to be incorporated into elite varieties with a better speed and precision than conventional methods (Sinha et al. 2021). Integration of high-throughput techniques and informatic tools has accelerated the rate of genetic gain. This has ensured that improved traits are expressed in farmer fields where they matter most. At the same time, advances in systems biology and functional genomics have deepened our understanding of complex plant responses to stress. This has in turn helped in designing resilient crops at both physiological and molecular levels (Kole et al. 2015; Shuhadah et al. 2022). Together, these innovations help maintain yield stability and provide hope for sustaining global food security in an era of intensifying climate challenges.

This chapter reviews how modern genomic and molecular innovations reshape crop improvement, highlighting their contribution to building climate-resilient and sustainable farming systems. Further, the integration of sustainable practices with advances in high-throughput sequencing, genome editing, epigenomics, and bioinformatics has also been explored. It provides a holistic view of the progress, challenges, and prospects for genomics-driven agricultural resilience.

1.2 Genomics and Climate-Smart Agriculture

The advent of genomics has reshaped crop improvement strategies, moving beyond phenotype-driven selection towards precision-based breeding. Technologies like high-throughput sequencing and genome-wide association studies (GWAS) and the development of pan-genomes have enabled the discovery of allelic variants, single-nucleotide polymorphisms (SNPs), and quantitative trait loci (QTLs) associated with stress tolerance (Mangal et al. 2024; Naqvi et al. 2022). These genomic resources form the foundation for approaches such as marker-assisted selection, marker-assisted backcrossing, and GS. These methods are increasingly applied in cereals and legumes to speed up the incorporation of climate-resilient traits (Kole et al. 2015; Sinha et al. 2023). On the other hand, genome editing tools such as CRISPR-Cas are being integrated with traditional breeding pipelines to introduce or improve genes linked to stress adaptation (Varshney et al. 2020). In rice, for instance, genomic studies have identified allelic variants of the SUB1A gene that provide tolerance to submergence. These have been successfully incorporated into widely grown varieties using genomics-assisted breeding (Kole et al. 2015). Similarly, in wheat, genomic prediction models have improved the accuracy of selecting lines resilient to drought and heat stress. These examples demonstrate how genomics has become a central pillar in breeding climate-smart crops, ensuring food security and sustainability in the face of environmental instability (Sinha et al. 2023).

Despite its advantages, conventional breeding faces limitations as it is time-consuming, intensively resourced and yields low genetic gain over time (Mangal et al. 2024). Strong influence of environmental conditions limits phenotype-based selection, narrowing the germplasm genetic diversity (Krishna et al. 2023). In contrast, genomics-assisted breeding offers better precision, as selection decisions are guided by molecular information rather than influenced by environmental variation. Another major advantage lies in the reduced cost and time requirement (Consortium et al. 2021). By replacing a significant fraction of labour-intensive phenotyping with genotyping, GS makes the identification of elite parental lines faster, which results in promising crosses (Sinha et al. 2023). Combining genomic tools with speed breeding, where multiple crop generations are grown yearly under controlled conditions, significantly shortens the breeding process. This approach allows the development of new varieties in just a few years instead of several decades (Krishna et al. 2023). Thus, genomics transforms crop improvement into a faster, more data-driven process, overcoming the bottlenecks of conventional breeding and ensuring rapid response to emerging climatic threats.

While genomics offers a strong foundation, its ability to accurately predict traits improves greatly when combined with phenomics and systems biology. Modern phenotyping platforms that use imaging tools, drones, and spectral sensors allow scientists to track how plants respond to environmental stress in real time and with minute details (Kumar and Kumar 2021). When this phenotypic data is integrated with genomic datasets, it strengthens predictive models, making it possible to identify resilient crop lines even before stress is experienced in the field. This is especially important for complex traits such as drought tolerance or nutrient-use efficiency, where interactions between genes and the environment strongly shape plant performance. A systems biology perspective takes this one step ahead by linking multiple layers of biological information, such as metabolomics, transcriptomics, and proteomics, with genomics. In wheat, for instance, metabolomics-based breeding has identified chemical markers associated with heat tolerance. Similarly, multi-omic studies in rice have uncovered regulatory networks that control adaptation to salt stress (Shuhadah et al. 2022). The integration of these strategies enhances the identification of causal genes, allowing the modelling of plant systems holistically, while also combining resilience with yield and nutritional quality.

The future of climate-smart agriculture lies in consistently connecting genomics, phenomics, and systems biology into unified breeding pipelines. By combining GS with large-scale field testing in different locations, researchers can better understand how genes and the environment work together in crops like maize and rice (Kumar and Kumar 2021; Shuhadah et al. 2022). Similarly, integrating machine learning (ML) with big data analytics improves the precision in predictive breeding. This, in turn, supports large-scale development of climate-resilient crops for diverse farming regions.

1.3 High-Throughput Genomic Technologies

Next-generation sequencing (NGS) and whole-genome sequencing (WGS) have been instrumental in addressing challenges posed by climate change. Unlike conventional sequencing, NGS allows rapid, parallel sequencing of millions of DNA fragments, enabling detailed, cost-effective analysis of genome structure, genetic variation, and regulations (Sahu et al. 2020; Satam et al. 2023). When applied to crop plants, WGS has eased the identification of novel alleles, structural variants (SVs), and SNPs associated with adaptive traits (Table 1.1). These findings provide a foundation for breeding programmes targeting tolerance to various abiotic stressors such as salinity, drought, and heat (Chauhan and Singh 2025; Varshney et al. 2024). For example, high-quality reference genomes (e.g. maize) and resequencing projects help scientists find genes for stress tolerance, making crop breeding faster and more effective (Jiao et al. 2017; Michael and Vanburen 2005). By integrating WGS with advanced bioinformatic tools, the challenges in decoding functional genetic variation underlying complex traits have been overcome (Sahu et al. 2020; Satam et al. 2023).

The NGS and WGS do not just list genetic variations but have been serving the purpose of bridging the genotype-phenotype gap in climate-resilient crops. For instance, resequencing rice, maize, and wheat germplasm has revealed genomic regions for drought tolerance, heat response, and nutrient use, helping identify key genes and QTLs essential for agriculture (Huang et al. 2012; Varshney et al. 2024). The WGS datasets have also provided valuable resources for developing pan-genomes, which capture variable stress-adaptive genes that may have been missed in single-reference assemblies (Gao et al. 2019). Importantly, NGS has been instrumental in advancing forward genetic approaches such as MutMap, QTL-seq, and

Table 1.1 Classification of genomic approaches for climate-smart crop development

Approaches	Core concept	Key techniques	Primary application
High-throughput genomics	Large-scale genome analysis for trait discovery	NGS, SNP genotyping, GWAS, pan-genomics	Identification of climate-resilient genes
Genome editing	Precise modification of genomic sequences	CRISPR/Cas, base editing, prime editing	Improvement of stress tolerance and yield stability
Epigenomics	Study of heritable gene regulation beyond DNA sequence	DNA methylation, histone modification	Understanding stress memory and epigenetic inheritance
Bioinformatics and big data analytics	Computational integration of multi-omic data	AI/ML pipelines, databases, predictive modelling	Trait prediction, gene network analysis, breeding optimization
Precision genomic integration	Synergistic use of genomics with precision agriculture tools	Phenomics, remote sensing, data-driven farming	Targeted deployment of climate-smart strategies

MutChromSeq, allowing rapid discovery of causal mutations conferring climate resilience, further accelerating breeding cycles (Abe et al. 2012; Sahu et al. 2020). Together, these advancements underscore how NGS and WGS empower breeders to move beyond phenotypic selection towards precise genome-informed strategies for developing crop varieties tailored to withstand the intensifying challenges of climate change (Varshney et al. 2024). Thus, NGS and WGS function as leading tools in using genomics to help crops adapt to climate change, paving the way for advanced methods like pan-genomics, GWAS, and multi-omic studies.

Pan-genomics has rapidly evolved from an idea to a practical tool in climate-smart breeding. It captures the full spectrum of intraspecific variation, including SVs (Table 1.1). These variations are often invisible to single-reference approaches yet underlie many stress adaptation traits (Hameed et al. 2022; Shang et al. 2022; Zhao et al. 2018). Graph and super-pan-genomes now enable breeders to map complex SVs linked to environmental resilience, improve GWAS power, and connect non-reference haplotypes to adaptive phenotypes across diverse environments (Hameed et al. 2022; Shang et al. 2022). In rice, pan-genome studies of both cultivated and wild types have found unique DNA patterns and structural changes linked to traits like flood tolerance, heat tolerance, seed shattering, and plant shape. These findings help scientists quickly find useful genes for adapting rice to climate change (Hameed et al. 2022; Shang et al. 2022; Song et al. 2020). Similar gains are evident in barley, where a 20-genome pan-genome exposed extensive gene content and SV diversity across domesticated and wild germplasm. This provides a scaffold to rediscover stress-relevant variation for marginal environments (Gui et al. 2022). In oilseed rape (*Brassica napus*), a species-level pan-genome clarified how SVs shape ecotype differentiation (e.g. winter vs. spring types) and agronomic adaptation, informing selection under cold, long-season, and other shifting climatic regimes (Song et al. 2020). Across cereals, the maize (pan-*Zea*) and pan-genomes have uncovered large reservoirs of non-reference genes/SVs and provided trait-to-variant links across altitudinal, aridity, and temperature gradients relevant to climate resilience (Guo et al. 2025; Ruperao et al. 2021).

In heat-tolerant pearl millet, a graph pan-genome captured >400 K SVs and implicated ER-related pathways and RWP-RK transcription factors (TFs) in thermal resilience, illustrating how graph models surface gene–environment mechanisms usable in selection or editing (Yan et al. 2023). Legume resources are expanding too; for instance, soybean pan-genomes, including a large PanSoy panel and 3D genome analyses, connect SVs and higher-order chromatin variation to expression changes that can be exploited for stress adaptation (Ni et al. 2023; Torkamaneh et al. 2021). For horticultural crops like tomato, pan-genome studies found thousands of non-reference genes and regulatory variants under selection, showing how this approach can bring back useful lost traits for breeding (Gao et al. 2019). As methods consolidate, high-level syntheses highlight that plant graph pan-genomics improves variant representation, gene discovery, and portability of findings across environments. These features are key for building climate-resilient ideotypes from standing diversity and wild relatives (Du et al. 2025; Hameed et al. 2022; Hu et al. 2025).

Genotyping-by-sequencing (GBS) has emerged as a powerful, cost-effective, high-throughput tool that captures the genetic diversity necessary for climate resilience. The GBS enables the rapid discovery of thousands to millions of SNP markers, even without a reference genome, by using simpler sequencing methods that look at only parts of the DNA (Pareek and Smoczynski 2011; Scheben et al. 2017). This capability has been harnessed to exhibit adaptive variation in wild common bean populations, revealing SNPs linked to drought-prone environments across Mesoamerica and offering breeders natural alleles for tolerance breeding (Cortés and Blair 2018). In rice pre-breeding, GBS has powered the genotyping of multi-parent populations such as MAGIC and NAM, dramatically increasing marker density, e.g. identifying over 79,000 SNPs tied to cadmium accumulation, drought response, and submergence tolerance (Bandillo et al. 2013). Across diverse species, GBS has enabled mapping climate adaptation traits in watermelon (drought and salinity tolerance), *Sorghum* (agro-climatic characteristics), and cotton (salt tolerance), enabling trait-linked marker discovery to speed up breeding (Xu et al. 2021). The GBS integrated with GWAS allows precise linkage of natural genetic variation to adaptive phenotypes, a combination crucial for breeding climate-resilient cultivars. The GWAS utilizes natural genetic mixes in identifying major and minor effect loci without the limitations of bi-parental crosses (Korte and Farlow 2013). For instance, in *Sorghum*, GBS identified about 265,000 SNPs across nearly 1000 worldwide accessions, enabling GWAS to map agro-climatic traits like inflorescence architecture and plant height, highlighting its adaptation across climates (Morris et al. 2013).

In maize, multi-locus GWAS under drought and normal conditions identified genes that control yield and flowering, directly helping link genetic markers to how crops perform under stress (Amadu et al. 2025). In chickpea, GWAS using GBS-based SNPs has found drought-resilient genes across different environments. This helps improve marker-assisted breeding in legumes under climate stress. Together, GBS and GWAS provide a strong system for climate-smart breeding. They allow quick marker discovery, testing in many environments, and predicting how genes and environments interact. This also supports GS models that capture complex, climate-dependent traits needed for the future (Crossa et al. 2017).

Transcriptomics has been the first and most widely deployed omics layer for decoding plant responses to climate-related stresses. It captures genome-wide changes in gene expression, regulatory networks, alternative splicing, and the dynamics of noncoding RNA. These processes underlie phenotypic plasticity. Transcriptome studies across model and crop species have repeatedly identified conserved stress-responsive modules, such as ABA signalling, heat-shock factors, ROS scavenging, and osmo-protectant biosynthesis. They have also revealed species- and ecotype-specific transcriptional strategies that breeders can target (Aslam et al. 2024; Kumar et al. 2024a, b, c; Xu et al. 2023). High-resolution time-series and tissue-specific RNA sequencing experiments have allowed researchers to distinguish early sensing events from later acclimation programmes. These studies have also helped prioritize candidate regulators for functional validation, such as TFs and splicing factors. These regulators control tolerance to drought, heat, and flooding (Baldoni 2022). Comparative transcriptomics, contrasting tolerant vs sensitive genotypes or wild relatives, has accelerated allele discovery and marker development, providing molecular entry points for marker-assisted selection, genomic prediction, and targeted editing (Li et al. 2022a, b; Roychowdhury et al. 2023).

Proteomics complements transcriptomics by revealing how stress alters protein abundance, post-translational modifications, and complex formation, determining biochemical activity and cellular resilience, but is not fully predicted by mRNA levels. Mass spectrometry-based studies in cereals and legumes have identified proteome signatures associated with heat, drought, and salinity tolerance. These include chaperones, antioxidant enzymes, metabolic enzymes, and transporters. Such proteins serve as functional biomarkers and, in some cases, as direct targets for trait improvement (Halder et al. 2022; Yan et al. 2022). Proteomic datasets also reveal the stability or turnover of key proteins under stress. This knowledge enables breeders and physiologists to prioritize robust alleles and design screens for protein-level traits. Examples include photosynthetic complex stability or root-system enzyme activity, which correlate with field performance (Nouraei et al. 2024; Sen et al. 2024).

Metabolomics provides deeper insights by quantitatively profiling small-molecule effectors, such as osmolytes, antioxidants, compatible solutes, phytohormones, and specialized metabolites, which act as proximate drivers of stress tolerance and yield stability under environmental extremes. Integrative metabolome–transcriptome and proteome studies have mapped how gene expression and protein network changes rewire central carbon, amino acid, and secondary metabolite pathways during stress. These studies have also identified biochemical markers, such as specific amino acids, polyols, and phenylpropanoids. Such markers can predict tolerance and reveal trade-offs with quality traits (Głuchowska and Zieniuk 2025; Wang et al. 2025). Importantly, multi-omic pipelines now combine transcriptomics, proteomics, and metabolomics with ML or network models. These approaches are being used to (i) prioritize causal genes, (ii) derive biochemical selection indices that are portable across environments, and (iii) improve genomic prediction models for complex, G × E-sensitive traits. Together, they help transform molecular insights into breeding decisions for climate resilience (Amin et al. 2025; Fan et al. 2025).

1.4 Genome Editing and Advanced Breeding Approaches

Genome editing and advanced breeding strategies reshape crop improvement by enabling precise, predictable, and accelerated genetic modifications (Gao 2021). Although conventional breeding remains valuable, it is often limited by long timelines, linkage drag, and environmental uncertainties. Genome editing platforms such as CRISPR/Cas systems, base editors, and prime editors have transformed precision breeding by allowing targeted modifications without introducing foreign DNA (Li et al. 2024). When integrated with GS, speed breeding, and high-throughput phenotyping, these approaches form a comprehensive framework for developing climate-resilient, pest-tolerant, and resource-efficient crops (Varshney et al. 2005). The ability to edit multiple genes simultaneously, together with rapid generation advancement, significantly shortens breeding cycles and enhances the efficiency of trait stacking (Li et al. 2024). Future progress will likely involve integrating genome editing tools with multi-omic datasets and ML approaches, enabling predictive breeding that accelerates genetic gains and strengthens global food and nutritional security.

CRISPR/Cas systems have transformed plant breeding by offering precise, efficient, and cost-effective alternatives to earlier tools such as zinc finger nucleases and transcription activator-like effector nucleases (Ahmad 2023). Their ease of design, high specificity, and multiplexing ability make them especially effective for increasing yield, enhancing stress tolerance, providing pathogen resistance, and improving nutritional content (Table 1.1). Variants like CRISPR/Cas12 and CRISPR/Cas13 expand functionality to RNA targeting, broadening applications in functional genomics and trait engineering (Li et al. 2024). High-fidelity Cas enzymes, along with new-generation base and prime editors, have further increased accuracy, while reducing off-target effects. Notable examples include CRISPR/Cas-mediated knockout of cytokinin oxidase (CKX) genes to boost grain yields in rice and maize (Mandal et al. 2022). Targeted modifications of susceptibility genes such as *MLO* in wheat and soybean or *LGS1* in sorghum have also been done to confer resistance to pathogens and parasitic plants (Garcia-Ruiz et al. 2021). Nonetheless, residual off-target activity and regulatory restrictions still pose challenges for large-scale agricultural use.

Speed breeding accelerates crop life cycles by employing controlled environments with extended photoperiods, optimized temperature regimes, and precise nutrient supply (Watson et al. 2018). This approach can reduce breeding timelines from several years to a few months, helping desirable traits and genome editing results move quickly into elite crops (Watson et al. 2018). When combined with tools like high-throughput phenotyping, genotyping, and GS, trait transfer and variety development are more efficient (Potts et al. 2023). Speed breeding allows several crop cycles each year, speeding up the creation of climate-resilient, high-yielding varieties, especially in long-duration crops like wheat, barley, and chickpea (Potts et al. 2023; Watson et al. 2018).

Integrating genome editing with GS offers a pathway to accelerate crop improvement. The GS utilizes genome-wide markers and predictive algorithms to estimate breeding values for complex traits, while genome editing enables precise modification of key loci to optimize these traits (Kumar et al. 2024a, b, c). Progress in computational biology and ML has strengthened GS, making genotype-to-phenotype predictions more accurate while lowering breeding costs and speeding up variety development. Together, these approaches provide a forward-looking framework for achieving high genetic gains and developing resilient, high-performing crop varieties.

1.5 Epigenomics and Its Role in Stress Adaptations

Epigenomics and regulatory genomics explore heritable changes in gene expression that occur without altering the underlying DNA sequence (Table 1.1). These mechanisms include DNA methylation, histone modifications, chromatin remodelling, and regulatory RNAs, all acting as flexible modulators of genome activity (Kumar and Rani 2023). Fine-tuning transcriptional programmes enables plants to adapt to developmental signals, environmental fluctuations, and stress conditions (Abdulraheem et al. 2024). Recent advances in sequencing-based epigenomic profiling, chromatin accessibility mapping, and high-resolution imaging have revealed complex regulatory networks shaping traits such as stress tolerance, yield, and flowering behaviour. Importantly, certain epigenetic modifications provide rapid, reversible responses, while others are stably inherited, forming the basis of stress memory (Kumar and Rani 2023). For example, drought, salinity, or pathogen exposure can establish persistent epigenetic states that prime plants for quicker and stronger responses to future challenges. This growing understanding has given the concept of epigenetic breeding, where beneficial epigenomic states are identified or engineered to enhance crop resilience and productivity under climate variability (Kumar 2018; Gahlaut et al. 2020). Integrating epigenomic insights with genome editing technologies represents a powerful frontier for next-generation crop improvement.

Chromatin remodelling, DNA methylation, and small RNAs form interconnected regulatory networks controlling stress-responsive gene expression (Table 1.1). Chromatin remodelling complexes reposition nucleosomes and modify histones, altering accessibility to transcriptional machinery (Abdulraheem et al. 2024). DNA methylation provides a longer-term regulatory framework, while small RNAs mediate sequence-specific transcriptional or post-transcriptional silencing. For instance, drought and salinity tolerance in rice and maize are partly regulated by small RNAs that modulate hormone signalling and osmo-protectant biosynthesis, supported by methylation-dependent reinforcement of gene expression states (Kumar and Rani 2023; Kumar 2018). Such multi-layered regulatory control highlights potential targets for selecting or engineering epigenomic states to develop climate-resilient crops.

Plant responses to climatic stress are profoundly influenced by epigenetic mechanisms that regulate gene activity without altering DNA sequences. For example, vernalization in *Arabidopsis thaliana* triggers histone methylation-mediated silencing of the floral repressor *FLOWERING LOCUS C* (Sharma et al. 2020). This process enables flowering after prolonged cold exposure. The mechanism is evolutionarily conserved and is also evident in cereals. In rice, drought endurance correlates with dynamic DNA methylation patterns across genes associated with root development and water-use efficiency (Abdulraheem et al. 2024). A comparison of drought-tolerant and drought-sensitive rice genotypes revealed that differential methylation in gene promoters and exonic regions modulates expression of key drought-responsive loci (Wang et al. 2016; Zheng et al. 2013). Further methylome sequencing in rice has revealed temporary changes in DNA methylation that act as a drought memory, enabling the plant to cope better with water stress (Kou et al.

2022). During heat stress, it has been observed that maintaining active histone marks such as H3K4me3 at stress-inducible loci establishes transcriptional memory. In *Arabidopsis*, higher H3K4 methylation at heat-responsive genes, controlled by heat-shock TFs, helps these genes switch back on more quickly during repeated heat stress, improving heat tolerance (Jin et al. 2024). Epigenetic regulation plays a key role in salinity tolerance as well. In crops such as maize and *Brassica*, changes in DNA methylation and histone acetylation influence the activity of ion transporters and antioxidant enzymes, helping the plant maintain ionic balance under salt stress (Roy and Soni 2022). These examples highlight how epigenomic variation shapes traits essential for climate adaptation. This knowledge drives the concept of epigenomic breeding, where selection or introduction of favourable epigenetic states can be done to create more resilient crop varieties. Epigenetic insights, along with genome editing and breeding technologies, can accelerate the development of climate-smart crops.

1.6 Bioinformatics and Big Data Analytics

The rapid growth of genomic, transcriptomic, proteomic, metabolomic, and epigenomic datasets in crop research has created a pressing need for advanced bioinformatics and big data analytics (Vasupalli et al. 2024). These frameworks integrate diverse datasets to unveil regulatory networks, predict trait performance, and support precision breeding strategies. High-throughput computational pipelines, artificial intelligence (AI), ML, and robust databases are now essential for handling large multi-omic datasets and guiding data-based crop improvement (Table 1.1) (Yetgin 2025). Computational tools offer rationalized workflows for handling large-scale datasets generated through multi-omic platforms. Specialized tools enable merging multi-omic data to construct system-level plant growth models, stress adaptation, and yield determination (Vasupalli et al. 2024). To handle the scale and complexity of big datasets, researchers are now turning to cloud computing and high-performance computing platforms, which provide the power needed for scalable analyses. AI and ML are transforming plant breeding by helping scientists to predict traits crucial for climate resilience (Yetgin 2025). Unlike traditional methods that struggle with complex gene–environment interactions, these tools can analyse massive datasets and reveal patterns linked to stress tolerance and yield. Techniques like random forests, support vector machines, gradient boosting, and deep neural networks are widely used to identify key genomic regions and markers for traits such as drought, heat, and salinity tolerance (Table 1.1) (Yetgin 2025). AI could potentially highlight the hidden relationships in the data and make it possible to predict complex traits more accurately and speed up the selection of breeding targets (Kumar et al. 2024a, b, c).

Deep learning has transformed high-throughput phenotyping through image analysis, remote sensing, and hyperspectral imaging. Computer vision models can automatically detect minor changes in leaf wilting, canopy architecture, chlorophyll fluorescence, or root growth under stress, generating real-time and scalable phenotypic data (Mochida et al. 2019). Such image-based predictions create powerful tools for tracking crop performance across diverse field environments when employed with remote sensing and drone-based monitoring. Integrating ML with crop growth models and environmental data helps in predicting how specific genotypes might perform under future stress conditions (Alimagham et al. 2025). These insights aid the selection of resilient varieties and management strategies tailored to local environments. Furthermore, deep learning tools can now predict the efficiency and specificity of CRISPR-Cas targeting sites. This reduces off-target risks and makes gene editing more accurate (Li et al. 2024). When coupled with multi-omic data like transcriptomics, proteomics, metabolomics, and epigenomics, AI models generate sharper predictions, identifying the key genes and pathways that drive resilience (Vasupalli et al. 2024; Yetgin 2025).

Databases and decision support tools are essential for managing and interpreting big genomic and trait data. Resources like Gramene, Ensembl Plants, and ICIS give researchers access to annotated genomes, QTLs, markers, and trait information (Kumar et al. 2024a, b, c). Inclusion of visualization tools, predictive models, and bioinformatic pipelines in many of these resources helps identify candidate genes and govern marker-assisted breeding (Nusrat et al. 2019). Cloud platforms further make collaboration easier by allowing real-time data sharing and AI-driven analysis.

1.7 Integrating Genomics with Precision Practices for Climate-Smart Agriculture

To build a climate-resilient and sustainable agricultural system, integrating genomics, precision agriculture, and soil health management becomes essential (Gao 2021; Li et al. 2024; Varshney et al. 2005). Genomic tools, such as GS, high-throughput sequencing, and trait-based gene discovery, allow the breeding of stress-resistant and resource-efficient crop varieties, directly reducing dependence on agrochemicals and enhancing nutrient-use efficiency (Adenekan 2024). Simultaneously, precision agriculture technologies such as GPS-guided equipment, drones, remote sensing, IoT sensors, and AI-driven modelling optimize site-specific applications of water, fertilizers, and inputs, boosting yields while minimizing environmental impact (Table 1.1) (Vullaganti et al. 2025). Together, these genomic and precision tools support site-specific nutrient management, improving soil fertility and crop performance and fostering sustainable intensification of production (Bello 2024; Vullaganti et al. 2025). The integration enables farmers to tailor management strategies to crops' genetic profile and the spatial variability of soil conditions, enhancing resource-use efficiency while maintaining or improving productivity (Table 1.1).

However, the success of these integrated innovations hinges equally on socio-economic dynamics and farmer adoption. Studies on adopting sustainable agriculture demonstrate that higher education, access to information and advisory services, and perceptions shaped by social norms and peer networks significantly enhance farmers' inclination to adopt new practices (Madureira et al. 2022; Rizzo et al.

2024). Conversely, barriers such as the complexity of technology, perceived risk, lack of extension support, and financial or resource constraints hinder adoption even when the environmental and productivity benefits are clear (Jones-garcia and Krishna 2021; Rizzo et al. 2024). Therefore, ensuring fair access to training, affordable tools, and support systems is vital for small resource-limited farmers to adopt genomics-based precision agriculture.

1.8 Challenges and Future Perspectives

Despite rapid technical progress, genomics-based breeding still faces substantial practical limits that slow the delivery of climate-resilient cultivars. High upfront costs for sequencing, genotyping, and building and maintaining reference and training populations make GS and pan-genome projects resource-intensive. Additional expenses for high-quality phenotyping and supporting infrastructure further limit accessibility for under-resourced breeding programmes and smallholder contexts. These economic and logistical barriers are well documented and continue to constrain scalability (Ansari et al. 2024; Burrow et al. 2021). Phenotypic complexity poses an additional challenge, as critical agronomic traits are controlled by many small-effect loci and are influenced by strong genotype × environment interactions. Predictive accuracy is reduced unless extensive multi-environment datasets and sophisticated G × E models are available. This creates statistical and operational bottlenecks (Cooper et al. 2025; Werner et al. 2025). The phenotyping gap remains acute, while genotyping costs have fallen, and reliable, affordable, high-throughput phenotyping (field and on-farm) is still a significant constraint, limiting training data for prediction models and thereby slowing genetic gain (Sankaran et al. 2025; Werner et al. 2025). Additional challenges include the heavy computational requirements and the bioinformatic expertise needed to assemble pan-genomes and run multi-omic integration pipelines. Long-term data management is also required. Furthermore, concerns about unintended outcomes, such as loss of diversity or trade-offs from rapid selection, call for careful planning and continuous monitoring (Li et al. 2022a, b).

The ethical, legal, and biosafety concerns make genome editing and new breeding methods subject to complex regulations. These rules differ greatly between countries, creating a patchwork system. This heterogeneity leads to market uncertainty and research friction for developers and farmers alike (Rabuma et al. 2024; Schiemann et al. 2020). Biosafety considerations, including off-target edits, pleiotropy, ecological interactions, and the limits of comparative risk assessments for complex engineered traits, require case-by-case assessment and sometimes new risk assessment approaches for products arising from cutting-edge methods (Eckerstorfer et al. 2021; El-mounadi et al. 2020). Intellectual property rules, access to germplasm and data, and benefit-sharing arrangements affect who gains from genomic innovations; without inclusive governance and attention to farmers' rights, new varieties and tools risk increasing inequities rather than alleviating them (Idris et al. 2023). Societal acceptability and trust depend on transparent engagement, clear labelling, and informed policy decisions. Robust biosafety governance is also essential to balance innovation with precaution. These factors call for multidisciplinary policy dialogue and capacity building among regulators and extension systems (Slokenberga et al. 2022).

Looking ahead, a pragmatic roadmap to next-generation crops combines multi-omics/pan-omics, advanced predictive modelling (including AI/ML), synthetic biology, and decentralized participatory breeding to accelerate rational trait discovery and deployment while managing risk (Fan et al. 2025; Mahmood et al. 2022). Integration of genomics with transcriptomics, proteomics, metabolomics, epigenomics, microbiome profiling, and spatial/single-cell omics, combined in pan-omic atlases, will improve causal inference from genotype to phenotype and help unravel complex, environment-sensitive traits (Amin et al. 2025). Coupling these multi-layer data with ML models and field-scale envirotyping promises better G × E prediction and more robust selection strategies, while synthetic biology and precision editing can translate validated mechanisms into targeted, resource-efficient trait improvements (Mohamedikbal et al. 2025; Zhang et al. 2025a, b). To ensure an equitable future, sustained investment is needed in open data infrastructure, interoperable standards, and proportionate regulatory frameworks alongside targeted capacity building for breeders and regulators in low-resource settings. Finally, participatory approaches that centre farmer priorities and livelihoods are essential; otherwise, the technical promise will not translate into broad social benefit (Bradbury et al. 2025; Zhang et al. 2025a, b).

1.9 Conclusion

The past two decades has seen remarkable advances in plant genomics, ranging from NGS and pan-genomics to genome editing, epigenomic engineering, and multi-omic integration. These innovations are reshaping crop improvement strategies. They allow scientists to pinpoint genetic variations, edit important genes, and understand how plants adapt to stress at a broader level. Methods like GBS and GWAS have sped up the search for genes behind complex traits, while pan-genomics has uncovered hidden diversity not seen in standard reference genomes. Genome editing platforms such as CRISPR-Cas, complemented by RNA-guided alternatives like TIGR-Tas, offer flexible and efficient tools for trait engineering across species. Epigenomics, at the same time, adds a new layer by revealing reversible, heritable modifications that regulate stress memory and phenotypic plasticity. Combined with other genomic tools, these advances go beyond traditional breeding and provide scalable, data-driven ways to improve yield, resilience, and resource use in a changing climate. Genomics, supported by AI and advanced breeding, can develop crops that thrive under climate stress while sustaining yields. By integrating these innovations with eco-friendly practices like reduced chemical use, nutrient recycling, and biodiversity conservation, we can build resilient agriculture systems. Successful implementation will rely on collaboration, open data sharing, and fair access to ensure benefits reach both smallholder and commercial farmers globally.

Acknowledgements This work was supported by the seed-money grant provided by Goa University to VKY (GU/D-RDRM/Seed Money/SBSB/VKY/7074/2025/374) and to SUB (GU/D-RDRM/Seed Money/SBSB/SB/12520/2026/671).

Declaration Authors declare the use of AI-assisted tools to enhance the clarity and language of the manuscript.

References

Abdulraheem MI, Xiong Y, Moshood AY, Cadenas-Pliego G, Zhang H, Hu J (2024) Mechanisms of plant epigenetic regulation in response to plant stress: recent discoveries and implications. Plants 13(2):163. https://doi.org/10.3390/PLANTS13020163

Abe A, Kosugi S, Yoshida K, Natsume S, Takagi H, Kanzaki H, Matsumura H, Yoshida K, Mitsuoka C, Tamiru M, Innan H, Cano L, Kamoun S, Terauchi R (2012) Genome sequencing reveals agronomically important loci in rice using MutMap. Nat Biotechnol 30(2):174–179. https://doi.org/10.1038/nbt.2095

Adenekan TK (2024) Advancing sustainable agriculture through genomics-based crop improvement strategies. November

Ahmad M (2023) Plant breeding advancements with "CRISPR-Cas" genome editing technologies will assist future food security. Front Plant Sci 14:1133036. https://doi.org/10.3389/FPLS.2023.1133036/XML

Alimagham S, van Loon MP, Ramirez-Villegas J, Berghuijs HNC, Rosenstock TS, van Ittersum MK (2025) Integrating crop models and machine learning for projecting climate change impacts on crops in data-limited environments. Agric Syst 228:104367. https://doi.org/10.1016/J.AGSY.2025.104367

Alkorta I, Epelde L, Garbisu C (2017) Environmental parameters altered by climate change affect the activity of soil microorganisms involved in bioremediation. FEMS:1–9. https://doi.org/10.1093/femsle/fnx200

Amadu MK, Beyene Y, Chaikam V, Tongoona PB, Danquah EY (2025) Genome—wide association mapping and genomic prediction analyses reveal the genetic architecture of grain yield and agronomic traits under drought and optimum conditions in maize. BMC Plant Biol. https://doi.org/10.1186/s12870-025-06135-3

Amin A, Zaman W, Park S (2025) Harnessing multi-omics and predictive modeling for climate-resilient crop breeding: from genomes to fields. Genes 16

Ansari S, Hossein-zadeh NG, Shadparvar AA (2024) Genomic predictions under different genetic architectures are impacted by mating designs. Vet Anim Sci 25(June):100373. https://doi.org/10.1016/j.vas.2024.100373

Arnell NW, Lowe JA, Challinor A, Osborn T (2019) Global and regional impacts of climate change at different levels of global temperature increase. Climatic Chnage:377–391

Aslam N, Li Q, Bashir S, Yuan L, Qiao L, Li W (2024) Integrated review of transcriptomic and proteomic studies to understand molecular mechanisms of Rice's response to environmental stresses. Biology 13

Asseng S, Foster I, Turners NC (2011) The impact of temperature variability on wheat yields. Glob Change Biol 5:997–1012. https://doi.org/10.1111/j.1365-2486.2010.02262.x

Baldoni E (2022) Improving drought tolerance: can comparative transcriptomics support strategic rice breeding? Plant Stress 3:100058. https://doi.org/10.1016/j.stress.2022.100058

Bandillo N, Raghavan C, Muyco PA, Anna M, Sevilla L, Lobina IT, Dilla-ermita CJ, Tung C, Mccouch S, Thomson M, Mauleon R, Singh RK, Gregorio G, Redoña E, Leung H (2013) Multi-parent advanced generation inter-cross (MAGIC) populations in rice: progress and potential for genetics research and breeding. Rice:1–15

Bello A (2024) Genomics and precision agriculture: applications of genetic technologies in farming. 7(3):2–3. https://doi.org/10.35841/aabb-7.3.208

Bibi F, Rahman A (2023) An overview of climate change impacts on agriculture and their mitigation strategies. MDPI:1–15

Bradbury A, Clapp O, Biacsi A, Kuo P, Gaju O, Hayta S, Zhu J, Lambing C (2025) Integrating genome editing with omics, artificial intelligence, and advanced farming technologies to increase crop productivity. Plant Commun 6(7):101386. https://doi.org/10.1016/j.xplc.2025.101386

Burrow HM, Mrode R, Mwai AO, Coffey MP, Hayes BJ (2021) Challenges and opportunities in applying genomic selection to ruminants owned by smallholder farmers. Agriculture (Basel):1–15

Chauhan D, Singh D (2025) Role of next-generation sequencing technology in crop improvement. April 2023. https://doi.org/10.1201/9781003361862-14

Consortium RD, Fugeray-scarbel A, Bastien C, Dupont-nivet M, Dupont-nivet M (2021) Why and how to switch to genomic selection: lessons from plant and animal breeding experience. Front Genet 12(July):1–9. https://doi.org/10.3389/fgene.2021.629737

Cooper M, Tomura S, Wilkinson MJ, Powell O, Messina CD (2025) Breeding perspectives on tackling trait genome-to-phenome (G2P) dimensionality using ensemble-based genomic prediction. Theor Appl Genet 138(7):1–24. https://doi.org/10.1007/s00122-025-04960-6

Cortés AJ, Blair MW (2018) Genotyping by sequencing and genome—environment associations in wild common bean predict widespread divergent adaptation to drought. Front Plant Sci 9(February):1–13. https://doi.org/10.3389/fpls.2018.00128

Crossa J, Pérez-rodríguez P, Cuevas J, Montesinos-lópez O, Jarquín D, Campos GDL, Burgueño J, González-camacho JM, Pérez-elizalde S, Beyene Y, Dreisigacker S, Singh R, Zhang X, Gowda M, Roorkiwal M, Rutkoski J, Varshney RK (2017) Genomic selection in plant breeding: methods, models, and perspectives. Trends Plant Sci 22(11):961–975. https://doi.org/10.1016/j.tplants.2017.08.011

Du Z, He J, Jiao W (2025) Plant graph-based pangenomics: techniques, applications, and challenges. aBIOTECH 6(2):361–376. https://doi.org/10.1007/s42994-025-00206-7

Dukes JS, Pontius J, Orwig D, Garnas JR, Rodgers VL, Brazee N, Cooke B, Theoharides KA, Stange EE, Harrington R, Ehrenfeld J, Gurevitch J, Lerdau M, Stinson K, Wick R, Ayres M (2009) Responses of insect pests, pathogens, and invasive plant species to climate change in the forests of northeastern North America: what can we predict? 1. Can J For Res 248:231–248. https://doi.org/10.1139/X08-171

Eckerstorfer MF, Grabowski M, Lener M, Engelhard M, Simon S, Dolezel M, Heissenberger A, Lüthi C (2021) Biosafety of genome editing applications in plant breeding: considerations for a focused case-specific risk assessment in the EU. BioTech 18:1–14

El-mounadi K, Morales-floriano ML, Garcia-ruiz H (2020) Principles, applications, and biosafety of plant genome editing using CRISPR-Cas9. Front Plant Sci 11(February):1–16. https://doi.org/10.3389/fpls.2020.00056

Fan B, Chen L, Chen L, Guo H (2025) Integrative multi-omics approaches for identifying and characterizing biological elements in crop traits: current Progress and future prospects. Int J Mol Sci 26:1466. https://doi.org/10.3390/ijms26041466

Gahlaut V, Zinta G, Jaiswal V, Kumar S (2020) Quantitative epigenetics: A new avenue for crop improvement. Epigenomes 4(4):25. https://doi.org/10.3390/epigenomes4040025

Gao C (2021) Genome engineering for crop improvement and future agriculture. Cell 184(6):1621–1635. https://doi.org/10.1016/J.CELL.2021.01.005

Gao L, Gonda I, Sun H, Ma Q, Bao K, Tieman DM, Burzynski-chang EA, Fish TL, Stromberg KA, Sacks GL, Thannhauser TW, Foolad MR, Diez MJ, Blanca J, Canizares J, Xu Y, Van Der Knaap E, Huang S, Klee HJ et al (2019) The tomato pan-genome uncovers new genes and a rare allele regulating fruit flavor. Nat Genet 51(June):1044

Garcia-Ruiz H, Szurek B, Van den Ackerveken G (2021) Stop helping pathogens: engineering plant susceptibility genes for durable resistance. Curr Opin Biotechnol 70:187–195. https://doi.org/10.1016/J.COPBIO.2021.05.005

Głuchowska A, Zieniuk B (2025) Unlocking plant resilience: metabolomic insights into abiotic stress tolerance in crops. Meta:1–32

Gui S, Wei W, Jiang C, Luo J, Chen L, Wu S, Li W, Wang Y, Li S, Yang N, Li Q, Fernie AR, Yan J (2022) A pan – Zea genome map for enhancing maize improvement. Genome Biol:1–22. https://doi.org/10.1186/s13059-022-02742-7

Guo D, Li Y, Lu H, Zhao Y, Kurata N, Wei X, Wang A, Wang Y, Zhan Q, Fan D, Zhou C, Lu Y, Tian Q, Weng Q, Feng Q, Huang T, Zhang L, Gu Z, Wang C, Wang Z (2025) A pangenome reference of wild and cultivated rice. Nature 642(June). https://doi.org/10.1038/s41586-025-08883-6

Halder T, Choudhary M, Liu H, Chen Y, Yan G, Siddique KHM (2022) Wheat proteomics for abiotic stress tolerance and root system architecture: current status and future prospects. Proteomes:1–36

Hameed A, Poznanski P, Nadolska-orczyk A, Orczyk W (2022) Graph pangenomes track genetic variants for crop improvement. Int J Mol Sci 23

Hu H, Zhao J, Thomas WJW, Edwards D, Batley J (2025) The role of pangenomics in orphan crop improvement. Nat Commun:1–9. https://doi.org/10.1038/s41467-024-55260-4

Huang X, Kurata N, Wei X, Wang Z, Wang A, Zhao Q, Zhao Y, Liu K, Dong G, Zhan Q, Li C, Fujiyama A, Toyoda A, Lu T, Feng Q, Qian Q, Li J (2012) A map of rice genome variation reveals the origin of cultivated rice. Nature 490(7421):497–501. https://doi.org/10.1038/nature11532

Idris SH, Sharleeza N, Jalaluddin MAT, Chang LW (2023) Ethical and legal implications of gene editing in plant breeding: a systematic literature review. J Zhejiang Univ-Sci B 24(12):1093–1105

Jiao Y, Peluso P, Shi J, Liang T, Stitzer MC, Wang B, Campbell MS, Stein JC, Wei X, Chin C, Guill K, Regulski M, Kumari S, Olson A, Gent J, Mccombie R, Schneider KL, Wolfgruber TK, May MR et al (2017) Improved maize reference genome with. Nat Publ Group 546(7659):524–527. https://doi.org/10.1038/nature22971

Jin Q, Chachar M, Ali A, Chachar Z, Zhang P (2024) Epigenetic regulation for heat stress adaptation in plants: new horizons for crop improvement under climate change. Agronomy:1–21

Jones-garcia E, Krishna VV (2021) Farmer adoption of sustainable intensification technologies in the maize systems of the Global South. A review. Agron Sustain Dev

Kole C, Muthamilarasan M, Henry R, Edwards D (2015) Application of genomics-assisted breeding for generation of climate resilient crops: progress and prospects. Front Plant Sci 6(August):1–16. https://doi.org/10.3389/fpls.2015.00563

Korte A, Farlow A (2013) The advantages and limitations of trait analysis with GWAS: a review. Plant Method:1–9

Kou S, Gu Q, Duan L, Liu G, Yuan P, Li H, Wu Z, Liu W (2022) Genome-wide bisulphite sequencing uncovered the contribution of DNA methylation to rice short-term drought memory formation. J Plant Growth Regul 41(7):2903–2917. https://doi.org/10.1007/s00344-021-10483-3

Krishna TPA, Veeramuthu D, Maharajan T, Soosaimanickam M (2023) The era of plant breeding: conventional breeding to genomics-assisted breeding for crop improvement. Curr Genomics:24–35. https://doi.org/10.2174/1389202924666230517115912

Kumar S (2018) Epigenomics of plant responses to environmental stress. Epigenomes 2018, Vol. 2, page 6, 2(1), 6. doi:https://doi.org/10.3390/EPIGENOMES2010006

Kumar R, Kumar A (2021) Omics technologies for sustainable agriculture and global food security II Decemberhttps://doi.org/10.1007/978-981-16-2956-3

Kumar M, Rani K (2023) Epigenomics in stress tolerance of plants under the climate change. Mol Biol Rep 50(7):6201–6216. https://doi.org/10.1007/S11033-023-08539-6

Kumar A, Chandran N, Paul P, Dhatt BK, Sandhu J, Irvin L, Oguro S, Shi Y, Zhang C, Walia H (2024a) Phenotypic and transcriptomic responses of diverse rice accessions to transient heat stress during early grain development. Front Plant Sci:1–14. https://doi.org/10.3389/fpls.2024.1429697

Kumar P, Paul RK, Roy HS, Yeasin M, Ajit, Paul AK (2024b) Big data analysis in computational biology and bioinformatics. Methods Mol Biol 2719:181–197. https://doi.org/10.1007/978-1-0716-3461-5_11

Kumar R, Das SP, Choudhury BU, Kumar A, Prakash NR, Verma R, Chakraborti M, Devi AG, Bhattacharjee B, Das R, Das B, Devi HL, Das B, Rawat S, Mishra VK (2024c) Advances in genomic tools for plant breeding: harnessing DNA molecular markers, genomic selection, and genome editing. Biol Res 57(1):1–23. https://doi.org/10.1186/S40659-024-00562-6

Li M, Li H, Sun A, Wang L, Ren C, Liu J, Gao X (2022a) Transcriptome analysis reveals genes in soybean. Front Genet:1–17. https://doi.org/10.3389/fgene.2022.1060529

Li W, Liu J, Zhang H, Liu Z, Wang Y, Xing L, He Q (2022b) Plant pan-genomics: recent advances, new challenges, and roads ahead. J Genet Genomics 49(9):833–846. https://doi.org/10.1016/j.jgg.2022.06.004

Li B, Sun C, Li J, Gao C (2024) Targeted genome-modification tools and their advanced applications in crop breeding. Nat Rev Genet 25(9):603–622. https://doi.org/10.1038/s41576-024-00720-2

Lobell DB, Bänziger M, Magorokosho C, Vivek B (2011) Nonlinear heat effects on African maize as evidenced by historical yield trials. Nat Clim Chang 1(March). https://doi.org/10.1038/nclimate1043

Madureira L, Marques C, Silva R (2022) Understanding farmers' adoption of sustainable agriculture innovations: A systematic literature review. Agronomy 12

Mahmood U, Li X, Fan Y, Chang W, Niu Y, Li J, Qu C, Lu K (2022) Multi-omics revolution to promote plant breeding efficiency. Front Plant Sci:1–20. https://doi.org/10.3389/fpls.2022.1062952

Mandal S, Ghorai M, Anand U, Roy D, Kant N, Mishra T, Mane AB, Jha NK, Lal MK, Tiwari RK, Kumar M, Radha, Ghosh A, Bhattacharjee R, Prоćków J, Dey A (2022) Cytokinins: A genetic target for increasing yield potential in the CRISPR era. Front Genet 13:883930. https://doi.org/10.3389/FGENE.2022.883930/XML

Mangal V, Kumar L, Kumar S, Saxena K, Sood S (2024) Triumphs of genomic-assisted breeding in crop improvement. Heliyon 10(15):e35513. https://doi.org/10.1016/j.heliyon.2024.e35513

Michael TP, Vanburen R (2005) Progress, challenges and the future of crop genomes. Curr Opin Plant Biol

Mochida K, Koda S, Inoue K, Hirayama T, Tanaka S, Nishii R, Melgani F (2019) Computer vision-based phenotyping for improvement of plant productivity: a machine learning perspective. GigaScience 8(1):1–12. https://doi.org/10.1093/GIGASCIENCE/GIY153

Mohamedikbal S, Al Mamun HA, Bestry MS, Batley J, Edwards D (2025) Integrating multi - omics and machine learning for disease resistance prediction in legumes. Theor Appl Genet 138(7):1–21. https://doi.org/10.1007/s00122-025-04948-2

Morris GP, Ramu P, Deshpande SP, Hash CT, Shah T, Upadhyaya HD (2013) Population genomic and genome-wide association studies of agroclimatic traits in sorghum. Proc Natl Acad Sci U S A 110(2). https://doi.org/10.1073/pnas.1215985110

Naqvi RZ, Siddiqui HA, Mahmood MA, Najeebullah S, Ehsan A, Azhar M, Farooq M, Amin I, Asad S, Mukhtar Z, Mansoor S, Asif M (2022) Smart breeding approaches in post-genomics era for developing climate-resilient food crops. Front Plant Sci:1–19. https://doi.org/10.3389/fpls.2022.972164

Ni L, Liu Y, Ma X, Liu T, Yang X, Wang Z, Liang Q, Liu S, Zhang M, Wang Z, Shen Y, Tian Z (2023) Pan - 3D genome analysis reveals structural and functional differentiation of soybean genomes. Genome Biol:1–26. https://doi.org/10.1186/s13059-023-02854-8

Nouraei S, Khan JM, Mia S, Yan G, Liu H, Turner NC (2024) Proteomic analysis of near-isogenic lines reveals key biomarkers on wheat chromosome 4B conferring drought tolerance. Plant Genome. https://doi.org/10.1002/tpg2.20343

Nusrat S, Harbig T, Gehlenborg N (2019) Tasks, techniques, and tools for genomic data visualization. Comput Graph Forum 38(3):781–805. https://doi.org/10.1111/CGF.13727;PAGE:STRING:ARTICLE/CHAPTER

Pachauri RK (2014). Climate change 2014 synthesis report

Pareek CS, Smoczynski R (2011) Sequencing technologies and genome sequencing. J Appl Genet:413–435. https://doi.org/10.1007/s13353-011-0057-x

Pareek A, Meena BM, Sharma S, Tetarwal ML, Kalyan RK, Meena BL (2017) Impact of climate change on insect pests and their management strategies (Issue November 2018)

Potts J, Jangra S, Michael VN, Wu X (2023) Speed breeding for crop improvement and food security. Crops 3(4):276–291. https://doi.org/10.3390/CROPS3040025

Rabuma T, Moronta-barrios F, Craig W, Craig W (2024) Navigating biosafety regulatory frameworks for genetic engineering in Africa: a focus on genome editing and gene drive technologies. Front Bioeng Biotechnol. https://doi.org/10.3389/fbioe.2024.1483279

Rizzo G, Migliore G, Schifani G, Vecchio R (2024) Key factors influencing farmers' adoption of sustainable innovations: a systematic literature review and research agenda. Org Agric:57–84. https://doi.org/10.1007/s13165-023-00440-7

Roy S, Soni P (2022) Unraveling the epigenetic landscape for salt tolerance in plants. Int J Plant Biol 13:443–462

Roychowdhury R, Das SP, Gupta A, Parihar P, Chandrasekhar K, Sarker U, Kumar A, Ramrao DP, Sudhakar C (2023) Multi-omics pipeline and omics-integration approach to Decipher plant's abiotic stress tolerance responses. Genes

Ruperao P, Thirunavukkarasu N, Gandham P, Selvanayagam S, Govindaraj M, Nebie B, Manyasa E, Deshpande SP, Rathore A (2021) Sorghum pan-genome explores the functional utility for genomic-assisted breeding to accelerate the genetic gain. Front Plant Sci 12(June):1–17. https://doi.org/10.3389/fpls.2021.666342

Sahu PK, Sao R, Mondal S, Vishwakarma G (2020) Approaches for identification and mapping of causal mutations in crop plants: a comprehensive review. Plants

Sankaran S, Singh KD, Sangjan W, Gore MA, Farmer EE, Mcnair M, Tolosa ST, Rife TW (2025) Affordable phenomics: expanding access to enhancing genetic gain in plant breeding. Plant Phenome J:1–25. https://doi.org/10.1002/ppj2.70034

Satam H, Joshi K, Mangrolia U, Waghoo S, Zaidi G, Rawool S, Thakare RP, Banday S, Mishra AK, Das G, Malonia SK (2023) Next-generation sequencing technology: current trends and advancements. Biology 12

Scheben A, Batley J, Edwards D (2017) Genotyping-by-sequencing approaches to characterize crop genomes: choosing the right tool for the right application. Plant Biotechnol J:149–161. https://doi.org/10.1111/pbi.12645

Schiemann J, Robienski J, Schleissing S, Spök A, Sprink T, Wilhelm RA (2020) Editorial: plant genome editing—policies and governance. Front Plant Sci 11(March):1–4. https://doi.org/10.3389/fpls.2020.00284

Sen A, Gumus T, Temel A, Ozturk I, Celik O (2024) Biochemical and proteomic analyses in drought-tolerant wheat. Plants

Shang L, Li X, He H, Yuan Q, Song Y, Wei Z, Hu M, Zhao F, Zhang C, Li Y, Gao H, Wang T, Liu X, Zhang H, Zhang Y, Cao S, Yu X, Zhang B, Zhang Y et al (2022) A super pan-genomic landscape of rice. Cell Res. https://doi.org/10.1038/s41422-022-00685-z

Sharma N, Geuten K, Giri BS, Varma A (2020) The molecular mechanism of vernalization in Arabidopsis and cereals: role of flowering Locus C and its homologs. Physiol Plant 170(3):373–383. https://doi.org/10.1111/PPL.13163

Shuhadah N, Saad M, Neik TX, Thomas WJW, Amas JC, Cantila AY, Craig RJ, Edwards D, Batley J (2022) Advancing designer crops for climate resilience through an integrated genomics approach. Curr Opin Plant Biol 67(Ml):102220. https://doi.org/10.1016/j.pbi.2022.102220

Sinha P, Singh VK, Bohra A, Kumar A, Reif JC, Varshney RK (2021) Genomics and breeding innovations for enhancing genetic gain for climate resilience and nutrition traits. Theor Appl Genet 134(6):1829–1843. https://doi.org/10.1007/s00122-021-03847-6

Sinha D, Maurya AK, Abdi G, Majeed M, Agarwal R, Mukherjee R, Ganguly S, Aziz R, Bhatia M (2023) Integrated genomic selection for accelerating breeding programs of climate-smart cereals. MDPI 14

Slokenberga, S., Minssen, T., & Nordberg, A. (2022). Governing, protecting, and regulating the future of genome editing

Song J, Guan Z, Hu J, Guo C, Yang Z, Wang S, Liu D, Wang B, Lu S, Zhou R, Xie W, Cheng Y, Zhang Y, Liu K, Yang Q, Chen L, Guo L (2020) Eight high-quality genomes reveal pan-genome architecture and ecotype differentiation of Brassica napus. Nat Plants 6(January). https://doi.org/10.1038/s41477-019-0577-7

Subedi B, Poudel A, Aryal S (2023) The impact of climate change on insect pest biology and ecology: implications for pest management strategies, crop production, and food security. J Agric Food Res 14(March):100733. https://doi.org/10.1016/j.jafr.2023.100733

Torkamaneh D, Lemay M-A, Belzile F (2021) The pan-genome of the cultivated soybean (PanSoy) reveals an extraordinarily conserved gene content. Plant Biotechnol J 13:1852–1862. https://doi.org/10.1111/pbi.13600

Varshney RK, Graner A, Sorrells ME (2005) Genomics-assisted breeding for crop improvement. Trends Plant Sci 10(12):621–630. https://doi.org/10.1016/J.TPLANTS.2005.10.004

Varshney RK, Sinha P, Singh VK, Kumar A, Zhang Q, Bennetzen JL (2020) 5Gs for crop genetic improvement. Curr Opin Plant Biol 56:190–196. https://doi.org/10.1016/j.pbi.2019.12.004

Varshney RK, Barmukh R, Bentley A, Nguyen HT (2024) Exploring the genomics of abiotic stress tolerance and crop resilience to climate change. Plant Genome:1–6. https://doi.org/10.1002/tpg2.20445

Vasupalli N, Bhat JA, Jain P, Sri T, Islam MA, Shivaraj SM, Singh SK, Deshmukh R, Sonah H, Lin X (2024) Omics big data for crop improvement: opportunities and challenges. Crop J 12(6):1517–1532. https://doi.org/10.1016/J.CJ.2024.10.007

Vullaganti N, Ram BG, Sun X (2025) Precision agriculture technologies for soil site-specific nutrient management: A comprehensive review. Artific Intell Agric 15(2):147–161. https://doi.org/10.1016/j.aiia.2025.02.001

Wang W, Qin Q, Sun F, Wang Y, Xu D, Li Z, Fu B (2016) Genome-wide differences in DNA methylation changes in two contrasting Rice genotypes in response to drought conditions. Front Plant Sci 7(November):1–13. https://doi.org/10.3389/fpls.2016.01675

Wang Y, Tong L, Liu H, Li B, Zhang R (2025) Integrated metabolome and transcriptome analysis of maize roots response to different degrees of drought stress. BMC Plant Biol 25

Watson A, Ghosh S, Williams MJ, Cuddy WS, Simmonds J, Rey MD, Asyraf Md Hatta M, Hinchliffe A, Steed A, Reynolds D, Adamski NM, Breakspear A, Korolev A, Rayner T, Dixon LE, Riaz A, Martin W, Ryan M, Edwards D et al (2018) Speed breeding is a powerful tool to accelerate crop research and breeding. Nat Plants 4(1):23–29. https://doi.org/10.1038/s41477-017-0083-8

Werner CR, Zaman-allah M, Assefa T, Cairns JE, Atlin GN (2025) Accelerating genetic on-farm sparse testing. Trends Plant Sci 30(1):17–20. https://doi.org/10.1016/j.tplants.2024.10.010

Xu P, Guo Q, Meng S, Zhang X, Xu Z, Guo W, Shen X (2021) Genome-wide association analysis reveals genetic variations and candidate genes associated with salt tolerance related traits in Gossypium hirsutum. BMC Genomics 22:1–14

Xu X, Flavio C, De Lima F, Vu LD, De Smet I (2023) When drought meets heat—a plant omics perspective. Front Plant Sci:1–9. https://doi.org/10.3389/fpls.2023.1250878

Yan S, Bhawal R, Yin Z, Thannhauser TW, Zhang S (2022) Recent advances in proteomics and metabolomics in plants. Molecular Horticulture. https://doi.org/10.1186/s43897-022-00038-9

Yan H, Sun M, Zhang Z, Jin Y, Zhang A, Lin C, Wu B, He M, Xu B, Wang J, Qin P, Mendieta JP, Nie G, Wang J, Jones CS, Feng G, Srivastava RK, Zhang X (2023) Pangenomic analysis identifies structural variation associated with heat tolerance in pearl millet. Nat Genet 55(March). https://doi.org/10.1038/s41588-023-01302-4

Yetgin A (2025) Revolutionizing multi-omics analysis with artificial intelligence and data processing. Quant Biol 13(3):e70002. https://doi.org/10.1002/QUB2.70002

Yuan X, Li S, Chen J, Yu H, Yang T, Wang C, Huang S, Chen H, Ao X (2024) Impacts of global climate change on agricultural production: a comprehensive review. MDPI 14

Zhang D, Xu F, Wang F, Le L, Pu L (2025a) Synthetic biology and artificial intelligence in crop improvement. Plant Commun. https://doi.org/10.1016/j.xplc.2024.101220

Zhang N, Tang L, Li S, Liu L, Gao M, Wang S, Chen D, Zhao Y, Zheng R, Soleymaniniya A, Zhang L, Wang W, Yang X, Ren Y, Sun C, Wilhelm M, Wang D, Li M, Chen F (2025b) Integration of multi-omics data accelerates molecular analysis of common wheat traits. Nat Commun. https://doi.org/10.1038/s41467-025-57550-x

Zhao Q, Feng Q, Lu H, Li Y, Wang A, Tian Q, Zhan Q, Lu Y, Zhang L, Huang T, Wang Y, Fan D, Zhao Y, Wang Z, Zhou C, Chen J, Zhu C, Li W, Weng Q et al (2018) Pan-genome analysis highlights the extent of genomic variation in cultivated and wild rice. Nat Genet 50(February). https://doi.org/10.1038/s41588-018-0041-z

Zheng X, Chen L, Li M, Lou Q, Xia H, Wang P, Li T (2013) Transgenerational variations in DNA methylation induced by drought stress in two rice varieties with distinguished difference to drought resistance. PLoS One 8(11):1–13

Potential of Crop Wild Relatives in Developing Abiotic Stress-Tolerant Crop Cultivars

2

Jatin Sharma, Deepak Bhamare, Arushi Arora, Priyanka Upadhyay, Pooja Kanwar Shekhawat, Anand Kumar, Amit Kumar, Shubham Sharma, and Dinesh Kumar Saini

Abstract

Abiotic stresses, such as extreme temperatures, drought, salinity, and other environmental adversities, significantly impact global food security by affecting crop yields. Crop wild relatives (CWRs) are plant taxa closely related to crops and serve as a valuable source of genetic diversity, which can help adapt crops to various abiotic stresses, particularly to meet the growing consumer demand amid the climate crisis. The historical evidence suggests that CWRs have been integral to crop improvement with successful examples in cereals, pulses, oilseeds, and other crops. Recent advances in genomics, including trait mapping and genome editing techniques, have successfully facilitated the identification and transfer of crucial traits from CWRs into modern crop varieties. Nevertheless, challenges such as the conservation of CWRs and the genetic complexities of integrating wild traits into cultivated varieties continue to be critical considerations in breed-

J. Sharma
Department of Plant Breeding and Genetics, Punjab Agricultural University, Ludhiana, Punjab, India

Department of Agriculture, Mandi, Himachal Pradesh, India

D. Bhamare
School of Agricultural Biotechnology, Punjab Agricultural University, Ludhiana, Punjab, India

A. Arora
Gurudev Singh Khush Institute of Genetics, Plant Breeding and Biotechnology, Punjab Agricultural University, Ludhiana, Punjab, India

P. Upadhyay
Department of Plant Breeding and Genetics, Dr. Rajendra Prasad Central Agricultural University, Pusa, Bihar, India

V. Gahlaut et al. (eds.), *Smart Crop Development*,
https://doi.org/10.1007/978-981-95-7680-7_2

ing programs. However, future direction toward an interdisciplinary approach combining advanced omic tools with effective strategies to ensure sustainable utilization of CWRs is an essential strategy to develop abiotic stress-resilient varieties. This chapter highlights the role of CWRs in enhancing abiotic stress tolerance, focusing on drought, salinity, and extreme temperatures in cereals, oilseeds, and pulses. It covers their genetic diversity, historical contributions, and potential in modern agriculture, along with advancements in genomics and breeding techniques. Challenges and future directions for utilizing CWRs are also discussed.

Keywords

Crop wild relatives · Climate change · Food security · Abiotic stresses

2.1 Introduction

Abiotic stress refers to the adverse effects of non-living components on biological organisms, such as plants. Plants face various abiotic stresses during their life cycle, including extreme temperatures, drought, salinity, and flooding. Climate change, pollution, and global warming exacerbate these stresses, significantly affecting both the quality and quantity of crop production (Bashir et al. 2019). Drought is a major constraint on plant growth, significantly reducing plant biomass and disrupting key physiological processes involved in growth, development, and metabolism (Brodersen et al. 2019). Yield losses from drought can range from 30% to 90%, depending on severity and crop growth stage (Fahad et al. 2017), and it remains a primary challenge to global food security, contributing to historical famines. As water supplies decrease and the global population grows, the effects of drought on farming intensify.

Salinity, another critical abiotic stress, results from the accumulation of water-soluble salts in soil, affecting soil performance, mainly in arid and semi-arid regions (Munns and Tester 2008). Approximately 6% of arable land is affected by soil

P. K. Shekhawat
Department of Genetics and Plant Breeding, Sardarkrushinagar Dantiwada Agricultural University, Dantiwada, Gujarat, India

A. Kumar
Faculty of Agriculture Sciences, GLA University, Mathura, Uttar Pradesh, India

A. Kumar
Department of Genetics and Plant Breeding, Govind Ballabh Pant University of Agriculture and Technology, Pantnagar, Uttarakhand, India

S. Sharma
Department of Plant Breeding and Genetics, Punjab Agricultural University, Ludhiana, Punjab, India

D. K. Saini (✉)
Department of Plant and Soil Science, Texas Tech University, Lubbock, TX, USA

salinization, with an annual increase of 1–2%, leading to significant yield reductions in important food crops such as maize, rice, and wheat (Munns and Tester 2008). Heat stress, caused by extreme temperatures, damages photosynthetic pigments, enzymes, and chlorophyll biosynthesis, while cold stress leads to frost damage and disrupts seed germination and metabolic processes (Mondal et al. 2023; Wang et al. 2024). Temperature stress, particularly from sudden temperature changes, markedly affects crops (Yang et al. 2021). Waterlogging stress, resulting from excess groundwater saturation, limits oxygen availability in the root zone, reducing carbohydrate and amino acid production and restricting plant growth, leading to yield losses ranging from 1% to 100%, depending on the crop and timing of waterlogging (Rocha et al. 2010; Kaur et al. 2020). The impact of abiotic stresses on crops, leading to altered physiological and metabolomic responses, growth reduction, and decreased grain yield and quality, is shown in Fig. 2.1.

Since, abiotic stresses are major challenges in modern agriculture, necessitating efforts to breed crops that can withstand harsh environmental conditions (Zhang et al. 2017). Crop wild relatives (CWRs) are wild species closely related to cultivated crops, sharing genetic similarities and evolving alongside them. While CWRs can naturally cross with crops, they remain undomesticated, making them valuable to plant breeders. Unlike cultivated plants, which are typically bred for optimal performance in controlled environments, CWRs have evolved to thrive in harsh conditions without human selection pressures (Renzi et al. 2022). These species

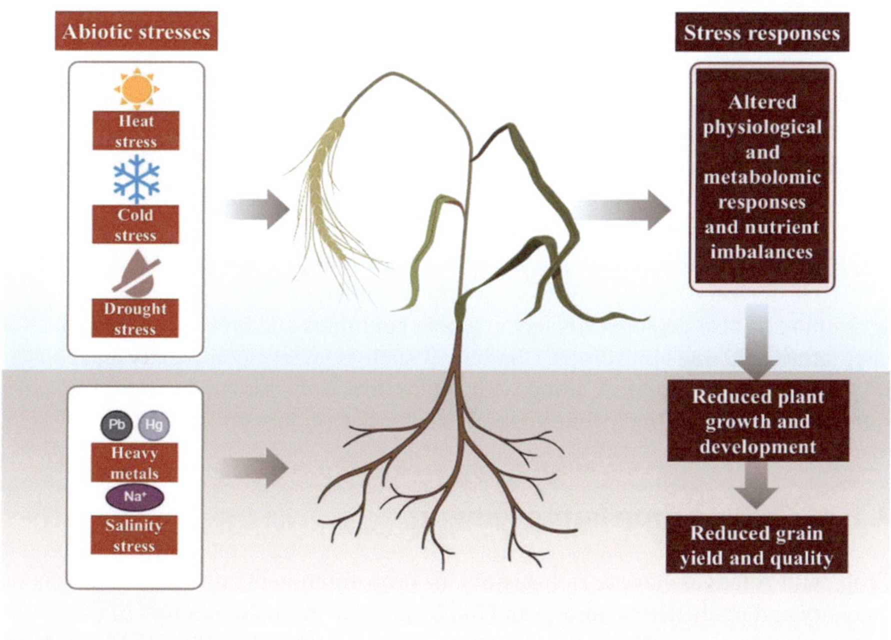

Fig. 2.1 Impact of abiotic stresses on crops, leading to altered physiological and metabolomic responses, reduced growth, and decreased grain yield and quality

possess unique genes and alleles, generally absent in modern crops, making them a valuable source of traits for improving drought, salinity, and temperature tolerance (Quezada-Martinez et al. 2021). By integrating CWR traits into elite cultivars, crop resilience to abiotic stress can be enhanced, ensuring stable and sustainable yields despite climate.

Crop wild relatives have long been utilized for crop improvement, contributing vital traits to modern cultivars. Nikolai Vavilov's research highlighted the importance of CWRs by identifying centers of origin for cultivated crops and emphasizing the need to protect their wild relatives. His work demonstrated the geographical connection between CWR hotspots and areas of agricultural diversity, underscoring the role of wild species in enhancing cultivated varieties (Maxted and Vincent 2021). These early studies laid the foundation for the ongoing integration of essential traits from CWRs into contemporary crops. The systematic use of CWRs in breeding programs began in the mid-twentieth century, notably with wild relatives of sugarcane. By the late 1960s, genes from CWRs were successfully integrated into wheat and barley to improve yield and disease resistance (Hajjar and Hodgkin 2007). The role of CWRs grew increasingly important in the 1970s and 1980s as research expanded to include more crops, helping address the narrow genetic base of modern cultivated plants (Hajjar and Hodgkin 2007; Kaur and Sangha 2023).

The use of new technologies, such as trait mapping, marker-assisted breeding, and CRISPR/Cas9 gene editing, has made accurate integration of favorable traits from CWRs into crops much easier. Some recent examples show how CWRs could be used as a source for improved abiotic stress tolerance in crops. As an example, tomato lines that show enhanced drought or salinity tolerance have been developed from the wild relative of tomato, *Solanum pennellii* (Albaladejo et al. 2017). Similarly, genes for heat tolerance have been introduced into the cultivated barley from the wild barley species *Hordeum spontaneum* (Ellis et al. 2000). Thus, the use of CWR in modern crops can provide tremendous benefits to increasing agricultural productivity, especially in cases where abiotic stresses are a problem.

This chapter discusses the role of CWRs in improving abiotic stress tolerance in major cereals, oilseeds, and pulses. The chapter highlights genetic diversity of CWRs, historical contributions to crop improvement, and their potential in modern agriculture. It also explores advancements in genomics and breeding techniques for integrating wild traits into crops. Challenges such as genetic complexity and conservation issues are addressed, along with future directions for enhancing CWR utilization through emerging technologies and interdisciplinary collaboration.

2.2 CWRs in Crop Improvement

Crop wild relatives have a rich history in crop improvement, enhancing genetic diversity and resilience to abiotic and biotic stresses. As wild relatives of cultivated crops, they serve as valuable genetic reservoirs, providing traits for modern cultivar enhancement. The following discussion focuses on the role of CWRs in cereal crops, pulses, and oilseeds, highlighting their historical use, current applications in abiotic stress tolerance, and advancements in genomics and breeding techniques.

2.2.1 CWRs in Cereal Crops

2.2.1.1 Historical Insights into the Role of CWRs in the Improvement of Cereals

The history of CWRs in cereal crops like wheat, rice, maize, and sorghum is closely tied to their evolution and domestication. These crops underwent whole-genome duplications followed by genome diversification, shaping their current genetic structures. Phylogenetic studies show that most cereal crops share a common ancestor, as evidenced by collinearity and synteny, revealing complex evolutionary relationships and genetic duplications. This knowledge underscores the potential of CWRs in enhancing cultivated cereals through beneficial traits, making an understanding of their evolutionary history essential for crop improvement (Thiel et al. 2009). Early use of CWRs began in the 1940s and 1950s when their genetic potential was recognized. Nikolai Vavilov's research identified the centers of origin for cultivated crops and emphasized the need to conserve their wild relatives. His work highlighted the overlap between CWR hotspots and regions of agricultural diversity, underscoring their role in improving cultivated species (Maxted and Vincent 2021). By the late 1960s, CWR genes were integrated into wheat and barley to enhance yield and disease resistance (Hajjar and Hodgkin 2007).

Such early efforts laid the foundation for the ongoing integration of crucial traits from CWRs into modern cultivars. However, the systematic utilization of CWRs in breeding programs began in the mid-twentieth century, with notable advancements in the 1970s and 1980s, when researchers began to recognize the potential of CWRs to combat the narrow genetic base of the present cultivated crops (Hajjar and Hodgkin 2007; Kaur and Sangha 2023). However, one of the most notable yet earliest examples of utilization of CWRs occurred in the development of disease-resistant varieties dating even back to the 1950s when the breeders successfully incorporated genes from wild relatives (such as *Aegilops* species) into the advanced breeding lines (Zair et al. 2017). These CWRs have played a crucial role in providing resistance to environmental stresses, several diseases, and pests in cereal crops. For instance, wild relatives of barley, such as *Hordeum spontaneum*, have been utilized in the development of varieties resistant to salinity and drought stress (Ellis et al. 2000). Similarly, certain genes from the wild relatives such as in wheat crops have been integrated successfully into the cultivated varieties to enhance resistance to the cereal cyst nematodes (Thormann et al. 2016; Zair et al. 2017). Incorporating such traits from wild rice species has also led to the development of varieties tolerant to abiotic stresses such as salinity and flooding (Kapazoglou et al. 2023). The growing recognition of CWRs has driven conservation efforts, including strategies like the Harlan and de Wet Inventory, which prioritize and identify CWRs for crop improvement (Jarvis et al. 2015). These initiatives aim to preserve CWR genetic diversity, ensuring breeders can access traits essential for developing resilient crop varieties (Redden et al. 2014; Jarvis et al. 2015).

The gene pools of major cereals like wheat, rice, and maize represent the genetic diversity essential for crop improvement. The primary gene pool includes cultivated crops and their wild relatives, the secondary comprises distantly related species capable of hybridizing with cultivated ones, and the tertiary encompasses genetically related species requiring specialized techniques for hybridization. The primary gene pool of wheat includes species such as *Triticum turgidum* and *Triticum dicoccoides*; for rice, *Oryza glaberrima* and *Oryza rufipogon*; and for maize, *Zea mays* ssp. *parviglumis* and *Zea mays* ssp. *mexicana* (Sharma et al. 2020; Varotto et al. 2022).

Crop wild relatives from the primary gene pool have significantly contributed to enhancing abiotic stress resistance in cereal crops throughout history. For instance, the *Oryza rufipogon* has been identified as a source of cold stress tolerance, with its genes successfully introgressed into cultivated rice. Similarly, the *Triticum dicoccoides* has contributed to drought tolerance, improving the performance of cultivated wheat under water stress. Likewise, the maize wild relative *Zea mays* ssp. *mexicana* exhibits cold stress tolerance and has been utilized to develop resilient maize varieties. In the secondary and tertiary gene pools, rice includes species such as *Oryza barthii*, *Oryza longistaminata*, *Oryza meridionalis*, and *Oryza punctata*; wheat includes *Aegilops tauschii*, *Aegilops speltoides*, *Aegilops longissima*, and *Aegilops uniaristata*; and maize includes *Zea diploperennis*, *Zea perennis*, *Tripsacum dactyloides*, and *Tripsacum laxum*. While these species harbor crucial genes for combating abiotic stresses, their limited crossability with cultivated crops poses significant challenges for conventional breeding methods (Hajjar and Hodgkin 2007; Maxted et al. 2010; Gonzalez-Segovia et al. 2019; Calfee et al. 2021).

Conventional methods like backcross breeding have transferred traits such as drought and flood tolerance from wild species like *Oryza rufipogon* in rice (Kumar et al. 2022). However, challenges such as poor agronomic performance, crossability barriers, and linkage drag persist (Bohra et al. 2022). Modern genomic techniques offer innovative solutions to overcome these barriers, enabling the successful transfer of valuable traits to crop varieties. The advances in high-throughput sequencing and genomic analysis have enabled efficient identification of beneficial alleles in CWRs, allowing breeders to incorporate valuable traits into cultivated varieties (Wendler et al. 2014; Gill et al. 2022). These tools have been instrumental in exploring CWRs in rice, revealing traits for improved grain quality and iron toxicity tolerance (Bierschenk et al. 2020).

These advancements in plant breeding, coupled with genomic and molecular biology innovations, have streamlined the selection and integration of desirable traits from CWRs into cultivated crops. Molecular markers and genomic tools now allow precise identification and incorporation of critical genes, revolutionizing crop improvement programs. For example, the gene *Gpc-B1* was incorporated from wild emmer wheat (*Triticum turgidum* L. *dicoccoides*) into cultivated wheat to enhance protein content in grains using these techniques (Iqbal et al. 2016). Overall, the historical and current utilization of CWRs in cereal crops reveals a trajectory of increasing usage and applications of these resources for enhancing genetic diversity, adaptability, and resilience, ultimately for the development of sustainable and robust crop varieties that can perform efficiently in diverse environments.

2.2.1.2 Current Status of CWR Usage in Cereals for Abiotic Stress Tolerance

Studies have highlighted the rich genetic diversity in CWRs from regions such as Central Asia and Africa, which serve as valuable sources of traits for enhancing abiotic stress tolerance in cultivated crops (Simon et al. 2020). For instance, wild relatives of wheat have been utilized in Morocco's durum wheat breeding program to improve resilience to climate change (Taghouti et al. 2023). Similarly, the phenotypic responses of CWR-derived lines have demonstrated their potential to enhance crop performance under abiotic stress. In rice, CWR-derived lines selected by farmers in the Mekong Delta have shown significant tolerance and robust performance under salt stress conditions (Tin et al. 2020).

The integration of CWRs has been instrumental in developing abiotic stress-resistant varieties in crops like rice and maize (Mammadov et al. 2018; Kapazoglou et al. 2023). Advanced omic technologies, including genomics, transcriptomics, proteomics, and metabolomics, have significantly improved our understanding of stress tolerance in plants (Toulotte et al. 2022; Bisht et al. 2023). The use of molecular markers has facilitated trait mapping, enabling breeders to select resilient lines more efficiently and effectively (Fleury et al. 2010). For instance, the rice CWR *Oryza rufipogon* has been utilized to transfer flood tolerance traits to cultivated species through marker-assisted selection (MAS), allowing precise tracking of desirable traits during crop improvement programs (Yang et al. 2021). Further, GWAS has emerged as a powerful tool for identifying QTLs associated with various traits (Saini et al. 2022). Afzal et al. (2019) demonstrated its successful application in uncovering alleles related to drought resistance in synthetic wheat, which can be transferred to advanced wheat lines. In addition to identifying beneficial alleles, GWAS also aids in understanding the complex genetic architecture of stress tolerance in plants (Reddy et al. 2023).

Genome editing tools like CRISPR/Cas9 have enhanced abiotic stress tolerance, such as editing the *OsDREB1* gene in rice for improved drought resilience. Similarly, Zhang et al. (2016) demonstrated CRISPR's potential in transferring drought-tolerant traits from CWRs to cultivated wheat to improve water scarcity resistance. Another approach to utilizing CWRs is speed breeding, which manipulates photoperiod and temperature to rapidly produce multiple generations per year. Though it requires years of inbreeding to fix genetic backgrounds, speed breeding has been successfully applied to crops like rice, wheat, sorghum, and barley. This technique not only accelerates the introgression of desirable alleles but also enables the selection against deleterious ones, helping develop stable, abiotic stress-resistant varieties. As a relatively recent technique, the development of new varieties using CWRs is still progressing (Tirnaz et al. 2022).

The transcriptomic approach complements these strategies by revealing gene expression profiles under stress, while proteomics provides insights into protein abundance and modifications, highlighting variations in plant proteomes under abiotic stress in CWR gene pools (Kosová et al. 2011; Hossain et al. 2022). Wang et al. (2017b) combined QTL mapping with transcriptomic techniques to identify candidate genes for salt tolerance from *Oryza rufipogon* at the seedling stage. They identified 1391 differentially expressed genes between the salt-tolerant and recipient parents. Additionally, they discovered four QTLs/candidate genes and identified the salt-tolerant line 9 L136 in the rice CWR *Oryza rufipogon*. These genomic regions can be used to develop salt-tolerant lines in cultivated rice.

Metabolomics provides valuable insights into biochemical changes in stressed plants, aiding the identification of metabolites like osmo-protectants and antioxidants that enhance stress tolerance. For instance, comparative studies in wheat have revealed distinct metabolic profiles in drought-sensitive and drought-tolerant genotypes, highlighting key stress-related mechanisms. Similarly, metabolomic analyses in crops like wheat have identified drought-tolerant genotypes (Guo et al. 2018). Integrating metabolomics with other omic approaches offers a comprehensive understanding of crop responses to stress, enabling the transfer of beneficial traits from CWRs to cultivated species.

Despite challenges like linkage drag and genetic distance, advancements in breeding techniques, such as marker-assisted breeding, have enabled the transfer of beneficial traits from CWRs with minimal retention of undesirable genes (Prohens et al. 2017; Mammadov et al. 2018). Crop wild relatives have been successfully used in crops like wheat, rice, and sorghum to develop cultivars resistant to various abiotic stresses (Table 2.1). Recently, the variety *Nachit* was developed from durum wheat in Morocco, demonstrating resilience to climate change-related challenges, including drought tolerance (Taghouti et al. 2023). Table 2.1 summarizes the traits transferred from respective CWRs to cultivated counterparts, along with the associated improvements.

2.2.2 CWRs in Oilseed Crops

2.2.2.1 Historical Insights into the Role of CWRs in the Improvement of Oilseed Crops

India is the largest importer of edible oils (15%), and its demand is expected to reach 40.9 Mt. by 2026. By 2050, it will need to produce 17.84 Mt. of edible oil (Mittal 2006). Key oilseeds like mustard-rapeseed, soybean, groundnut, sunflower,

Table 2.1 Examples of wild relatives used to enhance abiotic stress tolerance in cereals

Crop	CWRs	Traits targeted (abiotic stress)	References
Rice	*Oryza nivara* and *Oryza rufipogon*	Flood and drought tolerance	Bin Rahman and Zhang (2016)
Wheat	*Aegilops tauschii*	Drought tolerance	Yumurtaci (2015); Sohail et al. (2011)
Maize	*Zea mays* spp. *parviglumis*	Heat and drought tolerance	Maazou et al. (2017); Banda (2020)
Barley	*Hordeum spontaneum*	Salinity and drought tolerance	Ebrahim et al. (2020); Nevo and Chen (2010)
Pearl millet	*Pennisetum glaucum*	Heat and drought tolerance	Kapoor et al. (2022); Jacob et al. (2022)

and safflower suffer 17–94% yield losses due to abiotic stresses (Chugh et al. 2022). Some of these crops also face challenges in domestication, breeding, and low genetic diversity.

Unlike domesticated species, CWRs of oilseeds were never subjected to stringent anthropogenic selection pressure especially for yield-related traits (Quezada-Martinez et al. 2021). Instead, they faced the selection pressure of adverse environments to evolve and adjust accordingly to survive (Renzi et al. 2022). Therefore, CWRs serve as reservoirs of abiotic stress tolerance and resistance genes. However, many of these beneficial genes remain underexploited due to challenges such as differences in ploidy levels, pre- and post-fertilization incompatibilities, and the co-introgression of undesirable agronomic traits (linkage drag), which can negatively affect yield and crop performance (Bohra et al. 2022). Oilseed genetic diversity was historically enhanced through landraces, CWRs, and mutants. Molecular techniques have advanced MAS, QTL mapping, genotype analysis, and phylogenetic studies. However, they still cannot assess adaptive genetic diversity. Integrating biometry, population genetics, bioinformatics, and molecular innovations could enable sustainable improvements in oilseed crops to meet evolving climate and societal demands.

Brassica wild relatives and U's triangle species exhibit wide variability toward heat and drought tolerance (Kashyap et al. 2023). Various CWRs in *Brassica* include *Sinapis alba* L. (white mustard), *B. fruticulosa* Cirillo (twiggy turnip), *B. tournefortii* Gouan (African mustard), *Camelina sativa* L. (gold-of-pleasure), *Capsella bursa-pastoris* L. (shepherd's purse), *Crambe abyssinica* (Abyssinian mustard), *Diplotaxis erucoides* L. (white wall rocket), *D. muralis* L. (annual wall rocket), *D. tenuisiliqua* Delile (wall rocket), *Erucastrum cardaminoides* Webb ex Christ (dog mustard), *E. gallicum* Willd. (common dog mustard), *Lepidium sativum* L. (garden cress), etc. (Verma et al. 2023). Crop wild relatives are being targeted for abiotic stress regulation at multigenic levels in *B. juncea* (Singh et al. 2021; Kapazoglou et al. 2023; Kashyap et al. 2023). *S. alba* and *B. fruticulose* were identified as drought-tolerant relatives of Indian mustard (Kashyap et al. 2023). Genome-wide profiling of drought and re-watered leaves in *S. alba* showed differential gene expression, with 309 genes downregulated and 248 upregulated, linked to cell division and metabolic processes (Dong et al. 2012).

Various genotypes, including *B. juncea* (Pusa Jaikisan), *B. fruticulosa* (Spain), *B. tournefortii* (RBT 2003), *C. bursa-pastoris* (late), *C. sativa*, *C. abyssinica* (EC694145), *C. abyssinica* (EC400058), and *D. muralis*, exhibit drought resistance through enhanced germination percentage and proline content in leaves and roots under hypertonic PEG-6000 solution (Kashyap et al. 2023). Plants mainly rely on osmotic regulation to maintain metabolic activities during drought. In *B. napus*, groups 2 and 3 LEA proteins play key roles in drought tolerance (Koh et al. 2015). ABA biosynthesis and signal transduction are also critical, with farnesyl-transferase regulating ABA sensitivity. Suppression of farnesyl-transferase in *B. napus* enhances water retention and seed productivity under drought (Wang et al. 2005). High temperature negatively affects *Brassica* spp. yields from germination to seed ripening (Sandhu et al. 2019), but *S. alba* has been identified as heat-tolerant. Protoplast fusion of *S. alba* and *B. juncea* allowed seed setting at temperatures >38 °C (Kumari et al. 2018). Heat tolerance during germination was observed in *B. carinata* (PC-6), *B. tournefortii* (RBT 2002), *B. tournefortii* (Rawa), *D. gomez-campoi*, and *L. sativum* (Kashyap et al. 2023). Additionally, *B. napus* (GSC-6) showed both heat and drought tolerance. Among *Brassica* species, allotetraploids like *B. juncea*, *B. napus*, and *B. carinata* have higher salinity tolerance than their diploid parents (Ashraf et al. 2001; Kumar et al. 2009), with significant variability observed in intraspecific *B. napus* and *B. juncea* under hydroponic conditions (Yong et al. 2015; Singh et al. 2022). *B. juncea* var. *foliosa* has phytoremediation potential for thorium-contaminated soils (Zhou et al. 2016), and *B. rapa* subsp. *pekinensis* (Chinese cabbage) can accumulate heavy metals without showing symptoms (Xiong and Wang 2005).

Soybean was domesticated over 5000 years ago in China from its wild relative *Glycine soja* (Hymowitz 1970). Wild soybean shows greater genetic diversity and adaptability to abiotic stresses such as drought, salt stress, and alkalinity, making it valuable for improvement programs. *Glycine soja* readily crosses with cultivated soybean, producing fertile hybrids. The genus *Glycine* includes perennial species like *G. canescens*, *G. clandestina*, *G. latifolia*, and others, while the subgenus *Soja* includes *G. max* and its wild relative *G. soja* (Kim et al. 2010; Sherman Broyles et al. 2014; Zhou et al. 2015). *G. canescens* and *G. latifolia* have shown drought tolerance in studies, with lower epidermal conductance, higher relative water content, and enhanced osmotic adjustment compared to *G. max* genotypes (James et al. 2008). The *G. soja* genotype PI407155 also exhibited greater water use efficiency than *G. max* 'Essex' (Chen et al. 2006). Soybean CWRs are notable for salt tolerance, with *G. soja* exhibiting higher leaf tolerance to Cl^- toxicity, although it is susceptible to Na^+ accumulation (Luo et al. 2005). Salt tolerance in *G. soja* is due to mechanisms like sodium ion compartmentalization, cell membrane modifications, compatible solute synthesis, and enhanced antioxidant enzyme levels (Lu et al. 2008). Additionally, the wild soybean genotype Tongyu06311 shows salt tolerance through organic acid and amino acid metabolism with compatible solute accumulation (Yang et al. 2017).

Groundnut (*Arachis hypogaea* L.), an allotetraploid legume ($2n = 4x = 40$), is an oilseed crop that originated through hybridization of *A. duranensis* (AA) and *A. ipaensis* (BB), carrying wide genetic diversity (Bertioli et al. 2016). Its genome size (~2.7 Gb) is similar to the combined genomes of its wild progenitors (*A. duranensis* 1.25 Gb and *A. ipaensis* 1.56 Gb) (Samoluk et al. 2015). However, utilizing the genetic diversity of wild relatives is challenging due to ploidy differences and sterility barriers. Groundnuts face abiotic stress challenges, including drought, heat, and salinity. Heat shock factors regulate stress-responsive genes, such as heat shock proteins, dehydration-responsive element-binding proteins (DREBs), and abscisic acid-responsive element-binding proteins, essential for stress tolerance. DREB genes, particularly *PNDREB1*, are linked to improved transpiration efficiency, although their response to salinity remains unclear (Kapazoglou et al. 2023). Ethylene-responsive factor genes activate ABA and regulate drought, heat, cold, and salinity responses. Genes like *AhLea-3* and *AtNHX1* enhance salt tolerance, and overexpression of AtHDG11 improves both drought and salt tolerance (Asif et al. 2011; Kapazoglou et al. 2023). Wild groundnut species exhibit drought tolerance via two gene groups: (i) cell protectors and upstream regulators and (ii) regulatory genes like transcription factors (TFs) in the ABA pathway (Cason et al. 2020). Drought-responsive genes have been identified in species like *A. stenosperma* and *A. duranensis*, with traits such as reduced transpiration and high photosynthesis (Guimarães et al. 2012). Key drought-responsive candidate genes include expansin, nitrilase, NAC, and bZIP TFs (Gai et al. 2022). Additionally, wild genotypes such as *A. glabrata* and *A. paraguariensis* exhibit heat and cold tolerance, respectively (Nautiyal et al. 2008), with heat shock TFs identified in *A. duranensis* and *A. ipaensis* (Wang et al. 2017a). Nonspecific lipid transfer proteins in *A. duranensis* respond to salinity and cold stress (Song et al. 2020).

Sunflower is the second largest hybrid crop, after *Zea mays* L., and the fifth largest oilseed crop having an estimated value of US$20 billion globally (FAO 2016), contributing 10% of the world's edible plant-derived oil (Kleingartner 2004). Evidence of domestication was found in 4625 BP from samples at the Hayes site in middle Tennessee (Crites 1993), whereas molecular studies confirm that the Eastern United States (Mississippi River Valley) was the only independent origin for the domestication of cultivated sunflower (Blackman et al. 2011). Sunflower is severely affected by abiotic stresses like drought, salt stress, and waterlogging. Promising accessions have been identified in sunflowers during the last decade like AKSF-42-1, M-1026, 298-R, GMU-351, and hybrids like Laxmi-225, CO-2, CSFH-12205, DRSH-1, and KBSH-44 for drought tolerance (Yadav et al. 2022) as well as KBSH-44, Jwalamukhi (PSCL-5015), MLSFH-47 (AH-11-34), KBSH-41, KBSH-42, DRSF-108, PRORUN-09, Bhanu, KBSH-53, and Bhaskar for waterlogging (Boraiah et al. 2023).

Sunflower CWRs offer significant variability for combating abiotic stress (Ortiz 2015). Prescott-Allen and Prescott-Allen (1986) estimated that sunflower CWR contributes 25.9% to the crop's annual value, worth $185 million (Tyack and Dempewolf 2015). Key discoveries include the PET1 cytoplasm from *H. petiolaris*, cytoplasmic male sterility in wild prairie sunflower (*Helianthus petiolaris*) (Leclercq 1969), and restorer genes with recessive branching in *H. annuus* and *H. petiolaris* (Kinman 1970). The *Helianthus* genus is divided into four sections: annuals (*Helianthus*, *Agrestis*) and perennials (*Ciliares*, *Divaricati*) (Timme et al. 2007). The primary gene pool includes cultivated and wild species of *H. annuus* and winter sunflower (*H. winteri*), while the secondary pool includes *H. anomalus*, *H. paradoxus*, *H. petiolaris*, and *H. deserticola*, and the tertiary pool contains *H. hirsutus*, *H. tuberosus*, and *H. divaricatus* (Stebbins et al. 2013). Many *Helianthus* species thrive in salt-affected habitats, with *H. paradoxus* being a key candidate for salt tolerance due to its ability to dominate salt marshes (Seiler et al. 1981). Studies by Chandler and Jan (1984) found that *H. paradoxus* exhibited the highest salt tolerance, surviving at 1300 mM NaCl, while *H. debilis* showed salt tolerance similar to cultivated sunflower. Further evaluations (Miller 1995; Welch and Rieseberg 2002) demonstrated that *H. paradoxus* is five times more salt-tolerant than its progenitors. CDPK3, a calcium-dependent protein kinase in sunflower, maps to a salt tolerance

QTL on LG4 (Lexer et al. 2004), and *H. paradoxus* shows gene expression related to potassium and calcium transport for salinity adaptation (Edelist et al. 2009). *H. anomalus* is known for its drought tolerance and high oil content, making it valuable for breeding (Seiler et al. 2006). It has been used in breeding for improved water use efficiency and drought resilience (Baldini et al. 1993; Baldini and Vannozzi 1998). Other species like *H. deserticola*, *H. hirsutus*, *H. maximiliani*, and *H. tuberosus* are also recommended for drought-tolerant breeding (Skoric 2009).

Safflower (*Carthamus tinctorius* L.) is the only cultivated species in the *Carthamus* genus, which includes 25 species. Vavilov (1951) identified three centers of origin for cultivated safflower: India, Afghanistan, and Ethiopia. *C. palaestinus* and *C. oxyacanthus* are wild relatives of safflower that are crossable with the cultivated species (Ashri and Knowles 1960). Exploring genetic variation from both wild and cultivated types can help evaluate genetic diversity for abiotic stress tolerance in safflower. Drought is one of the most significant abiotic factors affecting safflower production. Gene introgression from wild relatives into the cultivated gene pool has enhanced genetic variation, enabling the identification of drought-tolerant genotypes with specific fatty acid profiles in an evaluation study of three interspecific crosses: *C. tinctorius* × *C. palaestinus*, *C. oxyacanthus* × *C. palaestinus*, and *C. oxyacanthus* × *C. palaestinus*. Crosses between *C. tinctorius* and *C. palaestinus* or *C. oxyacanthus* have the potential to improve drought tolerance without negatively affecting the resulting interspecific progenies (Shafiei Koij et al. 2019). *Carthamus oxyacantha* outperformed cultivated safflower in terms of drought tolerance in a comparative study (Bassiri et al. 1977). The high drought tolerance of this wild safflower species makes it a promising genetic resource for introgression breeding programs aimed at transferring tolerance genes to cultivated species (Majidi et al. 2011). While wild safflower species exhibit considerable genetic variation for adaptability and tolerance to abiotic factors such as salinity, transferring traits of interest from wild to cultivated species is constrained by chromosomal differences.

2.2.2.2 Current Status of CWR Usage for Abiotic Stress Tolerance in Oilseed Crops

In 2016, Brozynska et al. reviewed sequencing efforts for CWRs in oilseed crops, comparing them with their domesticated counterparts. Crop wild relatives exhibited higher heterozygosity, which may complicate genome assembly. Recently, the genomes of two wild diploid perennial *Brassica* species with C-genomes, *Brassica macrocarpa* and *Brassica cretica* from Greece, were sequenced (Golicz et al. 2016; Kioukis et al. 2020). These species, potential wild ancestors of *B. oleracea*, were assembled using *B. oleracea* as a reference genome (Branca et al. 2012). Expanding pangenomes and super-pangenomes will provide valuable genetic resources for crop improvement. Future research in *Brassica* genomics is expected to focus on whole-genome re-sequencing and pangenome studies to explore breeding histories and identify genes linked to abiotic stresses (Mohd Saad et al. 2021).

The availability of reference genomes in soybean (Haun et al. 2011) has facilitated the mining of economically important trait gene pools by generating draft assemblies of wild relatives (Sripathi et al. 2016). The development of SoySNP50K iSelect Illumina BeadChip and 50 K genome-wide SNPs has further promoted the study of wild relatives in soybean (Song et al. 2013). These genomic resources enable the characterization of genome content and structural variations (SVs) in CWRs, supporting large-scale evolutionary studies and pangenome comparisons. Pangenome studies in soybean have identified lineage-specific copy number variants that allow for the positive selection of stress tolerance and related traits (Li et al. 2014). Wild soybean accession PI483463 has been shown to contain a dominant gene for salt tolerance (Lee et al. 2009). Overexpression of the gene *GsWRKY20*, which encodes a WRKY-type TF from wild soybean (*Glycine soja*), enhanced salt and drought tolerance in transformed alfalfa (*Medicago sativa*). The transformed alfalfa exhibited lower relative membrane permeability, reduced MDA content, increased osmo-protectants, lower Na^+ accumulation, and higher K^+, along with morphological changes such as a thicker cuticle under salt treatments (Tang et al. 2014). Similarly, overexpression of the *GsJAZ2* gene, a novel JAZ family gene involved in jasmonate signaling, from *G. soja*, enhanced salt and alkalinity tolerance in *Arabidopsis*. *GsJAZ2* is a positive regulator for tolerance, as shown in GFP fusion protein studies (Zhu et al. 2012). Other genes such as *GmCHX1*, *GmSALT3*, *Ncl2*, *GmTIP2*, and *GsGST* from *G. soja* have also been demonstrated to be involved in salt tolerance (Guan et al. 2014; Ji et al. 2010; Lee et al. 2009; Qi et al. 2014; Zhang et al. 2017).

Genes and TFs in wild *Arachis* species often respond to multiple abiotic stresses. In *A. diogoi*, the gene *AdDjSKI* is induced by heat, salinity, drought, and osmotic stress and is linked to photosynthesis (Rampuria et al. 2018). NAC TFs involved in salt and drought responses were identified in *A. hypogaea* and its progenitors (Yuan et al. 2020). Valine-glutamine sequences related to environmental changes were found in *A. duranensis*, *A. ipaensis*, and *A. monticola*. Additionally, mTERF family genes, which aid plant acclimation to stress, were identified in *A. duranensis* and *A. ipaensis*. *A. glabrata* showed various TFs responding to drought, salt, and cold stress (Li et al. 2020). Transcriptomic analyses of wild and adaptable *A. glabrata* genotypes revealed several TFs expressed under drought, cold, and salt stress conditions (Zhao et al. 2021). In *A. stenosperma*, 4513 differentially expressed genes were identified under UV and dehydration stress, related to signaling and metabolism (Martins et al. 2022).

The NCBI database contains nucleotide sequences and expressed sequence tags for *Helianthus* species, covering 98% of the data. *Helianthus* is ideal for utilizing CWRs in studying abiotic stress tolerance. Baute (2015) genotyped 406 introgressed lines, including 351 stable and fertile lines from 11 CWR species. Recent efforts focus on generating extensive genomic sequences and assembling annotated genomes for *H. argophyllus*, *H. petiolaris*, and *H. maximiliani* (Kane et al. 2013). DNA sequence diversity in both cultivated and wild sunflowers has been analyzed using molecular markers (Mandel et al. 2011). Crosses between CWR species like *H. argophyllus*, *H. annuus*, and *H. petiolaris* with cultivated sunflower have increased molecular marker diversity and chromosomal rearrangements (Sujatha et al. 2008).

Wild safflower possesses valuable genes for stress tolerance, revealed through functional markers like ISSR (Sabzalian et al. 2009). A set of 75 RFLP- and 1412 PCR-based markers, developed from cDNA and genomic libraries, were mapped in *Carthamus tinctorius* and *Carthamus oxyacanthus* populations, representing the first major linkage analysis of *Carthamus* species (Mayerhofer et al. 2010). Table 2.2 provides some examples of how wild relatives have been utilized to enhance abiotic stress tolerance in oilseed crops.

2.2.3 CWRs in Pulses

2.2.3.1 Historical Insights into the Role of CWRs in the Improvement of Pulses

Pulses, including soybean, peanut, pea, chickpea, lentil, cowpea, lupin, and pigeon pea, have been grown alongside wheat and barley since the beginning of agriculture in the Near East (Kislev and Bar-Yosef 1988). Lentils, one of the oldest domesticated crops, were cultivated around 8,000 to 7,000 B.C.E. in the Fertile Crescent (modern Syria/Turkey) (Ljuština and Mikić 2010). Chickpeas were domesticated from *C. reticulatum* in southern Turkey and Syria (Kerem et al. 2007). In eastern India, *Cajanus cajanifolius* was once widespread but is now rare due to habitat loss (Fuller and Harvey 2006). Pea (*Pisum sativum*) was first domesticated in the Mediterranean about 10,000 years ago, leading to the development of numerous varieties (Daniel and Maria 2000). Furthermore, the genus *Cicer* comprises 9 annual and 35 perennial species. According to Harlan and de Wet (1971), *C. arietinum* is the only domesticated species among these 44.

The primary gene pool includes the cultivated chickpea, *C. arietinum*, and its wild relative, *C. reticulatum*, which can be easily crossed with cultivated chickpea due to their typical breeding characteristics. *C. echinospermum* forms the secondary gene pool. Although it can be crossed with cultivated species, the resulting hybrids and their improved offspring exhibit lower fertility. The remaining 6 annual and 35 perennial *Cicer* species constitute the tertiary gene pool. These species are difficult to cross with cultivated chickpea using traditional breeding methods, necessitating the use of modern techniques for gene transfer.

Chickpea CWRs provide significant genetic variation for incorporating desired traits, especially for abiotic stress tolerance. Chickpeas face multiple abiotic stresses, including drought, heat, cold, and salinity, which reduce production worldwide (Maphosa et al. 2020). Advances in genomics have clarified the genetic architecture of both wild and cultivated chickpeas, enabling the use of genomics-assisted breeding techniques like QTL mapping, GWAS, and QTL-seq (Varshney et al. 2021). These techniques aid in developing climate-resilient chickpea varieties by identifying donor sources for stress tolerance. Moreover, genomics facilitates the discovery of new genetic resources from wild plant populations and assists in their conservation (Brozynska et al. 2014).

Chickpea is traditionally grown in the spring across the Mediterranean region, but winter sowing can increase seed production due to a longer growing season and

Table 2.2 Examples of wild relatives used to enhance abiotic stress tolerance in oilseed crops

Crop	CWRs	Stress targeted	References
Brassica spp.	*Brassica tournefortii*	Drought and heat	Kashyap et al. (2023)
	Brassica fruticulosa	Drought	Kashyap et al. (2023)
	Camelina sativa		Purnamasari et al. (2019)
	Crambe abyssinica		Samarappuli et al. (2020)
	Capsella bursa-pastoris		Chen et al. (2007)
	Sinapis alba		Kumari et al. (2018)
	Lepidium sativum	Heat	Yadav et al. (2019)
Soybean	*Glycine soja*	Drought and salinity	Qi et al. (2014), Luo et al. (2005), and Lu et al. (2008)
	Glycine latifolia	Drought	James et al. (2008)
	Glycine canescens		
Groundnut	*Arachis dardani*	Drought	Cason (2008)
	Arachis diogoi		Rampuria et al. (2018)
	Arachis duranensis		Vinson et al. (2018); Song et al. (2020)
	Arachis glabrata		Zhao et al. (2021)
	Arachis magna		Brasileiro et al. (2015)
	Arachis diogoi	Heat	Rampuria et al. (2018)
	Arachis duranensis		Wang et al. (2017a)
	Arachis glabrata		Nautiyal et al. (2008)
	Arachis duranensis	Cold	Yuan et al. (2020)
	Arachis glabrata		Zhao et al. (2021)
	Arachis paraguariensis		Nautiyal et al. (2008)
	Arachis diogoi	Salinity	Rampuria et al. (2018)
	Arachis duranensis		Song et al. (2020)
	Arachis glabrata		Yuan et al. (2020)
	Arachis stenosperma	UV-light	Martins et al. (2022)
Sunflower	*Helianthus argophyllus*	Drought and heat	Hussain et al. (2016)
	Helianthus paradoxus	Salinity	Miller and Seiler (2003) and Lexer et al. (2003)
	Helianthus anomalus	Drought	Ludwig et al. (2004)
	Helianthus debilis Nutt. and *Helianthus divaricatus* L.	Multiple abiotic stresses	Kantar et al. (2015)
Safflower	*Carthamus oxyacanthus*	Drought	Majidi et al. (2011)
	Carthamus palaestinus		Shafiei Koij et al. (2019)
	Carthamus lanatus		Anjani et al. (2022)
	Carthamus turkestanicus		Anjani et al. (2022)

better moisture availability. However, freezing temperatures during winter can damage crops, particularly during flowering. An analysis of eight annual wild *Cicer* species found that *C. bijugum*, *C. reticulatum*, *C. echinospermum*, and *C. pinnatifidum* showed significant cold resistance, with some accessions scoring high for cold tolerance. In contrast, *C. chorassanicum*, *C. cuneatum*, and *C. yamashitae* were more vulnerable to cold stress. Soil salinity, a major threat to chickpea production, impacts various aspects of crop growth, and increasing salinity tolerance is crucial for agricultural productivity. Recent advances in RNA sequencing have enabled better profiling of stress-responsive genes, improving our understanding of how chickpea regulates transcription and translation under salt stress (Conesa et al. 2016; Shamloo-Dashtpagerdi et al. 2024). For instance, targeting NHX genes for modification could produce stress-tolerant crops, which are essential for improving agriculture (Parveen et al. 2023). Additionally, the introgression of the *CAMTA* gene into transgenic chickpea has shown enhanced photosynthetic activity under salinity stress.

The genus *Lens* includes seven species across four gene pools: the primary gene pool contains *L. culinaris*, *L. orientalis*, and *L. tomentosus*, and the secondary gene pool includes *L. odemensis* and *L. lamottei*, while *L. nigricans* and *L. ervoides* are in the tertiary and quaternary pools, respectively. *L. culinaris* ssp. *orientalis* is the closest wild progenitor of *L. culinaris* ssp. *culinaris*, while *L. nigricans* is the most distantly related. Viable hybrids have been reported between *L. culinaris* ssp. *orientalis* and *L. odemensis* (Ladizinsky et al. 1984). The genus is divided into two groups based on cross-ability: *L. culinaris* and *L. odemensis* form one group, while *L. ervoides* and *L. nigricans* form the other. Wild relatives of *L. culinaris* ssp. *orientalis*, *L. odemensis*, and *L. ervoides* have been used for crop improvement (Tripathi et al. 2024; Koul et al. 2017; Reddy et al. 2010; Wong et al. 2015).

Using 35 SSR markers, 258 alleles were identified across 278 genotypes, with genetic diversity ranging from 0.321 to 0.854 and polymorphism information contents from 0.299 to 0.836. Drought-tolerant genotypes clustered in group 6, while sensitive ones were in group 7 (Singh et al. 2016). Wild accessions showed higher allelic variation (151 alleles) than cultigens (114 alleles) for drought stress, with mean genetic diversity of 0.66 and 0.65, respectively (Hamwieh et al. 2009). Differential gene expression analysis of drought-treated genotypes identified 11,435 upregulated and 6934 downregulated transcripts, with 413 GO terms categorized into molecular function, cellular components, and biological processes (Singh et al. 2017b). Idrissi et al. (2016) evaluated the genetic diversity of 70 Mediterranean lentil landraces using 19 SSR markers and 7 AFLP primer combinations, assessing root and shoot traits and drought tolerance.

De novo transcriptome analysis of both genotypes identified 90,267–104,424 contigs with N50 values between 1755 and 1844 bp under heat stress and control conditions. The assembly yielded 194,178 high-quality SNPs, 141,050 microsatellites, and 7388 indels. Quantitative RT-PCR assessed the expression of ten genes (Singh et al. 2019). A total of 209,549 assembled transcripts were identified, with 161,809 containing coding regions, and 94,437 annotated. Differential gene expression analysis revealed 678 upregulated and 680 downregulated transcripts between tolerant and sensitive genotypes at the early reproductive stage, with 76 upregulated and 47 downregulated at the late reproductive stage under heat stress. The analysis identified 6852 genic SSRs and developed 4968 SSR primers for molecular mapping of heat-tolerant genes in lentils (Kumar et al. 2021). Seven miRNAs from six conserved families (miR408, miR172, miR169, miR167, miR166, and miR156) were identified, with miR408, miR169, and miR166 downregulated and miR156 upregulated in response to heat stress (Hosseini et al. 2021). Further, a GWAS using a mixed linear model found significant marker-trait associations on chromosomes 2 and 4, with candidate genes linked to potassium transporters associated with salt tolerance in related species (Dissanayake et al. 2021). Research on pea CWRs, including *Pisum sativum* subsp. *elatius*, *Pisum abyssinicum*, *Pisum fulvum*, and *P. sativum* subsp. *sativum*, has focused on their diversity and genetic relationships with cultivated lineages (Schreiber et al. 2019).

2.2.3.2 Current Status of CWR Usage for Abiotic Stress Tolerance in Pulses

Among the perennial accessions of *C. anatolicum*, *C. microphyllum*, *C. montbretii*, *C. oxydon*, *C. macracanthum*, *C. isauricum*, and *C. songaricum*, several exhibit drought stress resistance (Toker et al. 2007). Singh et al. (2021) reported that various annual *Cicer* species, including *C. bijugum*, *C. echinospermum*, *C. judaicum*, *C. pinnatifidum*, and *C. reticulatum*, as well as perennials like *C. microphyllum*, exhibit resistance to cold and chilling stress. A perennial accession of *C. isauricum* was found to withstand terminal drought in late summer and heat stress of 45 °C during the podding stage (Tekin et al. 2018). Srivastava et al. (2016) noted that *C. microphyllum* can tolerate saline stress, while annual wild accessions such as *C. reticulatum* and *C. pinnatifidum* can endure heat stress around 41.8 °C. Many wild *Cicer* species, particularly *C. reticulatum*, *C. echinospermum*, *C. bijugum*, and *C. pinnatifidum*, exhibit early flowering and maturity, enabling them to evade drought stress. Despite challenges in crossing, *C. echinospermum* has shown cold tolerance and increased yield in interspecific crosses (Knights et al. 2008). A crossable wild annual *C. reticulatum* (PI 489777) was used as a tolerant parent and *C. arietinum* (ICC 4958) as a sensitive parent to map QTLs for cold tolerance (Mugabe et al. 2019). Segregating populations from advanced backcrosses of *ICCV 95311* × *C. reticulatum* and *ICCV 95311* × *C. echinospermum* showed significant variation, with introgression lines exhibiting early flowering and other desirable traits (Choudhary et al. 2018). Chickpea lines from *ICCV 2* × *JG 11* showed two QTLs for salt tolerance on LGs05 and LGs07 (Pushpavalli et al. 2015).

Pigeon pea CWRs offer valuable genetic resources and are believed to have higher abiotic stress tolerance than the cultivated crop. The genus *Cajanus* consists of 3 clades—Indian, Australian, and Scarabaeoides—comprising 32–34 taxa. *Cajanus cajanifolius* is part of the secondary gene pool, and *C. acutifolius*, considered the crop's ancestor in the Indian clade, is a key species (Kassa et al. 2012). Wild relatives such as *C. latisepalus*, *C. cinereus*, *C. acutifolius*, *C. lanceolatus*, and *C. cajanifolius* are adapted to high temperatures, while *C. mollis*, *C. trinervius*, *C. confertiflorus*, and *C. platycarpus* are adapted to low temperatures. Despite its potential for crop diversification and climate resilience, pigeon pea has yet to undergo gene introgression studies, particularly for heat stress tolerance breeding.

Traditional breeding efforts have leveraged genomic resources to identify potential drought-tolerant genes, with 10 key genes related to U-box proteins, cation/H^+ antiporters, uncharacterized proteins, and universal stress protein A-like proteins found through expression analysis of 51 drought-responsive genes (Sinha et al. 2016). Comparative transcriptome analysis revealed significant species-specific variations in heat stress responses, with *C. cajanifolius*, *C. cajan*, *C. acutifolius*, and *C. scarabaeoides* showing thermotolerance. Gene expression profiling, histochemical assays, electrolyte leakage, and chlorophyll content analysis confirmed that *C. scarabaeoides* has adaptive heat tolerance traits. Time-series transcriptome analysis further explored the molecular basis of basal heat tolerance in *Cajanus* species (Ramakrishna et al. 2021). Additionally, soil salinization, common in dryland and irrigated agriculture, impedes crop development. Proteomic studies have revealed how post-translational modifications of stress-induced proteins enable plant adaptation to abiotic stresses. Wild relatives like *C. scarabaeoides*, *C. albicans*, and *C. platycarpus* exhibit a wide range of salinity tolerance, contributing to the diversity of pigeon pea's stress resilience. Similarly, *C. platycarpus*, *C. scarabaeoides*, and *C. sericeus* showed good salinity tolerance, aided by high levels of proline, amino acids, carbohydrates, and potassium ions. Genes providing salt tolerance in *C. albicans* can be introgressed into cultivated varieties (Choudary et al. 2011).

Previous studies have highlighted significant genetic diversity for key traits among lentil landraces and wild relatives, emphasizing the need for innovative breeding tools to incorporate this diversity into modern cultivars (Rajpal et al. 2023). The ICARDA gene bank holds 14,500 lentil accessions (Sharma 2019). The specific adaptations of cultivated lentils limit the direct use of exotic landraces in breeding programs. South Asian landraces are typically early-maturing, small-seeded, and red, while West Asian landraces are late-maturing, large-seeded, and yellow. Genes for yield traits like 100-seed weight and pods per plant have been identified in *Lens lamottei* and *Lens culinaris* ssp. *orientalis* (Ferguson et al. 1998; Singh et al. 2014). Recent advances include hybridization with *L. ervoides* via embryo rescue, introducing traits like disease resistance and stress tolerance (Tullu et al. 2013). Screening for heat stress tolerance in crop gene pools, combined with phenomic tools and NGS, can identify HS-tolerant genotypes. Techniques like genomic selection, GWAS, and MAS, as well as transgenic approaches, hold promise for developing heat-tolerant pea varieties (Devi et al. 2023). Table 2.3 provides examples of how wild relatives have been utilized to enhance abiotic stress tolerance in pulses.

2.3 Conservation of CWRs and Accessibility to Crop Researchers

Crop wild relatives hold immense genetic potential for abiotic stress adaptation but remain underutilized due to incomplete species documentation and inefficient management and monitoring.

Factors such as the replacement of landraces with high-yielding varieties, climate change, excessive fertilizer use, desertification, and overgrazing have led to genetic erosion, disrupting CWR diversity. Preserving, protecting, and effectively managing CWRs is crucial. Conservation efforts began in the 1970s at both national

Table 2.3 Examples of wild relatives used to enhance abiotic stress tolerance in pulses

Crop	CWRs	Stress targeted	References
Chickpea	*Cicer judaicum*	Cold and drought	Berger et al. (2012) Robertson et al. (1997)
	Cicer bijugum	Cold	Toker (2005)
	Cicer reticulatum, *Cicer echinospermum*, *Cicer reticulatum* Ladiz	Cold	Singh and Ocampo (1997); Mugabe et al. (2019)
	Cicer reticulatum	Drought, heat, and salinity	Canci and Toker (2009) Maliro et al. (2008)
	Cicer pinnatifidum	Drought and heat	Canci and Toker (2009)
	Cicer reticulatum	Drought	Moenga et al. (2020)
Alfalfa	*Medicago truncatula* and *Medicago ruthenica*	Drought, salt, and cold	Wang et al. (2021)
Cowpea	*Vigna trilobata* and *Vigna riukiensis*	Drought	Naito et al. (2022)
	Vigna stipulacea and *Vigna hainiana*	Heat	Van Zonneveld et al. (2020)
	Vigna luteola, *Vigna marina*, *Vigna vexillata*	Salt	Yoshida et al. (2020)
Groundnut	*Arachis diogoi* and *Arachis glabrata*	Drought	Zhao et al. (2021)
	Arachis diogoi	Heat	Rampuria et al. (2018)
	Arachis duranensis	Cold	Yuan et al. (2020)
	Arachis diogoi	Salinity	Rampuria et al. (2018)
Pigeon pea	*Cajanus cajanifolius*, *Cajanus scarabaeoides*	Heat	Khoury et al. (2015)
	Cajanus mollis	Cold	Khoury et al. (2015)
	Cajanus lineatus	Waterlogging and drought	Khoury et al. (2015)
	Cajanus acutifolius	Heat and drought	Khoury et al. (2015)
	Cajanus sericeus	Salinity	Srivastava et al. (2006)
	Cajanus platycarpus	Salinity	Srivastava et al. (2006)
	Cajanus albicans	Salinity	Subbarao et al. (1990)
Soybean	Wild accession W05	Salinity	Qi et al. (2014)
	Glycine soja, *Glycine tomentella*	Aluminum toxicity	Zeng et al. (2012)

(continued)

Table 2.3 (continued)

Crop	CWRs	Stress targeted	References
Lentil	*Lens culinaris*, *Lens nigricans*	Heat	Kumar et al. (2015); Gorim and Vandenberg (2017)
	Lens culinaris, *Lens orientalis*	Cold and salinity	Singh et al. (2017a)
	Lens odemensis, *Lens ervoides*	Drought	Omar et al. (2019); Gorim and Vandenberg (2017)

and international levels, leading to the establishment of the International Board for Plant Genetic Resources in 1975 (Engels and Thormann 2020). These efforts gained prominence with the inclusion of CWRs in the Aichi Biodiversity Targets and the Convention on Biological Diversity (CBD 2002). Today, conservation activities are integrated into the national missions of several countries.

There are two main strategies for conserving CWRs: in situ and ex situ. In situ conservation involves preserving CWRs in their natural habitats, including national parks, biosphere reserves, on-farm conservation areas, and gene sanctuaries (Kumar et al. 2019). Several countries have implemented in situ conservation initiatives. For instance, maize CWRs are conserved in Mexico's biosphere reserves, legumes and cereals in Turkey, and citrus and lychee relatives in Vietnam (FAO 2010). In 1981, India established a sanctuary for wild citrus relatives in the northeastern region (Hunter and Heywood 2011). More recently, Germany became the first European country to create a genetic reserve for four celery-related species (Frese et al. 2018). In situ conservation promotes genetic diversity by enabling natural crossing between related species. However, rapid environmental changes pose a significant challenge, as wild crops struggle to evolve quickly enough to adapt to such shifts. Another strategy for conserving CWRs is ex situ conservation, with gene banks serving as a cornerstone of this approach since its inception. Ex situ conservation involves protecting CWRs outside their native habitats using methods such as seed banks, field banks, DNA banks, and pollen banks (Kumar et al. 2019).

Seed gene banks are particularly effective, as most plants produce seeds that can tolerate the cold temperatures necessary for long-term storage. The largest seed bank globally is the Svalbard Global Seed Vault in the Arctic, governed by the Norwegian government (Asdal and Guarino 2018). India houses the Indian Seed Vault at Chang La Pass in Ladakh, the world's second largest seed bank (Ray and Bordolui 2021). Field banks, on the other hand, conserve crops or CWRs outside their natural habitats in dedicated areas. Botanical gardens also play a vital role in conserving CWRs through seeds, plants, or tissue cultures, with approximately 30% of global plant diversity preserved in such gardens (Mounce et al. 2017). Notable examples include the Kew Millennium Seed Bank in the United Kingdom, the RBG Sydney PlantBank, and China's Germplasm Bank of Wild Species (O'Donnell and Sharrock 2017).

DNA banks store complete genetic information of various CWRs, aiding in plant variation studies and characterization (Ebert et al. 2006). For species producing orthodox seeds or seeds with limited viability, pollen banks serve as a compact and efficient conservation method. Several databases support ex situ conservation efforts, such as Genesys, which documents four million species, with only 12% representing CWRs. The European Search Catalogue for Plant Genetic Resources includes 12.5% CWRs (Engels and Thormann 2020).

A complementary conservation approach combines in situ and ex situ strategies with careful planning to address gaps. While less tested, this method requires a comprehensive understanding of species biology, prioritization of CWRs, and assessment of threat levels. For instance, a study on tropical fruit trees concluded that both in situ and ex situ methods are essential for efficient conservation (Hunter and Changtragoon 2016). Despite various initiatives, the representation of CWRs in conservation systems remains low. A global study by Castaneda-Alvarez et al. (2016) found that of 1076 CWR-identified taxa, only 70.9% were included in gene banks, with 257 taxa represented by fewer than 10 accessions each.

To improve CWR conservation, it is crucial to document species comprehensively, assess threat statuses, and legalize or license their use. Adopting suitable conservation strategies and fostering collaborations at national, regional, and international levels are essential for effective management and preservation. The accessibility of CWRs to researchers is crucial for enhancing crop resilience to climate change. However, challenges such as limited ex situ collections, conservation gaps, and uncoordinated research efforts hinder their availability to breeders. Many CWRs are underrepresented in gene banks, which are vital for breeding programs (Maxted and Vincent 2021). Conservation gaps are exacerbated by climate change, habitat destruction, and other anthropogenic factors, further limiting their availability. Additionally, inadequate facilities and improper conservation practices hinder ex situ efforts, emphasizing the need for specialized protocols to improve utilization and conservation (Satori et al. 2021). While national seed collection programs have begun generating valuable characterization data, improved collaboration and systemic communication among researchers are essential to enhance access and utility (Eastwood et al. 2022).

2.4 Challenges and Future Directions for Use of CWRs in Crop Improvement

The potential of wild species as a source of novel genetic variation for crop improvement has been recognized since the early twentieth century. However, persistent challenges, including relatedness, compatibility, and crossability with cultivated counterparts, limit the production of fertile offspring. These barriers hinder the efficient integration of beneficial traits from CWRs into elite germplasm through conventional breeding approaches. For instance, in cotton, highly resistant sources against reniform nematode have been identified in wild diploid species such as *Gossypium somalense*, *G. longicalyx*, and *G. stocksii*. However, genetic incompatibility and ploidy differences hinder their use in breeding (Tirnaz et al. 2022). Environmental variations also influence trait selection, as traits beneficial in domesticated regions may not be advantageous in the wild. Moreover, introgressing wild-derived traits often causes linkage drag, negatively affecting yield and agronomic traits, as seen with efforts to incorporate disease resistance (Tirnaz et al. 2022).

The CWR species are adapted to survive in natural environments, serving as invaluable reservoirs of genetic diversity for plant breeders. This diversity evolves in nature under changing environmental conditions, a process halted when conserved in gene banks. Thus, conservation strategies must balance in situ methods with ex situ methods or combine both. In situ conservation offers advantages by maintaining genetic dynamism and facilitating continuous evolution under natural and human-driven selection pressures, enhancing resilience. However, it faces challenges like limited biological information, habitat destruction from civil strife, human settlement, and natural disasters. Ex situ conservation complements these efforts but is complex, as maintaining CWRs outside their natural adaptation zones can erode genetic variation over time. Ensuring optimal ecological conditions is critical for healthy seed and plant material propagation, highlighting the need for expertise in ex situ management.

Plant breeders often favor elite lines over CWRs due to the lower agronomic performance of elite × CWR progenies, which limits immediate genetic gains. However, pre-breeding strategies using DNA marker technology and QTL mapping have facilitated the efficient introgression of beneficial alleles from CWRs into elite lines. Chromosome segment substitution lines (CSSLs) are a successful tool in this regard, incorporating donor segments from CWRs into elite genomes to dissect complex traits and identify QTLs. Despite their value, the time and resources needed for CSSL development restrict their widespread use. To address these challenges, fast generation cycling systems using in vitro embryo culture can accelerate segregating generations from elite × CWR crosses. Alternatively, speed breeding, which manipulates photoperiod and temperature, offers a simpler, less labor-intensive method. Since genetic resources derived from a single CWR accession offer limited diversity, incorporating CWRs as founder parents in multi-parent breeding designs, such as multi-parent advanced generation intercross, can broaden genetic variation.

Advances in sequencing technologies have enabled the development of high-quality reference genomes for many crops, including CWRs like barley, rice, soybean, tomato, and wheat (Brozynska et al. 2016; Bohra et al. 2022). Sequencing and de novo assembly of CWR genomes facilitate gene characterization and targeted genome editing, significantly enhancing our ability to identify valuable genes/QTLs. Precise genome editing allows for the direct introduction of beneficial wild-derived genes into elite varieties, bypassing introgression challenges like linkage drag and reduced fertility (Bohra et al. 2022). As sequencing costs decrease and accuracy improves, the notion that a single reference genome can fully capture genetic diversity within a species has been challenged.

The increasing availability of high-quality genome sequencing data from diverse accessions (pangenomics) has highlighted the limitations of reference genome-based approaches in detecting SVs such as the presence-absence variation and copy number variation, which differentiate accessions. For breeders, this means that genomic segments introgressed from wild or exotic donors in CSSLs may carry genes absent from both the reference genome and the elite parent or vice versa. Pangenome studies in various crops, including rice, have revealed the significant role of SVs in crop domestication and breeding. For instance, a pangenome analysis of 66 rice accessions identified extensive presence-absence variations in genes linked to flowering time and other traits (Zhao et al. 2018). These findings underscore the potential of pangenome analysis to uncover genetic diversity lost or underrepresented in the domesticated gene pool. We have proposed a strategy demonstrating how the integration of CWRs with traditional and omics-assisted breeding approaches accelerates the development of abiotic stress-tolerant cultivars, as shown in Fig. 2.2.

Furthermore, systems biology can enhance the development of abiotic stress-tolerant cultivars by identifying key genes, regulatory networks, and metabolic pathways in CWRs. Through genomics, transcriptomics, and metabolomics, it enables the integration of stress-resilient traits from CWRs into cultivated crops (Saini 2024; Saini et al. 2024). Additionally, systems biology-based models can predict and optimize stress tolerance mechanisms, accelerating the breeding of resilient cultivars.

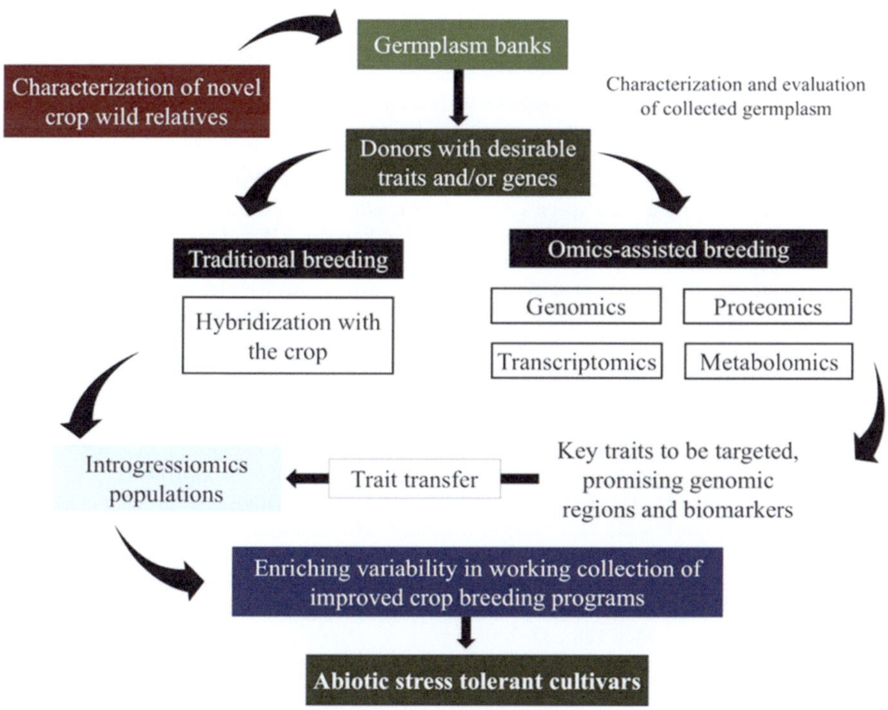

Fig. 2.2 Schematic representation of the utilization of crop wild relatives (CWRs) in crop improvement programs through traditional and omics-assisted breeding approaches to develop abiotic stress-tolerant cultivars

2.5 Conclusions

Unprecedented challenges like climate change, evolving abiotic and biotic stresses, and a growing population have driven breeders to turn toward CWRs. Since the domestication of crops, the potential for significant breeding advancements has plateaued, necessitating the identification and introduction of new gene sources for further crop improvement. Crop wild relatives, or the progenitors of modern crops, offer valuable solutions to overcome breeding bottlenecks and expand the gene pool. Unlike domesticated species, CWRs have not been subjected to selective pressure for yield traits but have evolved to withstand harsh environments, making them rich sources of abiotic stress tolerance genes. However, these genes remain underutilized due to issues such as varying ploidy levels, pre- and post-fertilization barriers, and undesirable yield-related alleles. While crop genetic diversity has historically been enhanced through landraces, CWRs, and mutants, advancements in molecular techniques, including trait mapping, map-based cloning, and MAS, have furthered the potential for improvement. The integration of biometry, population genetics, bioinformatics, and molecular tools holds promise for achieving sustainable crop improvements in response to changing climates and societal needs.

References

Afzal F, Li H, Gul A et al (2019) Genome-wide analyses reveal footprints of divergent selection and drought adaptive traits in synthetic-derived wheats. G3 Genes|Genomes|Genetics 9(6):1957–1973. https://doi.org/10.1534/g3.119.400010

Albaladejo I, Meco V, Plasencia F, Flores FB, Bolarin MC, Egea I (2017) Unravelling the strategies used by the wild tomato species *Solanum pennellii* to confront salt stress: from leaf anatomical adaptations to molecular responses. Environ Exp Bot 135:1–12

Anjani K, Ratnakumar P, Babu SS (2022) Photosynthetic characteristics in wild, cultivated species and interspecific inbred lines of safflower: photosynthetic characteristics in wild and cultivated species of safflower. J Oilseeds Res 39:26–33. https://doi.org/10.56739/jor.v39i1.137306

Asdal Å, Guarino L (2018) The Svalbard global seed vault: 10 years—1 million samples. Biopreserv Biobank 16(5):391–392

Ashraf M, Nazir N, McNeilly T (2001) Comparative salt tolerance of amphidiploid and diploid *Brassica* species. Plant Sci 160:683–689. https://doi.org/10.1016/S0168-9452(00)00449-0

Ashri A, Knowles PF (1960) Cytogenetics of safflower (*Carthamus L.*) species and their hybrids. Agron J 52:11–17. https://doi.org/10.2134/agronj1960.00021962005200010004x

Asif MA, Zafar Y, Iqbal J (2011) Enhanced expression of *AtNHX1* in transgenic groundnut (*Arachis hypogaea L.*) improves salt and drought tolerance. Mol Biotechnol 49:250–256. https://doi.org/10.1007/s12033-011-9399-1

Baldini M, Vannozzi GP (1998) Agronomic and physiological assessment of genotypic variation for drought tolerance in sunflower genotypes obtained from a cross between *H. annuus* and *H. argophyllus*. Agric Med 128:232–240

Baldini MF, Cecconi V, Vannozzi GP (1993) Influence of water deficit on gas exchange and dry matter accumulation in sunflower cultivars and wild species (*Helianthus argophyllus T. & G.*). Helia 16:1–10

Banda CM (2020) Genetic analysis for heat tolerance in tropical maize (*Zea mays*) (Doctoral dissertation, University of Zambia)

Bashir K, Matsui A, Rasheed S, Seki M (2019) Recent advances in the characterization of plant transcriptomes in response to drought, salinity, heat, and cold stress. F1000Res 8:658

Bassiri A, Khosh-Khui M, Rouhani I (1977) The influences of simulated moisture stress conditions and osmotic substrates on germination and growth of cultivated and wild safflowers. J Agric Sci 88:95–100. https://doi.org/10.1017/S0021859600033815

Baute GJ (2015) Genomics of sunflower improvement: from wild relatives to a global oil seed (Ph.D. dissertation, Univ. of British Columbia, Vancouver, BC). https://doi.org/10.14288/1.0166730

Berger JD, Kumar S, Nayyar H et al (2012) Temperature-stratified screening of chickpea (*Cicer arietinum L.*) genetic resource collections reveals very limited reproductive chilling tolerance compared to its annual wild relatives. Field Crops Res 126:119–129

Bertioli DJ, Cannon SB, Froenicke L et al (2016) The genome sequences of *Arachis duranensis* and *Arachis ipaensis*, the diploid ancestors of cultivated peanut. Nat Genet 48:438–446. https://doi.org/10.1038/ng.3517

Bierschenk B, Tagele MT, Ali B et al (2020) Evaluation of rice wild relatives as a source of traits for adaptation to iron toxicity and enhanced grain quality. PLoS One 15(1):e0223086

Bin Rahman AR, Zhang J (2016) Flood and drought tolerance in rice: opposite but may coexist. Food Energy Sec 5(2):76–88

Bisht A, Saini DK, Kaur B et al (2023) Multi-omics assisted breeding for biotic stress resistance in soybean. Mol Biol Rep 50:1–28

Blackman BK, Scascitelli MN, Kane MC et al (2011) Sunflower domestication alleles support a single domestication center in eastern North America. Proc Natl Acad Sci USA 108:14360–14365. https://doi.org/10.1073/pnas.1104853108

Bohra A, Kilian B, Sivasankar S et al (2022) Reap the crop wild relatives for breeding future crops. Trends Biotechnol 40:412–431. https://doi.org/10.1016/j.tibtech.2021.08.009

Boraiah KM, Basavaraj PS, Kakade VD et al (2023) Abiotic stress-tolerant crop varieties in India: status and a way forward. In: Updates in plant breeding. IntechOpen. https://doi.org/10.5772/intechopen.1001916

Branca F, Argento S, Alessandro T (2012) Assessing genetic reserves in Sicily (Italy): the *Brassica* wild relatives case study. In: Maxted N, Dulloo ME, Ford-Lloyd BV, Frese L, Iriondo JM, Pinheiro de Carvalho MAA (eds) Agrobiodiversity conservation: securing the diversity of crop wild relatives and landraces. CABI, Wallingford, pp 52–58. https://doi.org/10.1079/9781845938512.0052

Brasileiro AC, Morgante AC, Araujo AC et al (2015) Transcriptome profiling of wild Arachis from water-limited environments uncovers drought tolerance candidate genes. Plant Mol Biol Rep 33:1876–1892. https://doi.org/10.1007/s11105-015-0882-x

Brodersen CR, Roddy AB, Wason JW, McElrone AJ (2019) Functional status of xylem through time. Annu Rev Plant Biol 70:407–433

Brozynska M, Furtado A, Henry RJ (2016) Genomics of crop wild relatives: expanding the gene pool for crop improvement. Plant Biotechnol J 14:1070–1085. https://doi.org/10.1111/pbi.12454

Brozynska M, Omar ES, Furtado A et al (2014) Chloroplast genome of novel rice germplasm identified in northern Australia. Trop Plant Biol 7:111–120

Calfee E, Gates DM, Lorant A, Perkins MT, Coop G, Ross-Ibarra J (2021) Selective sorting of ancestral introgression in maize and teosinte along an elevational cline. PLoS Genet 17(10):e1009810. https://doi.org/10.1371/journal.pgen.1009810

Canci H, Toker C (2009) Evaluation of yield criteria for drought and heat resistance in chickpea (Cicer arietinum L.). J Agron Crop Sci 195(1):47–54

Cason JM (2008) Introgression pathway for drought tolerance in peanut (*Arachis hypogea* L.). Ph.D. Thesis. Texas A & M University, College Station, TX

Cason JM, Simpson CE, Rooney WL et al (2020) Drought-tolerant transcription factors identified in Arachis dardani and Arachis ipaënsis. Agrosyst Geosci Environ 3:e20069. https://doi.org/10.1002/agg2.20069

Castaneda-Alvarez NP, Khoury CK, Achicanoy HA et al (2016) Global conservation priorities for crop wild relatives. Nat Plants 2(4):1–6

CBD (2002) Global strategy for plant conservation. Secretariat of the Convention on Biological Diversity, Montreal, QC, p 48

Chandler JM, Jan CC (1984) Identification of salt-tolerant germplasm sources in the helianthus species. In: Agronomy abstracts. ASA, Madison, WI, p 61

Chen HF, Wang H, Li ZY (2007) Production and genetic analysis of partial hybrids in intertribal crosses between Brassica species (*B. rapa, B. napus*) and Capsella bursa-pastoris. Plant Cell Rep 26:1791–1800. https://doi.org/10.1007/s00299-007-0392-x

Chen Y, Chen P, de los Reyes BG (2006) Differential responses of the cultivated and wild species of soybean to dehydration stress. Crop Sci 46:2041–2046. https://doi.org/10.2135/cropsci2005.12.0466

Choudary AK, Sultana R, Pratap A, Nadarajan N, Jha UC (2011) Breeding for abiotic stress in pigeonpea. J Food Leg 24(3):165–174

Choudhary AK, Sultana R, Vales MI et al (2018) Integrated physiological and molecular approaches to improvement of abiotic stress tolerance in two pulse crops of the semi-arid tropics. Crop J 6(2):99–114

Chugh V, Kaur D, Purwar S (2022) Applications of molecular markers for developing abiotic-stress-resilient oilseed crops. Life 13(1):88. https://doi.org/10.3390/life13010088

Conesa A, Madrigal P, Tarazona S et al (2016) A survey of best practices for RNA-seq data analysis. Genome Biol 17:13

Crites GD (1993) Domesticated sunflower in fifth millennium B.P. temporal context: new evidence from middle Tennessee. Am Antiq 58:146–148. https://doi.org/10.2307/281459

Daniel Z, Maria H (2000) Domestication of plants in the Old World. Oxford University Press, New York, NY

Devi J, Sagar V, Mishra GP et al (2023) Heat stress tolerance in peas (*Pisum sativum* L.): current status and way forward. Front Plant Sci 13:1108276

Dissanayake R, Cogan NOI, Smith KF, Kaur S (2021) Application of genomics to understand salt tolerance in lentil. Genes 12(3):332. https://doi.org/10.3390/genes12030332

Dong CH, Li C, Yan XH et al (2012) Gene expression profiling of *Sinapis alba* leaves under drought stress and rewatering growth conditions with Illumina deep sequencing. Mol Biol Rep 39:5851–5857. https://doi.org/10.1007/s11033-011-1395-9

Eastwood RJ, Tambam B, Aboagye LM et al (2022) Adapting agriculture to climate change: a synopsis of coordinated national crop wild relative seed collecting programs across five continents. Plants 11(14):1840. https://doi.org/10.3390/plants11141840

Ebert AW, Karihaloo JL, Ferreira ME (2006) Opportunities, limitations and needs for DNA banks. In: De Vicente MC, Andersson MS (eds) DNA banks—providing novel options for genebanks? Topical reviews in agricultural biodiversity. International Plant Genetic Resources Institute, Rome, Italy, pp 61–68

Ebrahim F, Arzani A, Rahimmalek M et al (2020) Salinity tolerance of wild barley Hordeum vulgare ssp. spontaneum. Plant Breed 139(2):304–316

Edelist CX, Raffoux M, Falque C et al (2009) Differential expression of candidate salt tolerance genes in the halophyte Helianthus paradoxus and its glycophyte progenitors H. annuus and H. petiolaris (Asteraceae). Am J Bot 96:1830–1838. https://doi.org/10.3732/ajb.0900067

Ellis RP, Forster BP, Robinson D, Handley LL, Gordon DC, Russell JR, Powell W (2000) Wild barley: a source of genes for crop improvement in the 21st century? J Exp Bot 51:9–17. https://doi.org/10.1093/jexbot/51.342.9

Engels JM, Thormann I (2020) Main challenges and actions needed to improve conservation and sustainable use of our crop wild relatives. Plants 9(8):968

Fahad S, Bajwa AA, Nazir U, Anjum SA, Farooq A, Zohaib A, Sadia S, Nasim W, Adkins S, Saud S, Ihsan MZ, Alharby H, Wu C, Wang D, Huang J (2017) Crop production under drought and heat stress: plant responses and management options. Front Plant Sci 8:1147

FAO (2010) The second report on the state of the world's plant genetic resources for food and agriculture. Food and Agriculture Organization of the United Nations, Rome, Italy

FAO (2016) FAOSTAT, FAO Statistics Division, Rome. http://faostat3.fao.org/browse/E
Ferguson ME, Robertson LD, Ford-Lloyd BV, Newbury HJ, Maxted N (1998) Contrasting genetic variation amongst lentil landraces from different geographical origins. Euphytica 102:265–273
Fleury D, Jefferies SP, Kuchel H, Langridge P (2010) Genetic and genomic tools to improve drought tolerance in wheat. J Exp Bot 61(12):3211–3222. https://doi.org/10.1093/jxb/erq152
Frese L, Bönisch M, Herden T, Zander M, Friesen N (2018) In-situ-Erhaltung von Wild Selleriearten. Nat Landsch 50:155–163
Fuller DQ, Harvey EL (2006) The archaeobotany of Indian pulses: identification, processing, and evidence for cultivation. Environ Archaeol 11:219–246
Gai W, Sun H, Hu Y (2022) Genome-wide identification of membrane-bound fatty acid desaturase genes in three peanut species and their expression in Arachis hypogaea during drought stress. Genes 13:1718. https://doi.org/10.3390/genes13101718
Gill T, Gill SK, Saini DK et al (2022) A comprehensive review of high throughput phenotyping and machine learning for plant stress phenotyping. Phenomics:1–28
Golicz AA, Bayer PE, Barker GC et al (2016) The pangenome of an agronomically important crop plant Brassica oleracea. Nat Commun 7:13390. https://doi.org/10.1038/ncomms13390
Gonzalez-Segovia E, Perez-Limon S, Cintora-Martinez GC et al (2019) Characterization of introgression from the teosinte Zea mays ssp. mexicana to Mexican highland maize. PeerJ 7:e6815. https://doi.org/10.7717/peerj.6815
Gorim LY, Vandenberg A (2017) Evaluation of wild lentil species as genetic resources to improve drought tolerance in cultivated lentil. Front Plant Sci 8:1129
Guan R, Qu Y, Guo Y et al (2014) Salinity tolerance in soybean is modulated by natural variation in GmSALT 3. Plant J 80:937–950. https://doi.org/10.1111/tpj.12695
Guimarães PM, Brasileiro ACM, Morgante CV et al (2012) Global transcriptome analysis of two wild relatives of peanut under drought and fungi infection. BMC Genomics 13:387. https://doi.org/10.1186/1471-2164-13-387
Guo R, Shi L, Jiao Y et al (2018) Metabolic responses to drought stress in the tissues of drought-tolerant and drought-sensitive wheat genotype seedlings. AoB PLANTS 10(2). https://doi.org/10.1093/aobpla/ply016
Hajjar R, Hodgkin T (2007) The use of wild relatives in crop improvement: a survey of developments over the last 20 years. Euphytica 156(1–2):1–13. https://doi.org/10.1007/s10681-007-9363-0
Hamwieh A, Udupa SM, Sarker A, Jung C, Baum M (2009) Development of new microsatellite markers and their application in the analysis of genetic diversity in lentils. Breed Sci 59(1):77–86
Harlan JR, de Wet JM (1971) Toward a rational classification of cultivated plants. Taxon 20(4):509–517
Haun WJ, Hyten DL, Xu WW et al (2011) The composition and origins of genomic variation among individuals of the soybean reference cultivar Williams 82. Plant Physiol 155:645–655. https://doi.org/10.1104/pp.110.166736
Hossain A, Maitra S, Pramanick B et al (2022) Wild relatives of plants as sources for the development of abiotic stress tolerance in plants. In: Plant perspectives to global climate changes, pp 471–518
Hosseini SZ, Ismaili A, Nazarian Firouzabadi F, Fallahi H, Rezaeinejad A (2021) Identification of heat stress-responsive microRNAs in lentil (*Lens culinaris*) and their target genes. Cell Mol Res (Iran J Biol) 34(1):21–30
Hunter D, Changtragoon S (2016) Good practices for conservation and sustainable use of crop wild relatives of tropical fruit tree diversity. In: Tropical Fruit Tree Diversity, pp 127–139
Hunter D, Heywood V (2011) Crop wild relatives. A manual of *in-situ* conservation. Routledge, London, UK. 414p
Hussain MM, Rauf S, Riaz MA et al (2016) Determination of drought tolerance-related traits in *Helianthus argophyllus*, *Helianthus annuus* and their hybrids. Breed Sci 67:257–267. https://doi.org/10.1270/jsbbs.16095
Hymowitz T (1970) On the domestication of the soybean. Econ Bot 24:408–421. https://doi.org/10.1007/BF02860745

Idrissi O, Udupa SM, De Keyser E et al (2016) Identification of quantitative trait loci controlling root and shoot traits associated with drought tolerance in a lentil (*Lens culinaris* Medik.) recombinant inbred line population. Front Plant Sci 7

Iqbal M, Moakhar NP, Strenzke K et al (2016) Genetic improvement in grain yield and other traits of wheat grown in western Canada. Crop Sci 56(2):613–624. https://doi.org/10.2135/cropsci2015.06.0348

Jacob J, Sanjana P, Visarada KBR et al (2022) Seedling stage heat tolerance mechanisms in pearl millet (*Pennisetum glaucum* (L.) R. Br.). Russ J Plant Physiol 69(6):128

James AT, Lawn RJ, Cooper M (2008) Genotypic variation for drought stress response traits in soybean. I. Variation in soybean and wild *Glycine spp.* for epidermal conductance, osmotic potential, and relative water content. Aust J Agric Res 59:656–669. https://doi.org/10.1071/AR07159

Jarvis SG, Fielder H, Maxted N, Smart SM (2015) Distribution of crop wild relatives of conservation priority in the UK landscape. Biol Conserv 191:444–451. https://doi.org/10.1016/j.biocon.2015.07.039

Ji W, Zhu YM, Li Y et al (2010) Overexpression of a glutathione S-transferase gene, *GsGST*, from wild soybean (*Glycine soja*) enhances drought and salt tolerance in transgenic tobacco. Biotechnol Lett 32:1173–1179. https://doi.org/10.1007/s10529-010-0269-x

Kane NC, Burke JM, Marek L et al (2013) Sunflower genetic, genomic and ecological resources. Mol Ecol Resour 13:10–20. https://doi.org/10.1111/1755-0998.12023

Kantar MB, Sosa CC, Khoury CK et al (2015) Ecogeography and utility to plant breeding of the crop wild relatives of sunflower (*Helianthus annuus* L.). Front Plant Sci 6:841. https://doi.org/10.3389/fpls.2015.00841

Kapazoglou A, Gerakari M, Lazaridi E et al (2023) Crop wild relatives: a valuable source of tolerance to various abiotic stresses. Plants 12(2):328. https://doi.org/10.3390/plants12020328

Kapoor C, Singh SP, Sankar SM, Singh N (2022) Enhancing drought tolerance in pearl millet (*Pennisetum glaucum* L.): integrating traditional and omics approaches. Euphytica 218(7):104

Kashyap A, Kumari S, Garg P et al (2023) Indexing resilience to heat and drought stress in the wild relatives of rapeseed-mustard. Life (Basel) 13(3):738. https://doi.org/10.3390/life13030738

Kassa MT, Penmetsa RV, Carrasquilla-Garcia N et al (2012) Genetic patterns of domestication in pigeonpea (*Cajanus cajan* L. Millsp.) and wild *Cajanus* relatives. PLoS One 7(6):e39563

Kaur A, Sangha MK (2023) Crop wild relatives (CWRs) a genetic pool for crop improvement: a review. Agric Rev (Of). https://doi.org/10.18805/ag.r-2613

Kaur G, Singh G, Motavalli PP et al (2020) Impacts and management strategies for crop production in waterlogged or flooded soils: a review. Agron J 112(3):1475–1501

Kerem Z, Lev-Yadun S, Gopher A et al (2007) Chickpea domestication in the Neolithic Levant through the nutritional perspective. J Archaeol Sci 34:1289–1293

Khoury CK, Castañeda-Alvarez NP, Achicanoy HA et al (2015) Crop wild relatives of pigeonpea (*Cajanus cajan* L. Millsp.): distributions, ex situ conservation status, and potential genetic resources for abiotic stress tolerance. Biol Conserv 184:259–270

Kim MY, Lee S, Van K et al (2010) Whole genome sequencing and intensive analysis of the undomesticated soybean (*Glycine soja* Sieb. and Zucc.) genome. Proc Natl Acad Sci USA 107:22032–22037. https://doi.org/10.1073/pnas.1009526107

Kinman ML (1970) New developments in the USDA and state experiment station sunflower breeding programs. In: Proceedings of the 4th international sunflower conference, Memphis, TN. 23–25 June 1970. International Sunflower Association, Paris, pp 181–183

Kioukis A, Michalopoulou VA, Briers L et al (2020) Intraspecific diversification of the crop wild relative *Brassica cretica* Lam. using demographic model selection. BMC Genomics 21:48. https://doi.org/10.1186/s12864-019-6439-x

Kislev ME, Bar-Yosef O (1988) The legumes: the earliest domesticated plants in the Near East? Curr Anthropol 29:175–179

Kleingartner LW (2004) World outlook and future development of sunflower markets around the world. In: Seiler GJ (ed) Proceedings of the 16th international sunflower conference. 29 Aug–2 Sept 2004. International Sunflower Association, Paris, pp 69–77

Knights E, Southwell R, Schwinghamer M et al (2008) Resistance to *Phytophthora medicaginis* Hansen and Maxwell in wild *Cicer* species and its use in breeding root rot-resistant chickpea (*Cicer arietinum* L.). Aust J Agric Res 59:383–387

Koh J, Chen G, Yoo MJ et al (2015) Comparative proteomic analysis of *Brassica napus* in response to drought stress. J Proteome Res 14:3068–3081

Kosová K, Vítámvás P, Prášil IT et al (2011) Plant proteome changes under abiotic stress—contribution of proteomics studies to understanding plant stress response. J Proteomics 74(8):1301–1322

Koul PM, Sharma V, Rana M, Chahota RK, Kumar S, Sharma TR (2017) Analysis of genetic structure and interrelationships in lentil species using morphological and SSR markers. 3 Biotech 7:1–11

Kumar B, Dar A, Choudhary M et al (2019) Germplasm conservation and utilization in maize. National Workshop on Scientific Maize Cultivation in North East India, Aizawl, 5 March 2019

Kumar G, Purty RS, Sharma MP et al (2009) Physiological responses among brassica species under salinity stress show strong correlation with transcript abundance for SOS pathway-related genes. J Plant Physiol 166:507–520. https://doi.org/10.1016/j.jplph.2008.08.001

Kumar J, Gupta DS, Kesari R et al (2021) Comprehensive RNAseq analysis for identification of genes expressed under heat stress in lentil. Physiol Plant 173(4):1785–1807

Kumar S, Bhati J, Saha A et al (2022) Cerealestdb: a comprehensive resource for abiotic stress-responsive annotated ESTs with predicted genes, gene ontology, and metabolic pathways in major cereal crops. Front Genet 13. https://doi.org/10.3389/fgene.2022.842868

Kumar S, Rajendran K, Kumar J et al (2015) Current knowledge in lentil genomics and its application for crop improvement. Front Plant Sci 6:78

Kumari P, Bisht DS, Bhat SR (2018) Stable, fertile somatic hybrids between Sinapis alba and Brassica juncea show resistance to Alternaria brassicae and heat stress. Plant Cell Tissue Organ Cult 133:77–86

Ladizinsky G, Braun D, Goshen D, Muehlbauer FJ (1984) The biological species of the genus Lens L. Bot Gaz 145(2):253–261

Leclercq P (1969) Cytoplasmic male sterility in sunflower. Ann Amelior Plant 19:99–106

Lee JD, Shannon JG, Vuong TD et al (2009) Inheritance of salt tolerance in wild soybean (Glycine soja Sieb. and Zucc.) accession PI483463. J Hered 100:798–801. https://doi.org/10.1093/jhered/esp027

Lexer C, Lai Z, Rieseberg LH (2004) Candidate gene polymorphisms associated with salt tolerance in wild sunflower hybrids: implications for the origin of Helianthus paradoxus, a diploid hybrid species. New Phytol 161:225–233. https://doi.org/10.1046/j.1469-8137.2003.00925.x

Lexer C, Welch ME, Durphy JL (2003) Natural selection for salt tolerance quantitative trait loci (QTLs) in wild sunflower hybrids: implications for the origin of Helianthus paradoxus, a diploid hybrid species. Mol Ecol 12:1225–1235. https://doi.org/10.1046/j.1365-294X.2003.01803.x

Li L, Hu B, Li X (2020) Characterization of mTERF family in allotetraploid peanut and their expression levels in response to dehydration stress. Biotechnol Equip 34:1176–1187. https://doi.org/10.1080/13102818.2020.1825121

Li YH, Zhou G, Ma J et al (2014) *De novo* assembly of soybean wild relatives for pangenome analysis of diversity and agronomic traits. Nat Biotechnol 32:1045–1052. https://doi.org/10.1038/nbt.2979

Ljuština M, Mikić A (2010) Archaeological evidence for the domestication of lentil (Lens culinaris) and its distribution in Europe. J Lentil Res 4:26–29

Lu W, Bennett BD, Rabinowitz JD (2008) Analytical strategies for LC–MS based targeted metabolomics. J Chromatogr B Analyt Technol Biomed Life Sci 871:236–242. https://doi.org/10.1016/j.jchromb.2008.04.031

Ludwig F, Rosenthal DM, Johnston JA et al (2004) Selection on leaf ecophysiological traits in a desert hybrid Helianthus species and early generation hybrids. Evolution 58:2682–2692. https://doi.org/10.1111/j.0014-3820.2004.tb01621.x

Luo Q, Yu B, Liu Y (2005) Differential sensitivity to chloride and sodium ions in seedlings of Glycine max and G. soja under NaCl stress. J Plant Physiol 162:1003–1012. https://doi.org/10.1016/j.jplph.2004.11.008

Maazou ARS, Qiu J, Mu J, Liu Z (2017) Utilization of wild relatives for maize (Zea mays L.) improvement. African J Plant Sci 11(5):105–113

Majidi MM, Tavakoli V, Mirlohi A et al (2011) Wild safflower species (*Carthamus oxyacanthus Bieb.*): a possible source of drought tolerance for arid environments. Aust J Crop Sci 5:1055–1063

Maliro MFA, McNeil D, Redden B et al (2008) Sampling strategies and screening of chickpea (*Cicer arietinum L.*) germplasm for salt tolerance. Genet Resour Crop Evol 55:53–63

Mammadov J, Buyyarapu R, Guttikonda SK et al (2018) Wild relatives of maize, rice, cotton, and soybean: treasure troves for tolerance to biotic and abiotic stresses. Front Plant Sci 9. https://doi.org/10.3389/fpls.2018.00886

Mandel JR, Dechaine JM, Marek LF et al (2011) Genetic diversity and population structure in cultivated sunflower and a comparison to its wild progenitor, *Helianthus annuus* L. Theor Appl Genet 123:693–704. https://doi.org/10.1007/s00122-011-1619-3

Maphosa L, Richards MF, Norton SL, Nguyen GN (2020) Breeding for abiotic stress adaptation in chickpea (*Cicer arietinum* L.): a comprehensive review. Crop Breed Genet Genom 2:e200015

Martins ACQ, Mota APZ, Carvalho P (2022) Transcriptome responses of wild *Arachis* to UV-C exposure reveal genes involved in general plant defense and priming. Plants 11:408. https://doi.org/10.1186/s12870-020-02678-9

Maxted N, Kell SP, Toledo Á et al (2010) A global approach to crop wild relative conservation: securing the gene pool for food and agriculture. Kew Bull 65(4):561–576. https://doi.org/10.1007/s12225-011-9253-4

Maxted N, Vincent H (2021) Review of congruence between global crop wild relative hotspots and centres of crop origin/diversity. Genet Resour Crop Evol 68(4):1283–1297. https://doi.org/10.1007/s10722-021-01114-7

Mayerhofer R, Archibald C, Bowles V et al (2010) Development of molecular markers and linkage maps for the *Carthamus* species *C. tinctorius* and *C. oxyacanthus*. Genome 53:266–276. https://doi.org/10.1139/G10-002

Miller JF (1995) Inheritance of salt tolerance in sunflower. Helia 18:9–16

Miller JF, Seiler GJ (2003) Registration of five oilseed maintainer (HA 429-HA 433) sunflower germplasm lines. Crop Sci 43:2313–2314

Mittal S (2006) Structural shift in demand for food: projections for 2020. Working Paper No. 184; Indian Council for Research on International Economic Relations (ICRIER): New Delhi, India

Moenga SM, Gai Y, Carrasquilla-Garcia N et al (2020) Gene co-expression analysis reveals transcriptome divergence between wild and cultivated chickpea under drought stress. Plant J 104(5):1195–1214

Mohd Saad NS, Severn-Ellis AA, Pradhan A et al (2021) Genomics armed with diversity leads the way in *Brassica* improvement in a changing global environment. Front Genet 12:600789. https://doi.org/10.3389/fgene.2021.600789

Mondal S, Karmakar S, Panda D et al (2023) Crucial plant processes under heat stress and tolerance through heat shock proteins. Plant Stress 10:100227

Mounce R, Smith P, Brockington S (2017) Ex-situ conservation of plant diversity in the world's botanic gardens. Nat Plants 3(10):795–802

Mugabe D, Coyne CJ, Piaskowski J et al (2019) Quantitative trait loci for cold tolerance in chickpea. Crop Sci 59(2):573–582

Munns R, Tester M (2008) Mechanisms of salinity tolerance. Annu Rev Plant Biol 59:651–681. https://doi.org/10.1146/annurev.arplant.59.032607.092911

Naito K, Wakatake T, Shibata TF et al (2022) Genome sequence of 12 *Vigna* species as a knowledge base of stress tolerance and resistance. bioRxiv. https://doi.org/10.1101/2022.03

Nautiyal PC, Rajgopal K, Zala PV et al (2008) Evaluation of wild *Arachis* species for abiotic stress tolerance: I. Thermal stress and leaf water relations. Euphytica 159:43–57. https://doi.org/10.1007/s10681-007-9455-x

Nevo E, Chen G (2010) Drought and salt tolerances in wild relatives for wheat and barley improvement. Plant Cell Environ 33(4):670–685

O'Donnell K, Sharrock S (2017) The contribution of botanic gardens to ex-situ conservation through seed banking. Plant Divers 39(6):373–378

Omar GI, Saqer MM, Adwan GM (2019) Phylogenetic relationship among some species of the Genera Lens, Vicia, Lathyrus and Pisum (Leguminosae) in Palestine. Jordan J Biol Sci 12(3)

Ortiz R (2015) The importance of crop wild relatives, diversity, and genetic potential for adaptation to abiotic stress-prone environments. In: Redden R, Yadav SS, Maxted N, Dulloo ME, Guarino L, Smith P (eds) Crop wild relatives and climate change. Plant genetic resources and climate change. CAB International, Wallingford, pp 114–136. https://doi.org/10.1079/9781780641973.0114

Parveen K, Saddique MAB, Rehman SU et al (2023) Identification and characterization of salt stress-responsive NHX gene family in chickpea. Plant Stress 10:100266

Prescott-Allen CP, Prescott-Allen R (1986) The first resource: wild species in the North American economy. Yale Univ Press, New Haven, CT

Prohens J, Gramazio P, Plazas M et al (2017) Introgressiomics: a new approach for using crop wild relatives in breeding for adaptation to climate change. Euphytica 213(7). https://doi.org/10.1007/s10681-017-1938-9

Purnamasari MI, Erskine W, Croser JS et al (2019) Comparative reaction of Camelina sativa to Sclerotinia sclerotiorum and Leptosphaeria maculans. Plant Dis 103:2884–2892

Pushpavalli R, Krishnamurthy L, Thudi M et al (2015) Two key genomic regions harbour QTLs for salinity tolerance in ICCV 2 × JG 11 derived chickpea (Cicer arietinum L.) recombinant inbred lines. BMC Plant Biol 15:124

Qi X, Li MW, Xie M et al (2014) Identification of a novel salt tolerance gene in wild soybean by whole genome sequencing. Nat Commun 5:1–11. https://doi.org/10.1038/ncomms5340

Quezada-Martinez D, Addo Nyarko CP et al (2021) Using wild relatives and related species to build climate resilience in Brassica crops. Theor Appl Genet 134(6):1711–1728. https://doi.org/10.1007/s00122-021-03793-3

Rajpal VR, Singh A, Kathpalia R et al (2023) The prospects of gene introgression from crop wild relatives into cultivated lentil for climate change mitigation. Front Plant Sci 14:1127239

Ramakrishna G, Kaur P, Singh A et al (2021) Comparative transcriptome analyses revealed different heat stress responses in pigeonpea (Cajanus cajan) and its crop wild relatives. Plant Cell Rep 40:881–898

Rampuria S, Bag P, Rogan CJ et al (2018) Pathogen-induced AdDjSKI of the wild peanut, Arachis diogoi, potentiates tolerance of multiple stresses in E. coli and tobacco. Plant Sci Int J Exp Plant Biol 272:62–74. https://doi.org/10.1016/j.plantsci.2018.03.033

Ray J, Bordolui SK (2021) Role of seed banks in the conservation of plant diversity and ecological restoration. Res Rev J Environ Sci 3(2):1–16

Redden RJ, Yadav SPS, Maxted N et al (2014) Crop wild relatives and climate change. In: Plant genetic resources and climate change, pp 114–136. https://doi.org/10.1079/9781780641973.0114

Reddy MRK, Rathour R, Kumar N et al (2010) Cross-genera legume SSR markers for analysis of genetic diversity in Lens species. Plant Breed 129(5):514–518

Reddy SS, Saini DK, Singh GM et al (2023) Genome-wide association mapping of genomic regions associated with drought stress tolerance at seedling and reproductive stages in bread wheat. Front Plant Sci 14:1166439

Renzi JP, Coyne CJ, Berger J et al (2022) How could the use of crop wild relatives in breeding increase the adaptation of crops to marginal environments? Front Plant Sci 13:886162. https://doi.org/10.3389/fpls.2022.886162

Robertson LD, Ocampo B, Singh KB (1997) Morphological variation in wild annual Cicer species in comparison to the cultigen. Euphytica 95:309–319

Rocha M, Licausi F, Araujo W et al (2010) Glycolysis and the tricarboxylic acid cycle are linked by alanine aminotransferase during hypoxia induced by waterlogging of lotus japonicas. Plant Physiol 152:1501–1513

Sabzalian M, Mirlohi A, Saeidi G et al (2009) Genetic variation among populations of wild safflower, *Carthamus oxyacanthus*, analyzed by agro-morphological traits and ISSR markers. Genet Resour Crop Evol 56:1057–1064. https://doi.org/10.1007/s10722-009-9426-2

Saini DK (2024) Establishing the proper links: network-based techniques for comprehending gene functions in cereals. In: Saini DK, Kole C (eds) Omics and system biology approaches for delivering better cereals, pp 401–430

Saini DK, Chopra Y, Singh J et al (2022) Comprehensive evaluation of mapping complex traits in wheat using genome-wide association studies. Mol Breed 42:1

Saini DK, Kumar S, Saini K, Goyal S (2024) Omics and system biology for developing climate-ready cereals: an overview. In: Saini DK, Kole C (eds) Omics and system biology approaches for delivering better cereals, pp 401–430

Samarappuli D, Zanetti F, Berzuini S et al (2020) Crambe (Crambe abyssinica Hochst): a non-food oilseed crop with great potential: a review. Agronomy 10:1380. https://doi.org/10.3390/agronomy10091380

Samoluk SS, Robledo G, Podio M (2015) First insight into divergence, representation, and chromosome distribution of reverse transcriptase fragments from L1 retrotransposons in peanut and wild relative species. Genetica 143:113–125. https://doi.org/10.1007/s10709-015-9820-y

Sandhu SK, Pal L, Kaur J et al (2019) Genome-wide association studies for yield and its component traits under terminal heat stress in Indian mustard (*Brassica juncea* L.). Euphytica 215:1–15. https://doi.org/10.1007/s10681-019-2489-z

Satori D, Tovar C, Faruk A et al (2021) Prioritizing crop wild relatives to enhance agricultural resilience in sub-Saharan Africa under climate change. Plants People Planet 4(3):269–282. https://doi.org/10.1002/ppp3.10247

Schreiber M, Himmelbach A, Börner A, Mascher M (2019) Genetic diversity and relationship between domesticated rye and its wild relatives as revealed through genotyping-by-sequencing. Evol Appl 12(1):66–77

Seiler GJ, Cuk L, Rogers CE (1981) New and interesting distribution records for *Helianthus paradoxus* Heiser (Asteraceae). Southwest Nat 26:431–432

Seiler GJ, Gulya TJ, Marek L (2006) Exploration for wild *Helianthus* species from the desert southwestern USA for potential drought tolerance. Helia 29:1–10. https://doi.org/10.1515/helia-2023-0003

Shafiei Koij F, Majidi MM, Mirlohi A et al (2019) The use of wild relatives of safflower to increase genetic diversity for fatty acid composition and drought tolerance. Crop Sci 59:2109–2118. https://doi.org/10.2135/cropsci2019.01.0068

Shamloo-Dashtpagerdi R, Tanin MJ, Aliakbari M, Saini DK (2024) Unveiling the role of the *ERD15* gene in wheat's tolerance to combined drought and salinity stress: a meta-analysis of QTL and RNA-Seq data. Physiol Plant 176:e14570

Sharma S (2019) Utilizing the underutilized CWR: pigeonpea pre-breeding from discovery to delivery

Sharma SB, Sharma R, Govindaraj M et al (2020) Harnessing wild relatives of pearl millet for germplasm enhancement: challenges and opportunities. Crop Sci 61(1):177–200. https://doi.org/10.1002/csc2.20343

Sherman Broyles S, Bombarely A, Powell AF et al (2014) The wild side of a major crop: soybean's perennial cousins from Down Under. Am J Bot 101:1651–1665. https://doi.org/10.3732/ajb.1400121

Simon PW, Rolling W, Senalik D, Bolton A, Rahim MA, Mannan A, Shah AI (2020) Wild carrot diversity for new sources of abiotic stress tolerance to strengthen vegetable breeding in Bangladesh and Pakistan. Crop Sci 61:163–176. https://doi.org/10.1002/csc2.20333

Singh B, Sran AS, Sohi GS (2021) Innovative strategies to develop abiotic and biotic stress tolerance in mustard (*Brassicaceae*). In: Brassica breeding and biotechnology. IntechOpen, London, UK, pp 41–48. https://doi.org/10.5772/intechopen.95973

Singh D, Singh CK, Singh Tomar RS, Pal M (2017a) Genetics and molecular mapping of heat tolerance for seedling survival and pod set in lentil. Crop Sci 57(6):3059–3067

Singh D, Singh CK, Taunk J et al (2017b) Transcriptome analysis of lentil (*Lens culinaris* Medikus) in response to seedling drought stress. BMC Genom 18:1–20

Singh D, Singh CK, Taunk J, Jadon V, Pal M, Gaikwad K (2019) Genome-wide transcriptome analysis reveals vital role of heat responsive genes in regulatory mechanisms of lentil (Lens culinaris Medikus). Sci Rep 9:12976

Singh D, Singh CK, Tomar RSS et al (2016) Molecular assortment of Lens species with different adaptations to drought conditions using SSR markers. PLoS One 11:e0147213

Singh J, Singh V, Dutt V et al (2022) Insights into salt tolerance of mustard (Brassica juncea L. Czern & Coss): a metabolomics perspective. Environ Exp Bot 194:104760. https://doi.org/10.1007/s11738-019-2894-x

Singh KB, Ocampo B (1997) Exploitation of wild Cicer species for yield improvement in chickpea. Theor Appl Genet 95:418–423

Singh M, Bisht IS, Kumar S et al (2014) Global wild annual Lens collection: a potential resource for lentil genetic base broadening and yield enhancement. PLoS One 9:e107781

Sinha P, Pazhamala LT, Singh VK et al (2016) Identification and validation of selected universal stress protein domain containing drought-responsive genes in pigeonpea (Cajanus cajan L.). Front Plant Sci 6:1065

Skoric D (2009) Sunflower breeding for resistance to abiotic stresses. Helia 32:1–16. https://doi.org/10.2298/hel0950001s

Sohail Q, Inoue T, Tanaka H et al (2011) Applicability of Aegilops tauschii drought tolerance traits to breeding of hexaploid wheat. Breed Sci 61:347–357

Song Q, Hyten DL, Jia G et al (2013) Development and evaluation of SoySNP50K, a high-density genotyping array for soybean. PLoS One 8:e54985. https://doi.org/10.1371/journal.pone.0054985

Song X, Li E, Song H et al (2020) Genome-wide identification and characterization of nonspecific lipid transfer protein (nsLTP) genes in Arachis duranensis. Genomics 112:4332–4341. https://doi.org/10.1016/j.ygeno.2020.07.034

Sripathi VR, Buyyarapu R, Kumpatla SP et al (2016) Bioinformatics tools and genomic resources available in understanding the structure and function of Gossypium. Bioinform 231:233–258. https://doi.org/10.5772/64325

Srivastava N, Vadez V, Upadhyaya HD, Saxena KB (2006) Screening for intra and inter specific variability for salinity tolerance in pigeonpea (Cajanus cajan) and its related wild species. J SAT Agric Res 2(1):1–12

Srivastava R, Bajaj D, Malik A (2016) Transcriptome landscape of perennial wild Cicer microphyllum uncovers functionally relevant molecular tags regulating agronomic traits in chickpea. Sci Rep 6:1–17

Stebbins JC, Winchell CJ, Constable BB (2013) Helianthus winteri (Asteraceae), a new perennial species from the southern Sierra Nevada foothills, California. Aliso 31:19–24. https://doi.org/10.5642/aliso.20133101.04

Subbarao GV, Johansen C, Rao JK, Jana MK (1990) Salinity tolerance in F1 hybrids of pigeonpea and a tolerant wild relative. Crop Sci 30(4):785–788

Sujatha M, Prabakaran AJ, Dwivedi SL et al (2008) Cytomorphological and molecular diversity in backcross derived inbred lines of sunflower (Helianthus annuus L.). Genome 51:282–293. https://doi.org/10.1139/G08-008

Taghouti M, Bassi FM, Nasrellah N et al (2023) 'Nachit', a wild-relative-derived durum wheat resilient to climate change in Morocco. J Plant Regist 17:529–535. https://doi.org/10.1002/plr2.20292

Tang L, Cai H, Zhai H et al (2014) Overexpression of Glycine soja WRKY20 enhances both drought and salt tolerance in transgenic alfalfa (Medicago sativa L.). Plant Cell Tissue Organ Cult 118:77–86

Tekin M, Sari D, Catal M et al (2018) Eco-geographic distribution of Cicer isauricum PH Davis and threats to the species. Genet Resour Crop Evol 65:67–77

Thiel T, Graner A, Waugh R, Große I, Close TJ, Stein N (2009) Evidence and evolutionary analysis of ancient whole-genome duplication in barley predating the divergence from rice. BMC Evol Biol 9:209. https://doi.org/10.1186/1471-2148-9-20

Thormann I, Reeves PA, Reilley AA et al (2016) Geography of genetic structure in barley wild relative Hordeum vulgare subsp. spontaneum in Jordan. PLoS One 11:e0160745. https://doi.org/10.1371/journal.pone.0160745

Timme RE, Simpson BB, Linder CR (2007) High resolution phylogeny for helianthus (Asteraceae) using the 18S–26S ribosomal DNA external transcribed spacer. Am J Bot 94:1837–1852. 0.3732/ajb.94.11.1837

Tin HQ, Loi NH, Labarosa SJ et al (2020) Phenotypic response of farmer-selected CWR-derived rice lines to salt stress in the Mekong Delta. Crop Sci 61(1):201–218. https://doi.org/10.1002/csc2.20354

Tirnaz S, Zandberg J, Thomas WJ, Edwards D, Batley J (2022) Application of crop wild relatives in modern breeding: an overview of resources, experimental and computational methodologies. Front Plant Sci 13:4597

Toker C (2005) Preliminary screening and selection for cold tolerance in annual wild Cicer species. Genet Resour Crop Evol 52:1–5

Toker C, Canci H, Yildirim TOLGA (2007) Evaluation of perennial wild Cicer species for drought resistance. Genet Resour Crop Evol 54:1781–1786

Toulotte J, Pantazopoulou CK, Sanclemente MA, Voesenek LACJ, Sasidharan R (2022) Water stress resilient cereal crops: lessons from wild relatives. J Integr Plant Biol 64(2):412–430. https://doi.org/10.1111/jipb.13222

Tripathi K, Gupta D, Gore PG et al (2024) Exploitation of lentil wild relatives for genetic improvement. In: The lentil genome. Academic Press, pp 159–194

Tullu A, Bett K, Banniza S, Vail S, Vandenberg A (2013) Widening the genetic base of cultivated lentil through hybridization of Lens culinaris 'Eston' and L. ervoides accession IG 72815. Can J Plant Sci 93(6):1037–1047

Tyack N, Dempewolf H (2015) The economics of crop wild relatives under climate change. In: Redden R, Yadav SS, Maxted N, Dulloo ME, Guarino L, Smith P (eds) Crop wild relatives and climate change. Wiley, Hoboken, NJ, pp 281–291. https://doi.org/10.1002/9781118854396

Van Zonneveld M, Rakha M, Tan SY et al (2020) Mapping patterns of abiotic and biotic stress resilience uncovers conservation gaps and breeding potential of Vigna wild relatives. Sci Rep 10(1):2111

Varotto S, Krugman T, Cigliano RA et al (2022) Exploitation of epigenetic variation of crop wild relatives for crop improvement and agrobiodiversity preservation. Theor Appl Genet 135(11):3987–4003. https://doi.org/10.1007/s00122-022-04122-y

Varshney RK, Roorkiwal M, Sun S et al (2021) A chickpea genetic variation map based on the sequencing of 3,366 genomes. Nature 599(7886):622–627

Vavilov NI (1951) The origin, variation, Immunity and Breeding of Cultivated Plants. Ronald Press Company, New York, p 364

Verma S, Dubey N, Singh KH et al (2023) Utilization of crop wild relatives for biotic and abiotic stress management in Indian mustard [Brassica juncea (L.) Czern. & Coss.]. Front Plant Sci 14:1277922. https://doi.org/10.3389/fpls.2023.1277922

Vinson CC, Mota APZ, Oliveira TN et al (2018) Early responses to dehydration in contrasting wild Arachis species. PLoS One 13:e0198191. https://doi.org/10.1371/journal.pone.0198191

Wang L, de Kroon H, Bögemann GM et al (2005) Partial root drying effects on biomass production in *Brassica napus* and the significance of root responses. Plant Soil 276:313–326. https://doi.org/10.1007/s11104-005-5085-z

Wang P, Song H, Li C (2017a) Genome-wide dissection of the heat shock transcription factor family genes in *Arachis*. Front Plant Sci 8:106. https://doi.org/10.3389/fpls.2017.00106

Wang S, Cao M, Ma X et al (2017b) Integrated RNA sequencing and QTL mapping to identify candidate genes from *Oryza rufipogon* associated with salt tolerance at the seedling stage. Front Plant Sci 8

Wang T, Ren L, Li C et al (2021) The genome of a wild *Medicago* species provides insights into the tolerant mechanisms of legume forage to environmental stress. BMC Biol 19(1):96

Wang Y, Wang J, Sarwar R et al (2024) Research progress on the physiological response and molecular mechanism of cold response in plants. Front Plant Sci 15:1334913

Welch ME, Rieseberg LH (2002) Habitat divergence between a homoploid hybrid sunflower species, *Helianthus paradoxus* (Asteraceae) and its progenitors. Am J Bot 89:472–478. https://doi.org/10.3732/ajb.89.3.472

Wendler N, Mascher M, Nöh C et al (2014) Unlocking the secondary gene-pool of barley with next-generation sequencing. Plant Biotechnol J 12(8):1122–1131. https://doi.org/10.1111/pbi.12219

Wong MM, Gujaria-Verma N, Ramsay L et al (2015) Classification and characterization of species within the genus *Lens* using genotyping-by-sequencing (GBS). PLoS One 10(3):e0122025

Xiong ZT, Wang H (2005) Copper toxicity and bioaccumulation in Chinese cabbage (*Brassica pekinensis* Rupr.). Environ Toxicol 20:188–194. https://doi.org/10.1002/tox.20094

Yadav R, Choudhury PR, Koul AK et al (2022) Status of genetic resources in oilseed crops and their potential use for making India Atmanirbhar in edible oils. Indian J Plant Genet Resour. https://doi.org/10.5958/0976-1926.2022.00047.X

Yadav R, Nanjundan J, Gupta AK et al (2019) Novel source of biotic stress resistance identified from *Brassica* species and its wild relatives. PRO 36:195. https://doi.org/10.3390/proceedings2019036195

Yang DS, Zhang J, Li MX et al (2017) Metabolomics analysis reveals the salt tolerant mechanism in *Glycine soja*. J Plant Growth Regul 36:460–471. https://doi.org/10.1007/s11240-014-0463-y

Yang EY, Rajametov SN, Cho MC, Jeong HB, Chae WB (2021) Factors affecting tolerance to low night temperature differ by fruit types in tomato. Agriculture (Basel) 11

Yong HY, Wang C, Bancroft I et al (2015) Identification of a gene controlling variation in the salt tolerance of rapeseed (*Brassica napus* L.). Planta 242:313–326. https://doi.org/10.1007/s00425-015-2310-8

Yoshida J, Tomooka N, Yee Khaing T et al (2020) Unique responses of three highly salt-tolerant wild *Vigna* species against salt stress. Plant Prod Sci 23(1):114–128

Yuan C, Li C, Lu X (2020) Comprehensive genomic characterization of NAC transcription factor family and their response to salt and drought stress in peanut. BMC Plant Biol 20:454

Yumurtaci A (2015) Utilization of wild relatives of wheat, barley, maize and oat in developing abiotic and biotic stress tolerant new varieties. Emir J Food Agric 27(1):1. https://doi.org/10.9755/ejfa.v27i1.17852

Zair W, Maxted N, Amri A (2017) Setting conservation priorities for crop wild relatives in the Fertile Crescent. Genet Resour Crop Evol 65(3):855–863. https://doi.org/10.1007/s10722-017-0576-3

Zeng QY, Yang CY, Ma QB et al (2012) Identification of wild soybean miRNAs and their target genes responsive to aluminum stress. BMC Plant Biol 12:1–16

Zhang DY, Kumar M, Xu L, Wan Q, Huang YH, Xu ZL et al (2017) Genome-wide identification of Major Intrinsic Proteins in *Glycine soja* and characterization of GmTIP2;1 function under salt and water stress. Sci Rep 7:4106. https://doi.org/10.1038/s41598-017-04253-z

Zhang H, Mittal N, Leamy LJ, Barazani O, Song B (2016) Back into the wild—apply untapped genetic diversity of wild relatives for crop improvement. Evol Appl 10(1):5–24. https://doi.org/10.1111/eva.12434

Zhao C, He L, Xia H et al (2021) *De novo* full-length transcriptome analysis of *Arachis glabrata* provides insights into gene expression dynamics in response to biotic and abiotic stresses. Genomics 113:1579–1588. https://doi.org/10.1016/j.ygeno.2021.03.030

Zhao Q, Feng Q, Lu H et al (2018) Pan-genome analysis highlights the extent of genomic variation in cultivated and wild rice. Nat Genet 50(2):278–284

Zhou Z, Jiang Y, Wang Z et al (2015) Resequencing 302 wild and cultivated accessions identifies genes related to domestication and improvement in soybean. Nat Biotechnol 33:408–414. https://doi.org/10.1038/nbt.3096

Zhu D, Cai H, Luo X et al (2012) Over-expression of a novel JAZ family gene from *Glycine soja*, increases salt and alkali stress tolerance. Biochem Biophys Res Commun 426:273–279. https://doi.org/10.1016/j.bbrc.2012.08.086

Zhou S, Kai H, Zha Z et al (2016) Subcellular distribution and chemical forms of thorium in Brassica juncea var. foliosa. J Environ Radioact 157:60–66

Integrative Phenomics: Bridging Genomics and Stress Resilience in Crops

3

Navin Chander Gahtyari, Devender Sharma, Amit Kumar, Manoj Kumar, Prakash Ghasal, Ramesh Singh Pal, Nirmal Hedau, and Lakshmi Kant

Abstract

Advances in phenotyping include high-resolution trait measurement integrated with multi-omics, systems biology, and advanced analytics, thereby bridging the genotype–phenotype gap and giving rise to integrative phenomics. Stress resilience in crops is often quantitative, polygenic, and strongly influenced by genotype × environment interactions. Precise, multi-temporal phenotyping is therefore essential to translate genomic information into field performance. Integrative phenomics combines high-throughput sensing (RGB, thermal, multispectral, hyperspectral, chlorophyll fluorescence, LiDAR) deployed on platforms such as unmanned aerial vehicles, proximal sensors, gantries, and satellites to generate multidimensional trait datasets. Machine learning for large-scale data integration, together with genomic prediction, genome-wide association studies, and digital twin simulations, enables fusion of phenomic, genomic, and environmental datasets into predictive selection models. Applications across abiotic and biotic stresses show how canopy temperature, vegetation indices (NDVI), fluorescence metrics, and root phenotypes improve quantitative trait loci discovery and genomic selection accuracy, enabling earlier and resource-efficient identification of tolerant genotypes. Integration with proteomics and metabolomics links biochemical mechanisms to observable traits and supports functional validation. Cost-effective and accessible approaches include regional phenotyping hubs, adoption of open data standards (BrAPI, PhenoApp), low-cost sensor arrays, and participatory breeding that engages local programmes. Emerging technologies like robotics, IoT sensor networks, portable spectroscopy, and arti-

N. C. Gahtyari (✉) · D. Sharma · A. Kumar · M. Kumar · P. Ghasal · R. S. Pal · N. Hedau · L. Kant
ICAR-Vivekananda Parvatiya Krishi Anusandhan Sansthan (Viveknanda Institute of Hill Agriculture), Almora, Uttarakhand, India
e-mail: navin.gahtyari@icar.org.in

V. Gahlaut et al. (eds.), *Smart Crop Development*,
https://doi.org/10.1007/978-981-95-7680-7_3

ficial intelligence-driven analytics further strengthen the detective capabilities and adoption of the technology. Thus, breeding programmes must integrate high-throughput phenotyping tools along with advanced analytics for enhancing genetic gains for stress tolerance in breeding pipelines.

Keywords

Integrative phenomics · High-throughput phenotyping · Stress resilience · Multi-sensor imaging · Remote sensing

3.1 Introduction

Agricultural productivity in modern times is deeply challenged by the changing climate, resulting in increased temperature, erratic rainfall, drought, heat, salinity, and nutrient imbalances. Likewise, the distribution and severity of biotic stresses, including diseases and insect pests, are also continuously shifting (Mittler 2006; Chakraborty and Newton 2011). Traditional crop breeding has delivered significant improvements in stress tolerance. However, the underlying complex polygenic control of the stress response often limits its improvement (Collins et al. 2008; Mickelbart et al. 2015).

Advances in molecular genetics and multi-omics have uncovered numerous genes and pathways underlying stress responses. However, the translation of this knowledge into breeding has been slowed by the phenotyping bottleneck, that is, the accurate measuring of traits in multi-environmental field conditions (Furbank and Tester 2011; Fiorani and Schurr 2013). Conventional phenotyping is prone to subjectivity and error because it is mostly based on visual scoring and destructive sampling.

Hence, in modern times, phenomics has enabled a high-throughput, non-invasive, and quantitative assessment of plant traits at multiple scales (Furbank and Tester 2011; Fiorani and Schurr 2013; Araus and Cairns 2014). Integrative phenomics further combines imaging, spectral sensing, robotics, drones, and artificial intelligence (AI) with genomics and systems biology, generating data-rich insights into plant performance under stress (Tardieu et al. 2017). These advances not only enhance our ability to dissect complex stress tolerance traits but also accelerate their deployment in breeding pipelines. Let us examine how integrative phenomics can unlock abiotic stress resilience in crops, with emphasis on technological advances in the field and future prospects.

3.2 Climate Challenges and the Need for Phenomics

Agricultural systems worldwide are increasingly threatened by the combined pressures of population growth and resource limitations under the aegis of climate change. Global mean temperatures are projected to rise by 1.5–2.0 °C over the coming decades, accompanied by enhanced frequency of heat, flood, and drought (IPCC 2021). Concurrently, biotic stresses like diseases (fungal, bacterial, viral, etc.), insect pests, as well as weeds are being reshaped by changing environments. Warmer winters extend the survival of rust pathogens and insect vectors, whereas altered precipitation patterns create new ecological niches for pest outbreaks (Chakraborty and Newton 2011; Juroszek and von Tiedemann 2013). The convergence of these challenges poses a serious threat to global food security. Together, these biotic and abiotic stresses may responsible for up to 50% yield losses in major cereals under severe conditions (Mittler 2006; Fleury et al. 2010; Singh et al. 2016).

Traditional plant breeding has played a critical role in addressing both abiotic and biotic stresses, contributing varieties with enhanced tolerance or resistance (Allard 1999; Collins et al. 2008; Singh et al. 2016). However, the durability of resistance is increasingly challenged by the rapid evolution of pathogen races and insect biotypes (McDonald and Linde 2002). Similarly, tolerance to complex abiotic stresses often involves polygenic traits with strong genotype × environment (G × E) interactions, making them difficult to select through conventional methods (Cattivelli et al. 2008; Mickelbart et al. 2015). These limitations highlight the inadequacy of relying solely on conventional breeding strategies in the face of climate change.

Meanwhile, advances in molecular biology and genomics have expanded the knowledge base on stress-responsive genes, pathways, and regulatory networks (Mickelbart et al. 2015; Zandalinas et al. 2021). Yet, the major bottleneck lies not in genotyping but in phenotyping. Genomic data alone cannot drive the genetic gain without reliable, high-resolution trait data that capture how plants respond under realistic and variable environments (Furbank and Tester 2011; Fiorani and Schurr 2013).

This need has given rise to phenomics, a field dedicated to the large-scale, precise measurement of plant traits using advanced imaging and sensor technologies. Unlike conventional phenotyping, phenomics is capable of high-throughput, non-invasive, and real-time capturing of plant traits both under natural and artificial epiphytotic conditions. By bridging the gap between genetic potential and observable performance, phenomics provides the essential foundation for breeding crops resilient to biotic and abiotic stresses (Araus and Cairns 2014).

3.3 Bridging the Genotype–Phenotype Gap

Over the past two decades, advances in molecular biology, omics technologies (genomics, proteomics, etc.), and bioinformatics have revolutionized our understanding of plant stress responses. Genome sequencing, transcriptomics, metabolomics, and proteomics have identified thousands of genes, pathways, and regulatory networks linked to resilience against drought, salinity, heat, and biotic stresses (Mickelbart et al. 2015; Zandalinas et al. 2021). At the same time, the cost of genotyping has drastically reduced, from thousands of dollars per genome in the early 2000s to less than a few dollars per sample today with methods such as genotyping-by-sequencing (Varshney et al. 2014). This has made genotyping rapid, inexpensive, and routine in many breeding programmes. In contrast, phenotyping has lagged considerably, creating an imbalance often termed the phenotyping bottleneck (Furbank and Tester 2011). Without robust phenotypic data, the power of genomic tools such as quantitative trait loci (QTL) mapping, genome-wide association studies (GWAS), and genomic selection (GS) remains underutilized. The challenge is particularly acute for abiotic stress traits being highly quantitative and environmentally influenced and thus difficult to get assessed with conventional low-throughput methods. Bridging this gap between genetic potential and field performance remains one of the greatest barriers for accelerating genetic gain under climate change. Hence, the modern science of phenomics has gradually evolved, reflecting the shifting demands of crop improvement. Some key milestones in the evolution of phenomics include the following:

(a) Traditional phenotyping (early–mid-twentieth century): Based on visual scoring of morphological traits such as plant height, leaf colour, or disease severity. These methods culminated in major breeding achievements, including the Green Revolution, but were low-throughput, subjective, and highly environment-dependent (Allard 1999).
(b) Physiological phenotyping (mid–late twentieth century): Introduction of assays for photosynthesis, transpiration, water-use efficiency, and root depth provided insights into stress responses. However, these approaches were labour-intensive, destructive, and unsuitable for large breeding populations (Reynolds et al. 2012). Root phenotyping remains particularly challenging; alleviating this bottleneck has been a priority for automated platforms (Fiorani and Schurr 2013).
(c) High-throughput phenotyping (2000s): Advancements in imaging, robotics, and spectral sensors led to automated, non-invasive, and high-throughput measurements of morphological, physiological, and biochemical traits (Furbank and Tester 2011; Fiorani and Schurr 2013).
(d) Integrative phenomics (2010s–present): Integration of high-throughput platforms with molecular omics (genomics, transcriptomics, metabolomics) and computational modelling. This systems-level integration links molecular networks with whole-plant performance and accelerates genetic gains along with enriching breeding pipelines (Tardieu et al. 2017; Fiorani and Schurr 2013; Yang et al. 2020). An overview of the integrative phenomics workflow is presented in Fig. 3.1.

Advanced phenomics helps in capturing structural traits like canopy architecture, leaf area, and biomass accumulation. Enhancements in imaging and sensors had also helped in capturing physiological traits like transpiration efficiency, stomatal conductance, and photosynthetic performance. Together, these techniques along with the advanced phenotyping platforms help in capturing temporal dynamics like growth rates, stress onset, and recovery trajectories of the plants over time. When coupled with genomic information, such measurements enhance QTL detection, GWAS resolution, and the predictive accuracy of GS models (Crossa et al. 2017). Moreover, integration with transcriptomic, metabolomic, and proteomic datasets

Fig. 3.1 Conceptual workflow of integrative phenomics. Diverse imaging techniques (RGB, hyperspectral, thermal, chlorophyll fluorescence, LiDAR) are combined with environmental sensors, automation, and data analytics to capture structural, physiological, and temporal traits. When linked with genomics and other 'omics' datasets, these measurements bridge the genotype–phenotype gap and accelerate breeding for stress resilience

facilitates systems-level models, revealing the underlying regulatory networks of stress responses (Großkinsky et al. 2015).

A study at CIMMYT Mexico wheat fields involved capturing canopy temperature (CT) and normalized difference vegetation index (NDVI) from an aerial plane and using the generated datasets in genomic prediction models, thereby significantly improving selection accuracy for grain yield (Rutkoski et al. 2016). Such studies demonstrate the bridging capabilities of integrative phenomics between phenotype and genotype.

3.4 Technological Advances in Phenomics: Sensors, Remote Sensing, and UAV Platforms

Plant phenotyping traditionally had been a manual and labour-intensive process. However, the development of advanced sensors, imaging systems, and remote sensing platforms has transformed it into an automated and high-throughput technology. Modern phenomics leverages both controlled-environment platforms and field-based sensing systems to capture complex plant traits under diverse conditions (Fiorani and Schurr 2013).

3.4.1 Imaging-Based Technologies

Imaging forms the backbone of integrative phenomics. Unlike manual scoring, imaging captures plant dynamics in spatial and temporal domains with high precision by enabling automatable and non-invasive assessment of plant traits. Major sensor technologies and their trait applications are summarized in Table 3.1. Together, these techniques and platforms provide complementary insights into structural, physiological, and biochemical processes in plants.

3.4.1.1 RGB Imaging (400–700 nm)

This provides canopy architecture, greenness, and biomass estimates. High-throughput RGB imaging in controlled environments has been used to accurately estimate canopy traits, shoot biomass, etc. in various crops like wheat and barley under salinity stress conditions (Golzarian et al. 2011).

3.4.1.2 Thermal Imaging (8000–14,000 nm, Long-Wave Infrared)

This detects CT depression (CTD) linked to stomatal conductance and transpiration efficiency. Cooler canopies under stress indicate greater drought avoidance. Unmanned aerial vehicle (UAV)-based thermal imaging systems have been used to identify superior wheat genotypes under field drought conditions (Fiorani and Schurr 2013; Tardieu et al. 2017).

3.4.1.3 Hyperspectral Imaging (400–2500 nm)

This is capable of capturing reflectance across hundreds of bands ranging from visible spectrum to beyond red (infra-red) and short-wave infrared wavelengths of the electromagnetic spectrum. This minute detailing is capable of capturing pigment changes, photosynthetic efficiency, and structural alterations in the plants. It enables early pathogen detection in crops such as wheat and sugar beet before visible lesions appear (Mahlein et al. 2012).

Table 3.1 List of major sensor technologies used in plant phenomics

Sensor type	Traits measured	Key applications	References
RGB imaging	Projected shoot area, biomass (via image-derived models)	High-throughput estimation of shoot biomass in wheat and barley under salinity stress (controlled environment)	Golzarian et al. (2011)
Thermal imaging	CT, transpiration	Drought tolerance, stomatal conductance	Fiorani and Schurr (2013); Tardieu et al. (2017)
Hyperspectral imaging	Reflectance spectra, pigment composition, stress physiology	Early pathogen detection in crops	Mahlein et al. (2012)
Fluorescence imaging	Photosynthetic efficiency (PSII)	Heat, drought, pathogen stress	Baker (2008)
LiDAR/UAV 3D imaging	Plant height (high-throughput, heritable), canopy height dynamics	Correlated to biomass and flowering stage, detection of water stress impact, assessment of lodging	Madec et al. (2017)

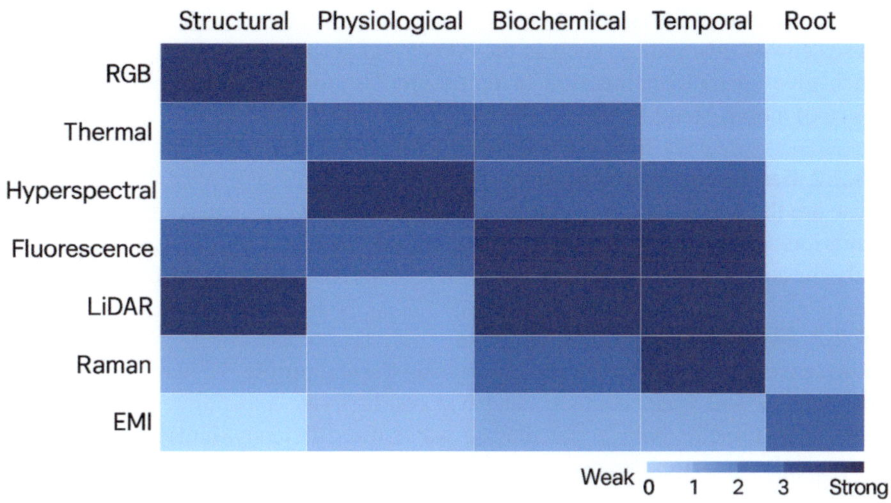

Fig. 3.2 Suitability matrix linking sensor modalities with measurable plant traits. Imaging platforms such as RGB, thermal, hyperspectral, fluorescence, and LiDAR vary in their capacity to quantify structural, physiological, and biochemical responses. Such matrices guide selection of appropriate sensors for stress resilience screening

3.4.1.4 Fluorescence Imaging

This is capable of capturing the re-emitted light from the chlorophyll molecule, which has not been utilized in the photosynthetic process. Thus, it gives spectral signatures especially for the photosystem II (PSII) (Fv/Fm) efficiency. The PSII efficiency is highly relevant as it is highly sensitive to abiotic and biotic stresses (Baker 2008).

3.4.1.5 3D Imaging and LiDAR

These can provide accurate measurements of plant and canopy heights and plant architecture in a 3D space. In wheat, UAV- and ground-based light detection and ranging (LiDAR) platforms had generated high-throughput data on wheat heights in well-watered and stress conditions, helping in predicting flowering, stress monitoring, and lodging assessment (Madec et al. 2017).

The suitability of different sensor modalities for measuring stress-related traits is summarized in Fig. 3.2.

3.4.2 Sensor-Based Physiological Monitoring

Small devices, which can be either handheld or mounted on small field instruments/vehicles, are very handy in the physiological monitoring of the crop plants. They are equipped with sensors capable of measuring light of specific wavelengths from the plants. Some of the important sensors include the following:

3.4.2.1 Spectral Sensors

These sensors give indices such as the NDVI and photochemical reflectance index, which give an indirect measure of chlorophyll status and photosynthetic performance of a crop plant.

3.4.2.2 Gas Exchange Sensors

These are the low-throughput instruments that measure CO_2 and water vapour concentration across the stomata (inside vs. outside of leaf). By doing so, they indirectly measure the photosynthetic rate and transpiration efficiency of a plant.

3.4.2.3 Soil and Root-Zone Sensors

These sensors assess moisture and nutrients in the soil profile. They measure parameters such as soil water content, electrical conductivity, and nitrate concentration. Such measurements help in calculating water-use efficiency, nutrient uptake efficiency, and many stress responses.

3.4.2.4 Handheld Devices

Small handheld devices like SPAD meters, infrared thermometers, and portable fluorometers provide cost-effective options for field phenotyping.

3.4.3 Remote Sensing and UAV Platforms

The use of UAVs and satellites for field phenotyping has given a giant leap to modern phenomics. Plot- to field-scale canopy stress responses are recorded by UAVs fitted with RGB, thermal, multispectral, or hyperspectral cameras (Yang et al. 2017). Remote sensing further offers the regional crop monitoring with coarser resolution (Araus and Cairns 2014), while proximal sensing with tractor- or gantry-mounted sensors provides high-resolution data in breeding nurseries (Deery et al. 2014). Table 3.2 provides a list of representative UAV and remote sensing platforms for field phenotyping. These methods provide a link between controlled-environment research and field-based G × E interactions (Fig. 3.3).

3.4.4 Integration with Data Analytics

The usage of advanced imaging and sensor platforms generates terabytes of data each cropping cycle. This high-throughput phenotyping generates massive datasets that require machine learning (ML), big data analytics, and predictive modelling for interpretation (Tardieu et al. 2017; Furbank and Tester 2011; Montesinos-López et al. 2023). A key hurdle is the integration of data coming from diverse platforms like imaging, sensors, and genomics, often in different formats. For management and visualization, these layers need sophisticated information systems and standardized workflows or pipelines (Tardieu et al. 2017). When properly mined, this

Table 3.2 UAV and remote sensing platforms in phenomics

Platform	Sensor payloads	Key applications	References
UAV (drone)	RGB, multispectral, thermal, hyperspectral	Large-scale field phenotyping, canopy stress analysis	Yang et al. (2017)
Tractor-mounted	Multispectral, fluorescence	Breeding nurseries, proximal sensing	Deery et al. (2014)
Remote sensing based	Multispectral, radar	Regional crop monitoring, stress mapping	Araus and Cairns (2014)

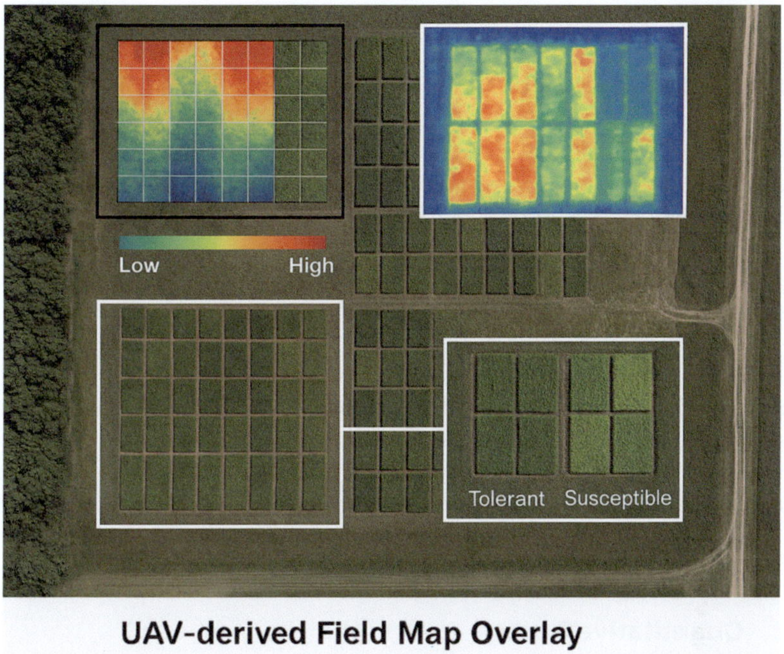

Fig. 3.3 UAV-based remote sensing of crop fields. High-resolution RGB, multispectral, or thermal images are overlaid into field maps, providing spatial overlays of CT, vegetation indices, and stress patterns. These maps allow breeders to visualize G × E interactions across large trial plots

data can significantly improve the breeding prediction by exposing G × E interactions (Tardieu et al. 2017; Montesinos-López et al. 2023).

Hence, the true potential of integrative phenomics can only be realized upon the effective data mining of these large datasets. By enabling earlier, more accurate, and environment-specific selection, this integration can serve as a crucial step in the breeding of climate-resilient crops.

3.5 Stress Biology in Plants: From Genetics to a Multidisciplinary Approach

Plants encounter a wide range of abiotic (drought, salinity, heat, nutrient deficiencies) and biotic (pathogens, pests, weeds) stresses that reduce productivity and pose a threat to food security. The traits governing these responses can be broadly classified either as qualitative or quantitative, depending on their genetic basis and inheritance patterns. Understanding this distinction is central to designing effective phenotyping and breeding strategies.

3.5.1 Qualitative Traits

Qualitative traits are typically controlled by one or a few major genes. They display discrete phenotypic classes such as resistant vs. susceptible. They often follow Mendelian inheritance patterns and are less influenced by environmental variation (Allard 1999). Examples include stem rust resistance in wheat conferred by *Sr31* (Singh et al. 2011) and bacterial blight resistance in rice conferred by *Xa21* (Song et al. 1995). Because of their simple inheritance, such traits have historically been easy to select using conventional phenotyping and marker-assisted selection. However, their durability is often limited, as pathogens and pests rapidly evolve to overcome single-gene resistance (McDonald and Linde 2002). Phenomics enhances the utility of qualitative traits by enabling rapid and objective-based large-scale screening. Imaging platforms can detect early infection symptoms, such as lesions, chlorosis, or necrosis, well before they are visible to the human eye, thus accelerating selection in breeding nurseries (Mahlein et al. 2012).

3.5.2 Quantitative Traits

Quantitative traits are often polygenic, with several minor-effect alleles contributing to a phenotype having continuous variation. Hence, they are strongly influenced by the interaction of genotype with environment (Collins et al. 2008). The high variation and environmental influence make their prediction difficult and pose a high complexity in selection. Key examples include yield stability under drought and heat, salinity tolerance in cereals, and partial resistance to foliar diseases such as spot blotch (*Bipolaris sorokiniana*), *Septoria tritici* blotch (*Zymoseptoria tritici*), and tan spot (*Pyrenophora tritici-repentis*) in wheat (Gahtyari et al. 2021; Roy et al. 2021; He et al. 2022). Traditional phenotyping of such traits through visual scoring of wilting, biomass reduction, or yield loss has been slow, subjective, and prone to environmental influence, limiting breeding progress (Reynolds and Langridge 2016). Modern integrative phenomics addresses these limitations by enabling high-throughput, objective, and precise trait measurement under field and controlled conditions. For example, thermal imaging quantifies CTD, which is highly correlated with stomatal conductance and water-use efficiency under drought conditions

(Fiorani and Schurr 2013; Tardieu et al. 2017). Hyperspectral reflectance detects subtle changes in pigment composition (e.g. chlorophylls, carotenoids), nutrient status, and stress-induced alterations in the plant system (Mahlein et al. 2012; Mahlein 2016). The UAV and remote sensing allow repeated, large-scale monitoring of stress responses across breeding nurseries (Araus and Cairns 2014). Root phenotyping involves conventional field phenotyping along with modern imaging approaches. Field methods like shovelomics are used for crown root scoring (Trachsel et al. 2011), while rhizotrons, 2D X-ray imaging, and X-ray computed tomography enable non-destructive analysis of root architecture (Fiorani and Schurr 2013). These approaches help in the development of root ideotypes, emphasizing deep and efficient root systems for capturing water and nutrients (Lynch 2013).

Thus, integrative phenomics turns phenotyping from a bottleneck to a driver of genetic gain for complex traits by capturing both temporal dynamics and spatial variation. This enhances QTL mapping, GWAS, and genomic prediction for stress resilience (Rutkoski et al. 2016; Crossa et al. 2017).

3.5.3 Agronomic Strategies and Phenomics Integration

Agronomic interventions remain a crucial approach to mitigate abiotic stresses, and their integration with phenomics platforms makes them more precise and data-driven. Real-time sensing tools such as UAV-based thermal imaging, hyperspectral reflectance, and canopy sensors allow the effectiveness of these strategies to be tracked and refined under field conditions.

3.5.3.1 Temperature-Related Stress Management

Adjusting sowing time helps crops to escape temperature extremes (both low and high temperature). In wheat, advancing sowing by 10–15 days in the Indo-Gangetic plains reduces terminal heat stress. Supplemental irrigation at anthesis stage reduces CT, whereas pre-frost irrigation stabilizes the canopy microclimate. Balanced application of nitrogen and potassium enhances tolerance, while foliar sprays of osmoregulators, anti-transpirants, and silicon further support stress resilience (Dubey et al. 2019; Adhikari et al. 2022). Phenomics tools, especially UAV thermal imaging, help to identify heat stress hotspots and guide decisions on optimal sowing windows and input management.

3.5.3.2 Water-Related Stress Management

Prolonged drought can be alleviated through supplemental irrigation, mulching, hydrogels, and in situ water harvesting. Anti-transpirants minimize transpiration losses, and drought-tolerant short-duration varieties synchronize crop maturity within the period of water availability. Similarly, ridge and furrow planting, raised beds, and efficient field drainage improve root aeration under situation of waterlogging, while split nitrogen and adequate potassium support in crop recovery (Bodner et al. 2015; Velmurugan et al. 2016). Phenomics allows rapid detection of canopy wilting, chlorophyll changes, and waterlogging symptoms, supporting timely corrective interventions.

3.5.3.3 Soil-Related Stress Management

Nutrient depletion, salinity, acidity, and heavy metal contamination remain major constraints. Application of slow-release fertilizers, balanced fertilization, and integrated nutrient management improve nutrient-use efficiency and soil health (Selim 2020; Rashid et al. 2021). Further, gypsum and organic amendments help in reclamation of saline soils, while biochar and compost immobilize toxic metals and restore soil fertility (Zeng et al. 2017; Saleem et al. 2024). High-throughput hyperspectral imaging and root phenotyping platforms enable early detection of nutrient deficiencies, salt stress, or heavy metal toxicity, facilitating site-specific remediation strategies.

3.5.3.4 Precision Agriculture and Integrated Nutrient Management

Precision agriculture applies sensors, IT, and site-specific machinery to optimize inputs under spatially variable soils (Gebbers and Adamchuk 2010). Integrated nutrient management complements this by blending organic amendments with judicious fertilizers to improve soil health and stress resilience (Selim 2020). Laser land levelling coupled with crop residue retention improves drainage and enhances water-use efficiency (Jat et al. 2006), while split nitrogen application, use of slow-release fertilizers, and adequate potassium strengthen crop recovery under both waterlogging and drought conditions (Nelson et al. 2011; Rashid et al. 2021). Phenomics links with precision agriculture–integrated nutrient management by mapping canopy nutrition, chlorophyll status, and stress signatures, enabling real-time adjustment of nutrient timing and placement, and guiding resilient, resource-efficient management.

3.5.4 Biochemical Foundations of Stress Resilience in Crops

Recent advances in integrative phenomics regarding linking genomics with high-throughput phenotyping are revealing how biochemical layers translate genetic variation into functional resilience. A recent pan-omics review emphasized the integration of pan-genomics, pan-transcriptomics, pan-proteomics, pan-metabolomics, and pan-phenomics to dissect core, dispensable, and unique biochemical components across diverse crop accessions, thereby identifying key determinants of stress resilience (Admas et al. 2025).

At the proteomic level, mass spectrometry (MS) technologies are mapping dynamic protein networks and post-translational modifications such as phosphorylation, ubiquitination, and acetylation, which fine-tune stress response pathways. Protein–protein interaction analyses, for instance, in drought-stressed rice, have identified regulatory hubs central to signalling cascades that may be targeted for molecular breeding (Amin et al. 2025; Choudhary et al. 2025).

Metabolomics complements proteomics by profiling small-molecule signatures that reflect the plant's physiological state under stress conditions. Advanced detection techniques (LC–MS, GC–MS, NMR) enable monitoring of shifts in osmoprotectants (e.g. proline, trehalose), antioxidants (e.g. phenolics, flavonoids), hormone precursors (e.g. ABA intermediates), and lipophilic compounds (e.g. fatty acids, phytosterols, VOCs), all of which contribute to protection and signalling under stress (Głuchowska et al. 2025).

Recent RNA-seq studies show that certain proteins, called transcription factors, help plants cope with stress. These transcription factors, which belong to different families, change how plant metabolism functions under stressful conditions. They also control the production of protective compounds such as flavonoids and alkaloids. These compounds make plants stronger and more resistant to harsh environments (Thingujam et al. 2025).

3.6 Phenomics for Abiotic Stress Resilience

It is estimated that around 50% crop yield is affected by abiotic stresses including drought, heat, salinity, nutrient deficiency, cold, and submergence in the vulnerable regions (Mittler 2006; Fleury et al. 2010; Singh et al. 2016). The advent of phenomics will help us in tackling these challenges with accurate high-throughput phenotyping tools for trait-based breeding.

3.6.1 Drought and Water Stress

Drone-based thermal imaging is a model widely applied for the evaluation of CTL (a trait strongly related to stomatal conductance and transpiration efficiency). These phenotyping measurements are complemented by root phenotyping techniques such as shovelomics (Trachsel et al. 2011) for rapid field scoring and advanced non-invasive imaging approaches (Fiorani and Schurr 2013). Ideotype concepts such as 'Steep, Cheap, and Deep' highlight the significance of deep and efficient root growth in enhancing drought adaptation (Lynch 2013).

3.6.2 Heat Stress

Heat tolerance can be assessed by chlorophyll fluorescence imaging, which measures the stability of PSII, a very sensitive biosensor to temperature-induced stress. Hyperspectral reflection approaches provide further information on pigment degradation and photosynthetic damage prior to their visual expression (Baker 2008; Mahlein et al. 2012; Mahlein 2016).

3.6.3 Salinity and Nutrient Stresses

Phenomics methods are also being used more and more in salinity and nutrient stresses. The UAV-borne multispectral indices, such as the NDVI and photochemical reflectance index, are used for high-throughput, non-destructive monitoring of canopy greenness and nitrogen status in cereals (Rutkoski et al. 2016; Varela et al. 2021). In contrast, hyperspectral reflectance has been effectively used to estimate leaf chlorophyll concentration and nitrogen content in maize across a range of field conditions (Yendrek et al. 2017).

3.6.4 Dynamic Stress Monitoring

Phenomics is essential for tracking recovery and growth dynamics in addition to stress detection. The UAV-based time-series imaging provides early indicators of yield performance under variable water availability by allowing continuous monitoring of canopy development and biomass accumulation in crops like sorghum (Varela et al. 2021).

3.7 Phenomics for Biotic Stress Resilience

Biotic stresses such as fungal, bacterial, and viral diseases, along with insect pests, are becoming more severe under changing climatic conditions (Chakraborty and Newton 2011). Traditional visual scoring methods often prove subjective and inconsistent, whereas phenomics technologies allow for early, objective, and quantitative detection of plant–pathogen interactions.

For early disease detection, hyperspectral reflectance has been shown to identify pathogen-induced physiological changes well before visible lesions develop (Mahlein et al. 2012; Mahlein 2016). Complementary tools such as thermal imaging detect localized canopy heating, and chlorophyll fluorescence captures photosynthetic decline during infection. In addition, automated image analysis enables precise quantification of lesion development over time (Bock et al. 2010), improving the evaluation of partial resistance traits such as slow rusting in wheat (Singh et al. 2016).

Beyond detection, phenomics contributes to breeding and epidemiology. High-resolution disease phenotyping strengthens QTL mapping, GWAS, and genomic prediction efforts aimed at developing durable resistance (Poland and Rutkoski 2016). When integrated with epidemiological models, these data also enhance the forecasting of disease outbreaks and risk management strategies (Pautasso et al. 2012).

Phenomics applications are also expanding to insect–plant interactions. Hyperspectral imaging has successfully detected aphid-induced stress signatures at the leaf level in cotton (Chen et al. 2018) and more recently in winter wheat through canopy reflectance combined with ML approaches (Skendžić et al. 2025). The

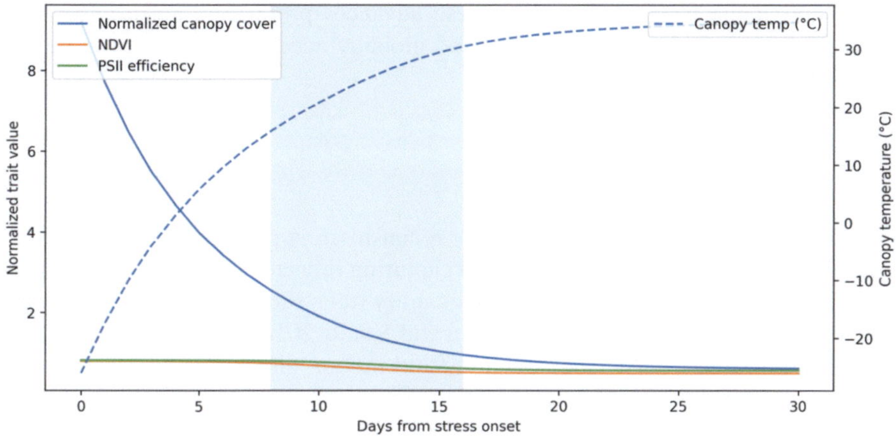

Fig. 3.4 Temporal trajectories of complementary phenotypes under drought stress. Normalized canopy cover (derived from RGB imaging), NDVI (from multispectral imaging), and chlorophyll fluorescence (PSII efficiency) each show distinct temporal patterns under drought. Together, these trajectories reveal the onset, severity, and recovery phases of stress, providing a multidimensional framework for assessing crop resilience

UAV-based surveys extend these capabilities to large-scale pest monitoring, while root phenotyping methods provide valuable insights into below-ground interactions with nematodes and soil insects (Trachsel et al. 2011; Lynch 2013).

3.7.1 Integration of Sensing Modalities for Stress Resilience

Beyond single traits, integrating multiple phenotyping modalities provides a dynamic view of plant stress responses. The RGB imaging captures canopy cover reduction, and multispectral indices such as NDVI track photosynthetic decline (Araus and Cairns 2014; Mahlein 2016), while chlorophyll fluorescence detects early losses in PSII efficiency (Baker 2008). Together, these complementary signals enable time-resolved monitoring of stress onset, intensity, and recovery, offering a robust framework to identify resilient genotypes (Furbank and Tester 2011; Rutkoski et al. 2016). Representative temporal trajectories of canopy cover (RGB), NDVI, and chlorophyll fluorescence under drought are shown in Fig. 3.4.

3.8 Emerging Technologies and AI Integration

The rapid evolution of sensing technologies, automation, and computational tools continues to expand the scope of plant integrative phenomics. Beyond traditional imaging and spectral platforms, several emerging technologies are reshaping how stress responses are measured and analysed. At the same time, the scale and complexity of integrative phenomics data demand advanced AI and big data analytics to derive actionable insights. Together, these advances promise to make integrative phenomics more powerful, predictive, and globally accessible. Some of those are discussed in the following sections.

3.8.1 LiDAR and 3D Imaging

LiDAR has become an important tool for reconstructing plant architecture in three dimensions. By emitting laser pulses and capturing reflected signals, LiDAR generates high-resolution canopy models that quantify traits such as plant height, canopy structure, and foliage distribution (Fiorani and Schurr 2013; Tardieu et al. 2017). In wheat, UAV-based LiDAR and structure-from-motion photogrammetry have been shown to estimate plant height accurately, with strong heritability and potential as proxies for biomass and stress responses (Madec et al. 2017). These approaches allow breeders to assess complex canopy traits quickly and objectively, rather than relying on manual scoring. In sorghum, UAV multispectral time-series imaging has captured canopy growth dynamics and enabled early yield prediction, further highlighting the potential of UAV platforms for large-scale breeding trials (Varela et al. 2021). In controlled environments, X-ray computed tomography and magnetic resonance imaging provide non-destructive visualization of root system architecture, a long-standing bottleneck in phenotyping (Mooney et al. 2012). These tools reveal deeper or more branched root systems that contribute to drought adaptation and resource capture.

3.8.2 Raman and Fluorescence Spectroscopy

Spectroscopic approaches are expanding phenomics into the biochemical domain by enabling non-destructive detection of metabolites and pigments associated with stress responses. For instance, Raman spectroscopy detects molecular vibrations to generate a biochemical 'fingerprint' of plant tissues. It enables non-destructive monitoring of metabolites such as carotenoids, phenolics, lignin, and nitrates, which are key indicators of stress resilience. Portable Raman leaf-clip sensors have demonstrated early, in vivo detection of nitrogen deficiency and drought responses in crops (Gupta et al. 2020), while broader applications include rapid diagnostics of both abiotic stresses and pathogen infections (Payne and Kurouski 2021). The fluorescence spectroscopy provides real-time information on photosynthetic pigments and chlorophyll status and is highly sensitive to stress-induced changes in PSII efficiency. Fluorescence imaging has been widely applied to detect early reductions in PSII activity under heat and other stresses, supporting the identification of stress-resilient genotypes in crops (Baker 2008; Fiorani and Schurr 2013; Mahlein et al. 2012).

Raman and fluorescence spectroscopy together provide valuable biochemical and physiological insights that complement imaging-based structural and thermal traits. These methods, by connecting pigment composition and metabolite dynamics with visible stress symptoms, enable the identification of genotypes that sustain functional stability under abiotic and biotic stresses.

3.8.3 Robotics and Automation

Automation is transforming controlled-environment phenotyping platforms. Conveyor belt systems automatically transport plants to imaging stations, enabling continuous monitoring of large populations (Granier et al. 2006). Robotic gantries and sensor arms ensure precise, repeatable data acquisition across experiments (Fiorani and Schurr 2013; Tardieu et al. 2017). In field settings, autonomous ground vehicles equipped with multispectral, hyperspectral, or LiDAR sensors are increasingly being used to collect plot-level data with minimal labour input (Tardieu et al. 2017). These technologies reduce costs, enhance throughput, and improve the reproducibility of phenomics experiments. Globally, several dedicated automated phenomics facilities have been established, ranging from controlled-environment conveyor platforms to field-based robotic systems (Table 3.3). These facilities serve as benchmarks for technological innovation and provide critical data for breeding pipelines.

3.8.4 Internet of Things and Portable Sensors

To complement high-end platforms, portable devices are increasingly used for scalable and affordable phenotyping. Chlorophyll fluorometers and infrared thermometers provide rapid, non-destructive measurements of photosynthetic efficiency and CT, traits directly linked to stress adaptation (Baker 2008). Smartphone-based imaging applications, often integrated with cloud-based ML, are now widely used to estimate canopy cover, disease severity, and biomass in crops (Dhanaraju et al. 2022; Röckel et al. 2022). Beyond individual devices, integrative sensor networks and field-deployable platforms combine soil and plant sensors to continuously monitor water availability, nutrient status, and microclimate (Fiorani and Schurr 2013; Tardieu et al. 2017). At ICRISAT, the LeasyScan facility integrates outdoor 3D laser scanning with lysimetric gravimetric scales, enabling high-throughput monitoring of canopy structure and transpiration efficiency under drought in crops including pearl millet, cowpea, and peanut (Vadez et al. 2015). By making stress-related trait measurement affordable, flexible, and field-ready, Internet of Things (IoT) and portable sensors are increasingly supporting integrative phenomics for breeding programmes in diverse environments.

3.8.5 Artificial Intelligence and Big Data Analytics

High-throughput phenotyping generates vast and complex datasets that require advanced computational tools for meaningful interpretation. The AI and ML

Table 3.3 List of selected automated high-throughput phenomics facilities across the globe

Facility (short name)	Institution/location	Core platform type	Key capabilities/ focus traits	References
The Plant Accelerator (APPF)	University of Adelaide, Australia	Greenhouse conveyor (LemnaTec)	High-throughput monitoring of growth dynamics, stress responses, and water-use efficiency	Furbank and Tester (2011)
Bellwether Facility	Danforth Plant Science Center, USA	Modular greenhouse imaging	Time-series imaging for growth and stress trait dynamics	Fahlgren et al. (2015)
Jülich Plant Phenotyping Center	Forschungszentrum Jülich, Germany	Greenhouse/ controlled-environment systems with conveyor and robotic imaging	High-throughput shoot and root phenotyping, growth dynamics, 3D architecture, non-invasive stress physiology	Fiorani and Schurr (2013)
ICRISAT LeasyScan	ICRISAT, Hyderabad, India	Outdoor 3D scanning + lysimetry	Canopy structure and transpiration phenotyping in drought-resilient cereals and legumes	Vadez et al. (2015)
Nanaji Deshmukh Plant Phenomics Centre	ICAR-IARI, New Delhi, India	High-throughput Scanalyzer greenhouse	RGB/NIR/IR/ hyperspectral imaging with automated weighing/watering and climate control	Singh et al. (2023)
PHENOPSIS	INRA, Montpellier, France	Automated growth-chamber system with pot weighing, irrigation, and digital imaging	High-throughput phenotyping of *Arabidopsis* under controlled soil water deficit; reproducible stress response curves	Granier et al. (2006)
LemnaTec Scanalyzer (The Plant Accelerator)	Univ. of Adelaide, Australia	Greenhouse RGB 2D/3D imaging	Biomass estimation in wheat and barley under salinity; model using projected area and age	Golzarian et al. (2011)

approaches automate trait extraction from imaging and spectral data, enabling rapid classification of phenotypic variation and stress responses (Montesinos-López et al. 2023; Ubbens and Stavness 2017). For example, deep learning applied to plant images has enabled tasks such as leaf counting, mutant classification, and age regression, providing a basis for UAV phenotyping applications (Ubbens and Stavness 2017). Beyond trait extraction, predictive models such as random forests, support vector machines, and neural networks can forecast yield and stress resilience from integrated datasets. Deep learning has also been applied to multi-trait genomic prediction, enabling breeders to prioritize superior genotypes earlier in the cycle (Montesinos-López et al. 2023). Big data frameworks further support this integration by combining phenotypic, genomic, and environmental datasets across multiple sites and seasons, strengthening the discovery of G × E interactions (Tardieu et al. 2017). Open-source initiatives such as BrAPI and mobile phenotyping tools like PhenoApp provide standardized pipelines and formats, ensuring interoperability and global sharing of phenomics data (Röckel et al. 2022). By linking AI-driven analytics with field-scale phenomics, breeders gain predictive models that enhance selection efficiency, reduce reliance on subjective scoring, and accelerate the development of stress-resilient cultivars.

3.9 Case Studies in Stress Resilience

Although integrative phenomics has shown considerable promise for crop improvement, its practical use in breeding pipelines is still rather limited. Even so, examples from recent work make it clear that these approaches can deliver real benefits. In cereals, for instance, imaging and spectral tools have been applied to monitor CT, biomass production, and photosynthetic activity, traits that serve as reliable indicators of drought and heat resilience. In legumes and root crops, phenomics has been applied to pick up stress symptoms early and to evaluate traits such as water-use efficiency, nutrient uptake, and resistance to diseases. More recently, researchers have begun extending these methods to underutilized crops such as sorghum, chickpea, quinoa, and cassava. Work in these crops highlights the potential of phenomics not only to improve resilience but also to broaden the range of species considered in breeding programmes. Examples of such applications across different crop groups are summarized in Table 3.4.

3.10 Translational Phenomics: Opportunities and Challenges for Breeding Pipelines

Integrated phenomics is the pressing need of modern times for translating the breeding pipeline into climate-resilient cultivars. However, the technology had largely been confined to the large and well-funded breeding programmes. However, the technology has immense potential for integration into the regular breeding programmes, even for the small breeding centres, as it has generated a lot of key insights into the stress biology of plants. Integrative phenomics adds value to the breeding programme in a variety of ways. Precision in phenotyping enhances the resolution capabilities of mapping techniques like QTL mapping and GWAS, leading to the identification of reliable and robust markers. This is truer in the case of many abiotic

Table 3.4 Selected applications of integrative phenomics in major and underutilized crops

Crop/stress	Phenomics approach	Key outcomes/applications	References
Wheat/drought and heat	Field-based thermal and multispectral imaging (CT, NDVI)	CT and NDVI as secondary traits improved accuracy of pedigree and GS for grain yield	Rutkoski et al. (2016)
Rice/nutrient stress (NPK)	Terrestrial hyperspectral imaging and deep learning	Accurate classification of N, P, and K deficiency patterns using transformer models	Wang et al. (2024)
Maize/drought and nutrient stress	Leaf hyperspectral reflectance (500–2400 nm) and PLSR modelling	Accurate, high-throughput prediction of leaf chlorophyll, nitrogen content, SLA, Vmax, sucrose, and other physiological/biochemical traits in the field	Yendrek et al. (2017)
Sorghum/biomass/drought	UAV multispectral time-series imaging + ML	Prediction of above-ground biomass from growth dynamics; early- and mid-season canopy traits improved yield prediction	Varela et al. (2021)
Chickpea/drought and heat	GWAS with phenotyping (root and yield traits)	Marker–trait associations for root traits and drought/heat adaptation	Thudi et al. (2017)
Quinoa/salinity	PlantArray functional phenotyping (gravimetric lysimetry)	Continuous monitoring of transpiration and growth; revealed contrasting water-use strategies under salinity	Jaramillo Roman et al. (2021)
Camelina sativa/germplasm variation	Trait phenotyping (seed, oil, morphology)	Characterization of 41 accessions; raw trait data provided for breeding use	Hotton et al. (2020)
Cassava/viral disease (CBSD)	Multispectral imaging + ML for disease detection	Early, presymptomatic detection of cassava brown streak virus (95% accuracy at 28 dpi); potential use in other vegetatively propagated crops	Grieve et al. (2024)

stresses, which are largely influenced by the environment (Rutkoski et al. 2016; Tardieu et al. 2017).

Early and non-invasive detection of stress responses such as CTD (Araus and Cairns 2014; Vadez et al. 2015), photosynthetic efficiency (Baker 2008), or root traits (Trachsel et al. 2011; Lynch 2013) enables earlier selection in the breeding cycle, saving time and resources. Integration of large genomics and phenomics datasets, together with the training of ML models on large phenomics datasets, not only improves selection efficiency but also improves yield forecasting and stress resilience prediction (Ubbens and Stavness 2017). Capabilities of field-based phenomics are further enhanced by UAVs, imaging, and proximal sensors, providing robust phenotypic data, which helps in better partitioning of genotypic, environmental, and G × E interactions for better cultivar deployment (Madec et al. 2017; Varela et al. 2021; Tardieu et al. 2017). Being discussed the high advantages of integrative phenomics, there are some major challenges for the widespread adoption of the technology.

3.10.1 High Costs and Infrastructure Needs

Integrative phenomics requires controlled-environment facilities and high-resolution imaging platforms. This requires major investments in infrastructure and staff (Furbank and Tester 2011).

3.10.2 Data Management and Analytics

Big data to the tune of many terabytes is generated in each cropping cycle. Hence, robust ML models and breeding pipelines are required for the efficient utilization of the humongous data generated each cycle (Selby et al. 2019).

3.10.3 Human Capacity Development

Integrative phenomics requires advanced operational knowledge in the field of modern sensor, big data analysis, and integration of information with genomics and breeding pipelines. Hence, human capacity development and training are a major challenge and requirement in the field.

3.10.4 Accessibility and Wider Adoption

Due to the higher costs involved, utilization of integrative phenomics facilities is confined to advanced research centres. Affordable tools like handheld sensors, smartphone imaging, and IoT devices can enhance the reach of the technology (Dhanaraju et al. 2022; Röckel et al. 2022).

Several translational strategies are emerging to overcome the mentioned challenges. Advanced phenotyping infrastructure can be used as regional phenotyping hub and can be shared among the breeding community. This not only saves costs but also enhances accessibility of the technology. Meanwhile, open-source tools and standards such as BrAPI and PhenoApps are facilitating the global sharing of phenomics datasets and streamlining data formats (Selby et al. 2019; Röckel et al. 2022). Resource-limited breeding programmes can be benefitted by cost-effective technologies like low-cost UAVs, portable sensors, and smartphone-based apps (Dhanaraju et al. 2022; Röckel et al. 2022). Finally, integrative phenomics can help in participatory approach of plant breeding by connecting farmer-defined priorities with phenomics-driven trait discovery.

3.11 Future Prospects

The future of crop improvement will depend on the tools that can accelerate the genetic gains and have the ability to deliver climate stress-resilient cultivars. Integrative phenomics will be one of the core pillars to such a vision, having the ability to bridge the genotype–phenotype gap through advances in sensing, computation, and systems biology. 'Digital twins' are predictive crop simulations that can create virtual crops and farming systems, enabling projections under climate change scenarios. These models can be continuously updated with the phenomic and environmental data (Wang 2024). By simulating genotype × management × environment interactions under diverse climates, such models can guide ideotype design and breeding strategies (Hammer et al. 2020).

Similarly, AI and ML models are further evolving, which can transform phenomics datasets into predictive tools. Deep learning, random forests, and support vector machines have been utilized to predict yield and stress resilience from imaging and environmental data, thereby enhancing selection efficiency (Montesinos-López et al. 2023; Ubbens and Stavness 2017). The role of integrative phenomics in enabling climate-smart agriculture is illustrated in Fig. 3.5.

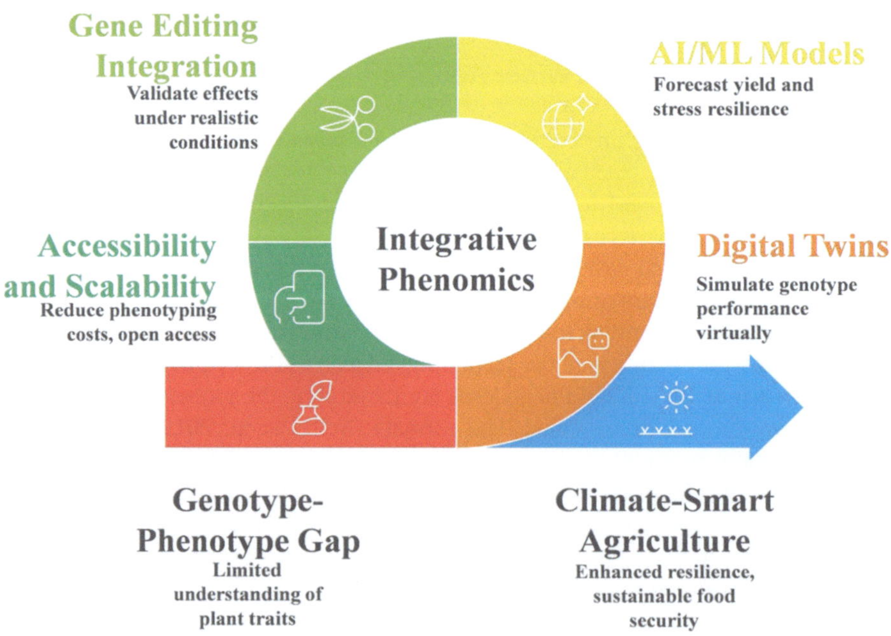

Fig. 3.5 Integrative phenomics framework for climate-smart agriculture. The model shows how advanced sensing platforms (imaging, UAVs, IoT devices), multi-omics datasets, gene editing, and AI converge within integrative phenomics to generate high-resolution trait information. When combined with breeding pipelines and agronomic management, these insights accelerate the development of stress-resilient cultivars and support sustainable, climate-smart cropping systems

Another frontier is the integration with gene editing and genomic tools. CRISPR/Cas-based editing can generate novel alleles for stress tolerance. Supplementally, integrative phenomics can validate the effects of these novel alleles under realistic stress conditions and can also detect potential pleiotropic impacts (Zhu et al. 2020). The integration of phenomics, multi-omics, and systems biology is essential for comprehending complex plant responses to a variety of stresses so as to understand the underlying molecular pathway in a better way (Zandalinas et al. 2021). The development of cost-effective sensors, portable devices, and IoT-based smartphone applications will help lower the cost of phenotyping and make it available to smaller breeding programmes. National breeding programmes and smallholder-focused research can benefit from shared platforms, open-source data standards, and training initiatives (Dhanaraju et al. 2022; Röckel et al. 2022; Selby et al. 2019).

3.12 Conclusions

A growing global population, resource constraints, and climate change make modern agriculture highly vulnerable to reduced productivity. Conventional breeding and molecular tools have delivered significant gains in the recent past, but are constrained by the phenotyping bottleneck. Genomic sciences had witnessed unprecedented growth, but phenotyping has lagged behind. The lag is more pressing in case of complex stress responses, often governed by the polygenes and heavily influenced by the environment. Thus, integrative phenomics has emerged as a transformative discipline to bridge this gap by enabling high-throughput, non-invasive, and dynamic measurement of plant traits under both controlled and field conditions.

Integrative phenomics uses advanced imaging, sensor platforms, UAVs, and remote sensing to give very accurate information about plant structure, function, and chemical processes. When combined with genomics, transcriptomics, and metabolomics, these datasets help us to understand how plants respond to stress on a systems level. This makes it easier to find novel traits and speeds up the breeding for climate resilience. Integrative phenomics has been successfully utilized in the major cereals like wheat, rice and maize, indicating its potential in maintaining food security. Conversely, usage in underutilized crops highlights its role in agricultural diversification and climate-smart farming systems. Emerging tools such as LiDAR, Raman spectroscopy, robotics, IoT-enabled devices, and smartphone-based platforms are expanding the scope of measurable traits and making integrative phenomics more accessible to breeding programmes. At the same time, advances in AI and big data analytics are transforming massive phenomics datasets into predictive breeding models. Moving forward, integration with digital twins, crop simulation models, and gene editing will further enhance the precision and efficiency of crop improvement.

For integrative phenomics to realize its full potential, translational efforts must address critical challenges of cost, scalability, data integration, and skilled human resources. Ensuring equitable access, even to the small-scale breeding programmes, will depend on the shared platforms, open-source solutions, and cost-effective technologies. In conclusion, integrative phenomics is slowly transforming from a new scientific frontier to a regular key feature of modern breeding programmes. Integrative phenomics can play a phenomenal role by developing climate-resilient varieties and securing global food and nutritional security in the twenty-first century by elucidating intricate stress-resilient traits.

References

Adhikari L, Baral R, Paudel D, Min D, Makaju SO, Poudel HP, Acharya JP, Missaoui AM (2022) Cold stress in plants: strategies to improve cold tolerance in forage species. Plant Stress 4:100081. https://doi.org/10.1016/j.stress.2022.100081

Admas T, Jiao S, Pan R, Zhang W (2025) Pan-omics insights into abiotic stress responses: bridging functional genomics and precision crop breeding. Funct Integr Genomics 25:128. https://doi.org/10.1007/s10142-025-01633-x

Allard RW (1999) Principles of plant breeding, 2nd edn. Wiley, New York

Amin A, Zaman W, Park S (2025) Harnessing multi-omics and predictive modeling for climate-resilient crop breeding: from genomes to fields. Genes 16:809. https://doi.org/10.3390/genes16070809

Araus JL, Cairns JE (2014) Field high-throughput phenotyping: the new crop breeding frontier. Trends Plant Sci 19(1):52–61. https://doi.org/10.1016/j.tplants.2013.09.008

Baker NR (2008) Chlorophyll fluorescence: a probe of photosynthesis in vivo. Annu Rev Plant Biol 59:89–113. https://doi.org/10.1146/annurev.arplant.59.032607.092759

Bock CH, Poole GH, Parker PE, Gottwald TR (2010) Plant disease severity estimated visually, by digital photography and image analysis, and by hyperspectral imaging. Crit Rev Plant Sci 29(2):59–107. https://doi.org/10.1080/07352681003617285

Bodner G, Nakhforoosh A, Kaul HP (2015) Management of crop water under drought: a review. Agron Sustain Dev 35(2):401–442. https://doi.org/10.1007/s13593-015-0283-4

Cattivelli L, Rizza F, Badeck F-W, Mazzucotelli E, Mastrangelo AM, Francia E, Marè C, Tondelli A, Stanca AM (2008) Drought tolerance improvement in crop plants: an integrated view from breeding to genomics. Field Crop Res 105:1–14. https://doi.org/10.1016/j.fcr.2007.07.004

Chakraborty S, Newton AC (2011) Climate change, plant diseases and food security: an overview. Plant Pathol 60(1):2–14. https://doi.org/10.1111/j.1365-3059.2010.02411.x

Chen T, Zeng R, Guo W, Hou X, Lan Y, Zhang L (2018) Detection of stress in cotton (*Gossypium hirsutum* L.) caused by aphids using leaf-level hyperspectral measurements. Sensors 18(9):2798. https://doi.org/10.3390/s18092798

Choudhary R, Ahmad F, Kaya C, Upadhyay SK, Muneer S, Kumar V, Meena M, Liu H, Upadhyaya H, Seth CS (2025) Decrypting proteomics, transcriptomics, genomics, and integrated omics for augmenting the abiotic, biotic, and climate change stress resilience in plants. J Plant Physiol 305:154430. https://doi.org/10.1016/j.jplph.2025.154430

Collins NC, Tardieu F, Tuberosa R (2008) Quantitative trait loci and crop performance under abiotic stress: where do we stand? Plant Physiol 147(2):469–486. https://doi.org/10.1104/pp.108.118117

Crossa J, Pérez-Rodríguez P, Cuevas J, Montesinos-López O, Jarquín D, de los Campos G, Burgueño J, González-Camacho JM, Pérez-Elizalde S, Beyene Y, Dreisigacker S, Singh R, Zhang X, Gowda M, Roorkiwal M, Rutkoski J, Varshney RK (2017) Genomic selection in plant breeding: methods, models, and perspectives. Trends Plant Sci 22(11):961–975. https://doi.org/10.1016/j.tplants.2017.08.011

Deery D, Jimenez-Berni J, Jones H, Sirault X, Furbank R (2014) Proximal remote sensing buggies and potential applications for field-based phenotyping. Agronomy 4(3):349–379. https://doi.org/10.3390/agronomy4030349

Dhanaraju MD, Krishnan P, Reddy DS, Bhadana VP, Padaria RN, Solanke AU, Bansal KC (2022) Internet of things (IoT) and mobile applications for crop phenotyping: opportunities and challenges. Curr Genomics 23:531–543. https://doi.org/10.2174/1389202923666220214110037

Dubey R, Pathak H, Singh S et al (2019) Impact of sowing dates on terminal heat tolerance of different wheat (Triticum aestivum L.) cultivars. Natl Acad Sci Lett-India 42:445–449. https://doi.org/10.1007/s40009-019-0786-7

Fahlgren N, Feldman M, Gehan MA, Wilson MS, Shyu C, Bryant DW, Hill ST, McEntee CJ, Warnasooriya SN, Kumar I, Ficor T, Turnipseed S, Gilbert KB, Brutnell TP, Carrington JC, Mockler TC, Baxter I (2015) A versatile phenotyping system and analytics platform reveals diverse temporal responses to water availability in *Setaria*. Mol Plant 8(10):1520–1535. https://doi.org/10.1016/j.molp.2015.06.005

Fiorani F, Schurr U (2013) Future scenarios for plant phenotyping. Annu Rev Plant Biol 64:267–291. https://doi.org/10.1146/annurev-arplant-050312-120137

Fleury D, Jefferies S, Kuchel H, Langridge P (2010) Genetic and genomic tools to improve drought tolerance in wheat. J Exp Bot 61(12):3211–3222. https://doi.org/10.1093/jxb/erq152

Furbank RT, Tester M (2011) Phenomics—technologies to relieve the phenotyping bottleneck. Trends Plant Sci 16(12):635–644. https://doi.org/10.1016/j.tplants.2011.09.005

Gahtyari NC, Roy C, He X, Roy KK, Reza MMA, Hakim MA, Malaker PK, Joshi AK, Singh PK (2021) Identification of QTLs for spot blotch resistance in two bi-parental mapping populations of wheat. Plants 10(5):973. https://doi.org/10.3390/plants10050973

Gebbers R, Adamchuk VI (2010) Precision agriculture and food security. Science 327(5967):828–831. https://doi.org/10.1126/science.1183899

Głuchowska A, Zieniuk B, Pawełkowicz M (2025) Unlocking plant resilience: metabolomic insights into abiotic stress tolerance in crops. Meta 15:384. https://doi.org/10.3390/metabo15060384

Golzarian MR, Frick RA, Rajendran K, Berger B, Roy S, Tester M, Lun DS (2011) Accurate inference of shoot biomass from high-throughput images of cereal plants. Plant Methods 7:2. https://doi.org/10.1186/1746-4811-7-2

Granier C, Aguirrezabal L, Chenu K, Cookson SJ, Dauzat M, Hamard P, Thioux JJ, Rolland G, Bouchier-Combaud S, Lebaudy A, Muller B, Simonneau T, Tardieu F (2006) PHENOPSIS, an automated platform for reproducible phenotyping of plant responses to soil water deficit in *Arabidopsis thaliana* permitted the identification of an accession with low sensitivity to soil water deficit. New Phytol 169(3):623–635. https://doi.org/10.1111/j.1469-8137.2005.01609.x

Grieve B, Duffy S, Dallas MM, Ascencio-Ibáñez JT, Alonso-Chavez V, Legg J, Hanley-Bowdoin L, Yin H (2024) Early detection of plant virus infection using multispectral imaging and machine learning. Plant Health Cases 2024:0010. https://doi.org/10.1079/planthealthcases.2024.0010

Großkinsky DK, Svensgaard J, Christensen S, Roitsch T (2015) Plant phenomics and the need for physiological phenotyping across scales to narrow the genotype-to-phenotype knowledge gap. J Exp Bot 66(18):5429–5440. https://doi.org/10.1093/jxb/erv345

Gupta S, Huang CH, Singh GP, Park BS, Chua NH, Ram RJ (2020) Portable Raman leaf-clip sensor for rapid detection of plant stress. Sci Rep 10:20206. https://doi.org/10.1038/s41598-020-76485-5

Hammer GL, McLean G, van Oosterom E, Chapman S, Zheng B, Wu A, Doherty A, Jordan D (2020) Designing crops for adaptation to the drought and high-temperature risks anticipated in future climates. Crop Sci 60(2):605–621. https://doi.org/10.1002/csc2.2011

He X, Gahtyari NC, Roy C, Dababat AA, Brar GS, Singh PK (2022) Globally important non-rust diseases of wheat. In: Reynolds MP, Braun H (eds) Wheat improvement: food security in a changing climate. Springer, Cham, pp 143–148. https://doi.org/10.1007/978-3-030-90673-3_15

Hotton SK, Kammerzell M, Chan R, Hernandez BT, Young HA, Tobias C, McKeon T, Brichta J, Thomson NJ, Thomson JG (2020) Phenotypic examination of *Camelina sativa* (L.) Crantz accessions from the USDA-ARS National Genetics Resource Program. *Plants* 9(5):642. https://doi.org/10.3390/plants9050642

IPCC (2021) Climate change 2021: the physical science basis. Contribution of working group I to the sixth assessment report of the intergovernmental panel on climate change. Cambridge University Press. https://doi.org/10.1017/9781009157896

Jaramillo Roman V, van de Zedde R, Peller J, Visser RGF, van der Linden CG, van Loo EN (2021) High-resolution analysis of growth and transpiration of quinoa under saline conditions. Front Plant Sci 12:634311. https://doi.org/10.3389/fpls.2021.634311

Jat ML, Chandna P, Gupta R, Sharma SK, Gill MA (2006) Laser land leveling: a precursor technology for resource conservation. Rice-Wheat Consortium Technical Bulletin Series, 7, p 48

Juroszek P, von Tiedemann A (2013) Climate change and potential future risks through wheat diseases: a review. Eur J Plant Pathol 136(1):21–33. https://doi.org/10.1007/s10658-012-0144-9

Lynch JP (2013) Steep, cheap and deep: an ideotype to optimize water and N acquisition by maize root systems. Ann Bot 112(2):347–357. https://doi.org/10.1093/aob/mcs293

Madec S, Baret F, de Solan B, Thomas S, Dutartre D, Jezequel S, Hemmerlé M, Colombeau G, Comar A (2017) High-throughput phenotyping of plant height: comparing unmanned aerial vehicles and ground LiDAR estimates. Front Plant Sci 8:2002. https://doi.org/10.3389/fpls.2017.02002

Mahlein AK (2016) Plant disease detection by imaging sensors - parallels and specific demands for precision agriculture and plant phenotyping. Plant Dis 100(2):241–251. https://doi.org/10.1094/PDIS-03-15-0340-FE

Mahlein A-K, Oerke E-C, Steiner U, Dehne H-W (2012) Recent advances in sensing plant diseases for precision crop protection. Eur J Plant Pathol 133(1):197–209. https://doi.org/10.1007/s10658-011-9878-z

McDonald BA, Linde C (2002) Pathogen population genetics, evolutionary potential, and durable resistance. Annu Rev Phytopathol 40:381–410. https://doi.org/10.1146/annurev.phyto.40.011402.113723

Mickelbart MV, Hasegawa PM, Bailey-Serres J (2015) Genetic mechanisms of abiotic stress tolerance that translate to crop yield stability. Nat Rev Genet 16(4):237–251. https://doi.org/10.1038/nrg3901

Mittler R (2006) Abiotic stress, the field environment and stress combination. Trends Plant Sci 11(1):15–19. https://doi.org/10.1016/j.tplants.2005.11.002

Montesinos-López A, Rivera C, Pinto F, Piñera F, Gonzalez D, Reynolds M, Pérez-Rodríguez P, Li H, Montesinos-López OA, Crossa J (2023) Multimodal deep learning methods enhance genomic prediction of wheat breeding. G3 Genes|Genomes|Genetics 13(5):jkad045. https://doi.org/10.1093/g3journal/jkad045

Mooney SJ, Pridmore TP, Helliwell J, Bennett MJ (2012) Developing X-ray computed tomography to non-invasively image 3-D root systems architecture in soil. Plant Soil 352(1–2):1–22. https://doi.org/10.1007/s11104-011-1039-9

Nelson KA, Scharf PC, Stevens WE, Burdick BA (2011) Rescue nitrogen applications for corn. Soil Sci Soc Am J 75(1):143–151. https://doi.org/10.2136/sssaj2010.0097

Pautasso M, Döring TF, Garbelotto M, Pellis L, Jeger MJ (2012) Impacts of climate change on plant diseases - opinions and trends. Eur J Plant Pathol 133(1):295–313. https://doi.org/10.1007/s10658-012-9936-1

Payne WZ, Kurouski D (2021) Raman-based diagnostics of biotic and abiotic stresses in plants: a review. Front Plant Sci 11:616672. https://doi.org/10.3389/fpls.2020.616672

Poland JA, Rutkoski JE (2016) Advances and challenges in genomic selection for disease resistance. Annu Rev Phytopathol 54:79–98. https://doi.org/10.1146/annurev-phyto-080615-100056

Rashid M, Hussain Q, Khan KS, Alwabel MI, Hayat R, Akmal M, Ijaz SS, Alvi S (2021) Carbon-based slow-release fertilizers for efficient nutrient management: synthesis, applications, and future research needs. J Soil Sci Plant Nutr 21(2):1144–1169. https://doi.org/10.1007/s42729-021-00490-0

Reynolds MP, Pask AJD, Mullan DM (2012) Physiological breeding II: a field guide to wheat phenotyping. CIMMYT, Mexico

Reynolds M, Langridge P (2016) Physiological breeding. Curr Opin Plant Biol 31:162–171. https://doi.org/10.1016/j.pbi.2016.04.005

Röckel F, Schreiber T, Schüler D, Kilias D, Junker A, Klukas C, Arend D, Scholz U, Oppermann M (2022) PhenoApp: a mobile tool for plant phenotyping to record field and greenhouse observations. F1000Res 11:12. https://doi.org/10.12688/f1000research.74239.2

Roy C, Gahtyari NC, He X, Mishra VK, Chand R, Joshi AK, Singh PK (2021) Dissecting quantitative trait loci for spot blotch resistance in South Asia using two wheat recombinant inbred line populations. Front Plant Sci 12:641324. https://doi.org/10.3389/fpls.2021.641324

Rutkoski J, Poland J, Mondal S, Autrique E, Pérez LG, Crossa J, Reynolds M, Singh R (2016) Canopy temperature and vegetation indices from high-throughput phenotyping improve accuracy of pedigree and genomic selection for grain yield in wheat. G3 Genes|Genomes|Genetics 6(9):2799–2808. https://doi.org/10.1534/g3.116.032888

Saleem A, Ur Rahim H, Khan U et al (2024) Organic materials amendments can improve NPK availability and maize growth by reducing heavy metals stress in calcareous soil. Int J Environ Sci Technol 21:2533–2546. https://doi.org/10.1007/s13762-023-05135-w

Selby P, Abbeloos R, Backlund JE, Basterrechea Salido M, Bauchet G, Benites-Alfaro OE, Birkett C, Calaminos VC, Carceller P, Cornut G, Covarrubias-Pazaran G et al (2019) BrAPI—an application programming interface for plant breeding applications. Bioinformatics 35(20):4147–4155. https://doi.org/10.1093/bioinformatics/btz190

Selim MM (2020) Introduction to the integrated nutrient management strategies and their contribution to yield and soil properties. Int J Agron 2020:2821678. https://doi.org/10.1155/2020/2821678

Singh B, Kumar S, Elangovan A, Vasht D, Arya S, Duc NT, Swami P, Pawar GS, Raju D, Krishna H, Sathee L, Dalal M, Sahoo RN, Chinnusamy V (2023) Phenomics based prediction of plant biomass and leaf area in wheat using machine learning approaches. Front Plant Sci 14:1214801. https://doi.org/10.3389/fpls.2023.1214801

Singh RP, Hodson DP, Huerta-Espino J, Jin Y, Njau P, Wanyera R, Herrera-Foessel SA, Ward RW (2011) The emergence of Ug99 races of the stem rust fungus is a threat to world wheat production. Annu Rev Phytopathol 49:465–481. https://doi.org/10.1146/annurev-phyto-072910-095423

Singh RP, Singh PK, Rutkoski J, Hodson DP, He X, Jørgensen LN, Hovmøller MS, Huerta-Espino J (2016) Disease impact on wheat yield potential and prospects of genetic control. Annu Rev Phytopathol 54:303–322. https://doi.org/10.1146/annurev-phyto-080615-095835

Skendžić S, Novak H, Zovko M, Pajač Živković I, Lešić V, Maričević M, Lemić D (2025) Hyperspectral canopy reflectance and machine learning for threshold-based classification of aphid-infested winter wheat. Remote Sens 17(5):929. https://doi.org/10.3390/rs17050929

Song Y, Wang L, Chen L, Kim S, Pi Y, Holsten T, Gardner J, Wang B, Zhai X, Zhu H, Fauquet C, Ronald P (1995) A receptor kinase-like protein encoded by the rice disease resistance gene, Xa21. Science 270(5243):1804–1806. https://doi.org/10.1126/science.270.5243.1804

Tardieu F, Cabrera-Bosquet L, Pridmore T, Bennett M (2017) Plant phenomics, from sensors to knowledge. Curr Biol 27(15):R770–R783. https://doi.org/10.1016/j.cub.2017.05.055

Thingujam D, Gouli S, Cooray SP, Chandran KB, Givens SB, Gandhimeyyan RV, Tan Z, Wang Y, Patam K, Greer SA et al (2025) Climate-resilient crops: integrating AI, multi-omics, and advanced phenotyping to address global agricultural and societal challenges. Plants 14:2699. https://doi.org/10.3390/plants14172699

Thudi M, Upadhyaya HD, Rathore A, Gaur PM, Krishnamurthy L et al (2017) Correction: genetic dissection of drought and heat tolerance in chickpea through genome-wide and candidate gene-based association mapping approaches. PLoS One 12(4):e0175609. https://doi.org/10.1371/journal.pone.0175609

Trachsel S, Kaeppler SM, Brown KM, Lynch JP (2011) Shovelomics: high throughput phenotyping of maize (*Zea mays* L.) root architecture in the field. Plant Soil 341(1–2):75–87. https://doi.org/10.1007/s11104-010-0623-8

Ubbens JR, Stavness I (2017) Deep plant phenomics: a deep learning platform for complex plant phenotyping tasks. Front Plant Sci 8:1190. https://doi.org/10.3389/fpls.2017.01190

Vadez V, Kholová J, Hummel G, Zhokhavets U, Gupta SK, Hash CT (2015) LeasyScan: a novel concept combining 3-D imaging and lysimetry for high-throughput phenotyping of traits controlling plant water budget. J Exp Bot 66(18):5581–5593. https://doi.org/10.1093/jxb/erv251

Varela S, Pederson T, Bernacchi CJ, Leakey ADB (2021) Understanding growth dynamics and yield prediction of sorghum using high temporal resolution UAV imagery time series and machine learning. Remote Sens 13:1763. https://doi.org/10.3390/rs13091763

Varshney RK, Terauchi R, McCouch SR (2014) Harvesting the promising fruits of genomics: applying genome sequencing technologies to crop breeding. PLoS Biol 12(6):e1001883. https://doi.org/10.1371/journal.pbio.1001883

Velmurugan A, Swarnam TP, Ambast SK, Kumar N (2016) Managing waterlogging and soil salinity with a permanent raised bed and furrow system in coastal lowlands of humid tropics. Agric Water Manag 168:56–67. https://doi.org/10.1016/j.agwat.2015.11.015

Wang L (2024) Digital twins in agriculture: a review of recent progress and open issues. Electronics 13(11):2209. https://doi.org/10.3390/electronics13112209

Wang J, Chu Y, Chen G, Zhao M, Wu J, Qu R, Wang Z (2024) Characterization and identification of NPK stress in rice using terrestrial hyperspectral images. Plant Phenomics 6:0197. https://doi.org/10.34133/plantphenomics.0197

Yang G, Liu J, Zhao C, Li Z, Huang Y, Yu H, Xu B, Yang X, Zhu D, Zhang X, Zhang R, Feng H, Zhao X, Li Z, Li H, Yang H (2017) Unmanned aerial vehicle remote sensing for field-based crop phenotyping: current status and perspectives. Front Plant Sci 8:1111. https://doi.org/10.3389/fpls.2017.01111

Yang W, Feng H, Zhang X, Zhang J, Doonan JH, Batchelor WD, Xiong L, Yan J (2020) Crop phenomics and high-throughput phenotyping: past decades, current challenges, and future perspectives. Mol Plant 13(2):187–214. https://doi.org/10.1016/j.molp.2020.01.008

Yendrek CR, Tomaz T, Montes CM, Cao Y, Morse AM, Brown PJ, McIntyre LM, Leakey ADB, Ainsworth EA (2017) High-throughput phenotyping of maize leaf physiological and biochemical traits using hyperspectral reflectance. Plant Physiol 173(2):614–626. https://doi.org/10.1104/pp.16.01447

Zandalinas SI, Fritschi FB, Mittler R (2021) Global warming, climate change, and environmental pollution: recipe for a multifactorial stress combination disaster. Trends Plant Sci 26(6):588–599. https://doi.org/10.1016/j.tplants.2021.02.011

Zeng G, Wan J, Huang D, Hu L, Huang C, Cheng M, Xue W, Gong X, Wang R, Jiang D (2017) Precipitation, adsorption and rhizosphere effect: the mechanisms for phosphate-induced Pb immobilization in soils—a review. J Hazard Mater 339:354–367. https://doi.org/10.1016/j.jhazmat.2017.06.035

Zhu H, Li C, Gao C (2020) Applications of CRISPR–Cas in agriculture and plant biotechnology. Nat Rev Mol Cell Biol 21:661–677. https://doi.org/10.1038/s41580-020-00288-9

Recent Advances in Understanding Heat Stress Responses in Major Pseudocereals

4

Priya Kumari, Aastha, Himanshi Gangwar, Vijay Gahlaut, and Vandana Jaiswal

Abstract

Pseudocereals such as amaranth, buckwheat, chia, and quinoa are increasingly recognized as climate-resilient and nutritionally rich crops with rising global market demand. India represents a significant centre of plant diversity, with landraces and wild relatives that provide important genetic resources for crop improvement. Pseudocereals also exhibit resilience to diverse environmental conditions, making them promising candidates for climate-resilient agriculture. However, heat stress (HS) remains a major problem affecting plant growth, reproductive development, yield, and grain nutrient quality. The rise in temperatures disrupts protein structure, impairs membrane integrity, generates reactive oxygen species, and causes hormonal imbalances, which lead to reduced pollen viability, seed set, and grain filling. This chapter first outlines the global production, yield, and market status of major pseudocereals, highlighting their growing economic relevance in diverse agro-ecological zones. Despite their nutritional

Authors Priya Kumari, Aastha and Himanshi Gangwar have equally contributed to this chapter.

P. Kumari · H. Gangwar · V. Jaiswal (✉)
Biotechnology Division, CSIR-Institute of Himalayan Bioresource Technology, Palampur, India

Academy of Scientific and Innovative Research (AcSIR), Ghaziabad, India
e-mail: vandana.jaiswal@csir.res.in

Aastha
Biotechnology Division, CSIR-Institute of Himalayan Bioresource Technology, Palampur, India

V. Gahlaut
University Center for Research and Development, Chandigarh University, Mohali, Punjab, India

V. Gahlaut et al. (eds.), *Smart Crop Development*,
https://doi.org/10.1007/978-981-95-7680-7_4

promise, pseudocereals are highly vulnerable to abiotic stresses, with HS posing a major challenge under global climate change. The chapter reviews molecular and physiological responses to HS in pseudocereals, emphasizing key signalling pathways involving reactive oxygen species, calcium, and phytohormones, as well as the role of transcription factors such as HSFs, HSPs, and TCPs. Advances in genomics, transcriptomics, proteomics, and metabolomics have accelerated the discovery of candidate genes and regulatory networks associated with thermotolerance. Furthermore, the chapter highlights that modern approaches provide promising avenues for developing heat-resilient pseudocereal cultivars. Together, these insights underscore the dual importance of pseudocereals as future-ready crops for global food and nutritional security and the urgent need for integrating molecular tools with breeding to ensure their sustainability under rising temperatures.

Keywords

Climate-resilient crops · CRISPR/Cas · Heat stress · Nutritional quality · Pseudocereals

4.1 Introduction

Pseudocereals are broadleaf (dicotyledonous) plants, botanically distinct from true cereals, which are grasses (monocots) (Schoenlechner and Bender 2020; Mankotia et al. 2025). They are excellent sources of high-quality protein, essential amino acids, dietary fibre, and bioactive phytochemicals such as saponins, polyphenols, phytosterols, phytosteroids, and betalains that contribute to potential health benefits (Martinez-Villaluenga et al. 2020; Mankotia et al. 2025). These compounds contribute to multiple health benefits such as reduced risk of diabetes, cancer, and cardiovascular diseases (Mir et al. 2018; Wanniarachchi et al. 2023). With their wide genetic diversity, they are increasingly recognized as potential future crops with the ability to adapt to a wide range of environmental conditions from tropical to temperate climates (Joshi et al. 2018, 2019). Among the most widely studied pseudocereals, amaranth, buckwheat, chia, and quinoa are gaining global attention for their nutritional value and resilience to abiotic stresses. Grain amaranth (*Amaranthus hypochondriacus*), a C4 plant native to Central and North America, is highly adaptable and valued for its nutrient-rich seeds and leaves (Joshi et al. 2018). Similarly, buckwheat (*Fagopyrum esculentum*) is widely cultivated across Asia and Europe and has gained importance as a gluten-free alternative to wheat-based diets. Chia (*Salvia hispanica* L.), native to Central and South America, is renowned for its omega-3 fatty acids, dietary fibre, and antioxidant compounds, making it a functional food with significant health-promoting properties. Quinoa (*Chenopodium quinoa* Willd.), referred to as the "golden grain" of the Andes, thrives at altitudes between 2000 and 4000 m and is known for its diverse secondary metabolites with antimicrobial, antioxidant, and anti-inflammatory properties (Lin et al. 2019; Angeli et al. 2020).

Despite their nutritional and agronomic advantages, pseudocereal productivity is threatened by rising global temperatures. Heat stress (HS) disrupts protein stability, impairs membrane integrity, induces reactive oxygen species (ROS), and causes hormonal imbalances, thereby affecting plant growth, reproductive development, and yield (Hasanuzzaman et al. 2013). To mitigate these effects, plants activate heat shock responses and stress signalling pathways that protect cellular processes by activating stress-responsive genes to maintain metabolic and developmental balance (Niu and Xiang 2018; Haider et al. 2021). Advances in genomics, transcriptomics, and gene editing technologies have accelerated the identification of stress-responsive genes such as heat shock transcription factors (HSFs) and transcription factors (TFs) in pseudocereals (Joshi et al. 2023). These advances offer new insights into key pathways and gene families associated with heat tolerance in these crops. For example, pan-genomic studies in pearl millet have identified genetic variations and TF families, such as RWP-RK, associated with improved thermotolerance (Yan et al. 2023). Their huge genetic diversity and less breeding histories further position them as promising candidates for rapid genetic improvement using conventional breeding, as well as modern tools such as chemical and physical mutagenesis, CRISPR/Cas-based gene editing, next-generation sequencing technologies, and genome-wide association study-based approaches (Wolter et al. 2019; Rodríguez et al. 2020; Yabe and Iwata 2020).

This book chapter explores the physiological effects of HS on pseudocereals, highlights recent progress in identifying heat-responsive genes and regulatory networks, and discusses the role of advanced genomic tools in developing heat-tolerant varieties. It further highlights on combining molecular advancements with breeding strategies to develop pseudocereals that have yield stability and nutritional quality and can grow well under rising temperatures. Altogether, understanding the molecular, genetic, and physiological responses of pseudocereals to HS lays the foundation for developing resilient crops. Integrating genomic insights with modern breeding strategies offers the potential to enhance heat tolerance and crop yield and preserve the nutritional quality of pseudocereals, ensuring their role in sustainable agriculture and global food security.

4.2 Global Status and Market Aspect of Major Pseudocereals

Pseudocereals such as amaranth, buckwheat, chia, and quinoa are increasingly recognized as high-value crops in global markets due to their rich nutritional composition, which includes carbohydrates, starch, dietary fibre, proteins, lipids, and essential micronutrients (Bender and Schoenlechner 2021).

4.2.1 Amaranth (*Amaranthus* spp.)

Amaranth is one of the oldest domesticated crops, valued as both a leafy vegetable and a grain rich in protein and micronutrients. It is cultivated as a minor crop across the Asia Pacific region including India, China, Manchuria, Nepal, Bhutan, Afghanistan, Indonesia, Japan, Thailand, and Israel. In India, its cultivation spans both hills and plain regions across states such as Jammu and Kashmir, Himachal Pradesh, Uttarakhand, Sikkim, Assam, Meghalaya, Arunachal Pradesh, Nagaland, Tripura, Jharkhand, Chhattisgarh, Maharashtra, Gujarat, Odisha, Karnataka, Kerala, and Tamil Nadu (Raiger et al. 2009). The global amaranth market was valued at USD 9.3 billion in 2024, and it is projected to expand at a CAGR of over 18.1% between 2025 and 2034 (Fig. 4.1a; https://www.gminsights.com/industry-analysis/amaranth-market).

4.2.2 Buckwheat (*Fagopyrum* spp.)

Buckwheat is valued for its gluten-free nature and high rutin content. Its genetic resources are mainly distributed around its centre of origin in southwestern China, following historical routes of migration and domestication. Common buckwheat (*Fagopyrum esculentum*) is now found across all continents, whereas Tartary buckwheat (*Fagopyrum tataricum*) remains primarily in mountainous regions of China and the Himalayas. Buckwheat resources include wild species, subspecies, landraces, and improved breeding varieties. Wild *Fagopyrum* species are restricted mainly to the areas of Yunnan and Sichuan provinces in China, while traditional landraces are cultivated in highland regions of China, Nepal, India, and Bhutan, regions often considered less economically developed (Suvorova and Zhou 2018). The global buckwheat production in 2023 reached 2.2 million tonnes (MT) with a yield of 1007.5 kg/ha, and the market reached USD 621.9 million in 2024, projected to grow at a CAGR of 3.1% from 2025 to 2034 (Fig. 4.1a, b; https://www.gminsights.com/industry-analysis/buckwheat-market).

4.2.3 Chia (*Salvia hispanica* L.)

Chia has gained popularity as a "superfood" due to its high omega-3 fatty acid, fibre, protein, and antioxidant content (Ayerza and Coates 2007). Native to southern Mexico and northern Guatemala, chia was domesticated around 4500 years ago for both food and medicinal purposes (Ayerza and Coates 2007, 2009; Anand et al. 2024). Today, it is cultivated in countries such as Australia, Bolivia, Colombia, Guatemala, Mexico, Peru, Paraguay, and Argentina, with Mexico serving as the largest production hub exporting to the United States, Japan, and Europe (Busilacchi et al. 2013; Alenbrant et al. 2014). Its cultivation is also expanding in Asia and

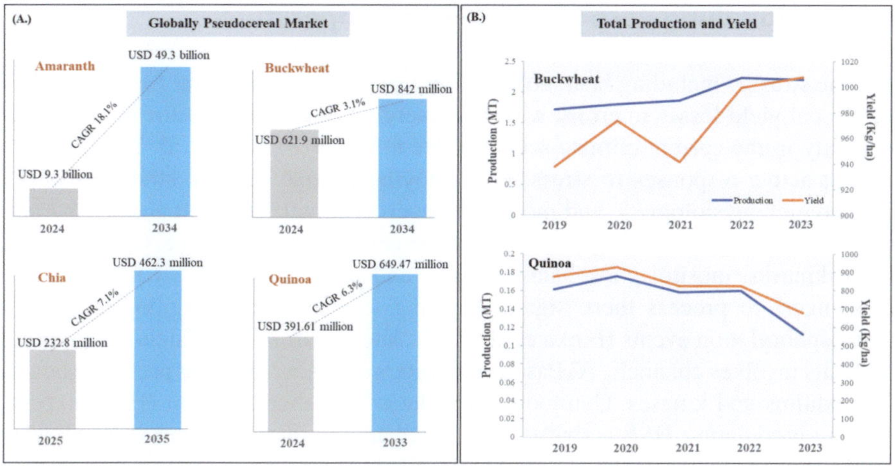

Fig. 4.1 Global market trends and production of major pseudocereals. (**a**) Market value of amaranth, buckwheat, chia, and quinoa for 2024–2035. (**b**) Total production in million tonnes (MT) and yield (kg/ha) of buckwheat and quinoa from 2019 to 2023. (Data sources: https://www.gminsights.com/industry-analysis/amaranth-market; https://www.gminsights.com/industry-analysis/buckwheat-market; https://www.futuremarketinsights.com/reports/chia-seed-market; https://www.marketgrowthreports.com/market-reports/quinoa-market-105179; FAO and national statistics)

Africa, with India reporting increased acreage in states including Uttar Pradesh, Rajasthan, Andhra Pradesh, and Karnataka (Anand et al. 2024). The global chia seed market is projected to grow from USD 232.8 million in 2025 to USD 462.3 million by 2035, registering a CAGR of 7.1%, driven by its functional food applications (Fig. 4.1a; https://www.futuremarketinsights.com/reports/chia-seed-market).

4.2.4 Quinoa (*Chenopodium quinoa* Willd)

Quinoa was originally cultivated in the Andean region, including Bolivia, Peru, Ecuador, Colombia, Argentina, and Chile, recognized by Vavilov as one of the primary centres of crop domestication (Lescano 1994; Jacobsen et al. 2003). Once a traditional Andean staple, quinoa cultivation has now expanded globally and is reported in over 95 countries, such as Tibet, Morocco, France, India, China, the United Kingdom, Sweden, Denmark, the Netherlands, and Italy (Bazile et al. 2013, 2016; Bhargava et al. 2006; Pulvento et al. 2012). The global production in 2023 reached 0.11 MT with a yield of 676.9 kg/ha. The market was valued at USD 391.61 million in 2024 and is expected to reach USD 649.47 million by 2033, growing at a CAGR of 6.3% from 2025 to 2033 (Fig. 4.1a, b; https://www.marketgrowthreports.com/market-reports/quinoa-market-105179).

4.3 Signalling Responses to High Temperature

Abiotic stresses including heat, cold, salinity, and drought, account for over half of the global yield losses in crops, with HS emerging as a critical factor limiting productivity in the current climate scenario (He et al. 2018; Janni et al. 2020). Plants exhibit active responses to stress by employing diverse defence strategies at the phenotypic, physiological, and molecular levels (Hasanuzzaman et al. 2013). Plant cells first sense stress through receptors in the cell wall or membrane. Calcium ions, MAP kinases, inositol phosphatases, ROS, and phytohormones act as secondary messengers to process these signals, leading to sequential phosphorylation and dephosphorylation events (Finka et al. 2012; Wang et al. 2016). Calcium signalling in plants involves channels, ATPases, antiporters, and Ca^{2+}-binding proteins such as calmodulins and kinases. Calmodulin regulates heat shock protein (HSP) expression by modulating HSF activity, thereby enhancing thermotolerance (Medvedev 2005; Zhao et al. 2021). On the other hand, HS-induced Ca^{2+} signalling is conveyed through RBOH proteins, leading to an apoplastic ROS burst. The ROS produced by RBOHs enter the cell via aquaporins and stimulate additional Ca^{2+} release through TPC1 channels, thereby modulating diverse signalling and stress response pathways (Mittler and Blumwald 2015; Zhu 2016). Excess ROS production activates HS-responsive pathways and stimulates the expression of stress-related and antioxidant proteins. This enhances the activity of key enzymes such as ascorbate peroxidase, catalase, and superoxide dismutase, which collectively mitigate oxidative damage (Suzuki 2023; Yang et al. 2023). In addition to ROS and calcium signalling, phytohormones serve as key regulators of plant responses to extreme temperatures (Torres et al. 2017). Hormones such as ABA, BRs, CKs, SA, JA, and ET orchestrate intricate signalling networks that activate stress-responsive TFs (Zandalinas et al. 2016; Iqbal et al. 2019; Balfagon et al. 2019; Prerostova et al. 2020; Huang et al. 2021; Luo et al. 2022). Among these, ABA is particularly important for heat acclimation, as it promotes osmotic adjustment through osmoprotectant accumulation, enhances antioxidant defence systems, and facilitates HSP synthesis to maintain protein stability (Kumar et al. 2012; Islam et al. 2018). Moreover, ABA activates antioxidant enzymes that detoxify excess ROS generated during HS, thereby safeguarding cellular homeostasis (Islam et al. 2018). Overall, these findings emphasize the integrated roles of hormones and ROS in regulating plant tolerance to temperature extremes. Their interaction extends beyond each other to include signalling molecules such as NO, Ca^{2+}, kinases, phosphatases, TFs, and diverse metabolites, collectively driving molecular, metabolic, and physiological adjustments that support acclimation to HS (Devireddy et al. 2021) (Fig. 4.2).

4.3.1 Heat Stress Response in Pseudocereals

Heat stress significantly impacts pseudocereals, triggering a range of molecular and physiological responses that enhance plant adaptation and survival. Specialized gene families and protein networks, including HSFs and TFs, play essential roles in

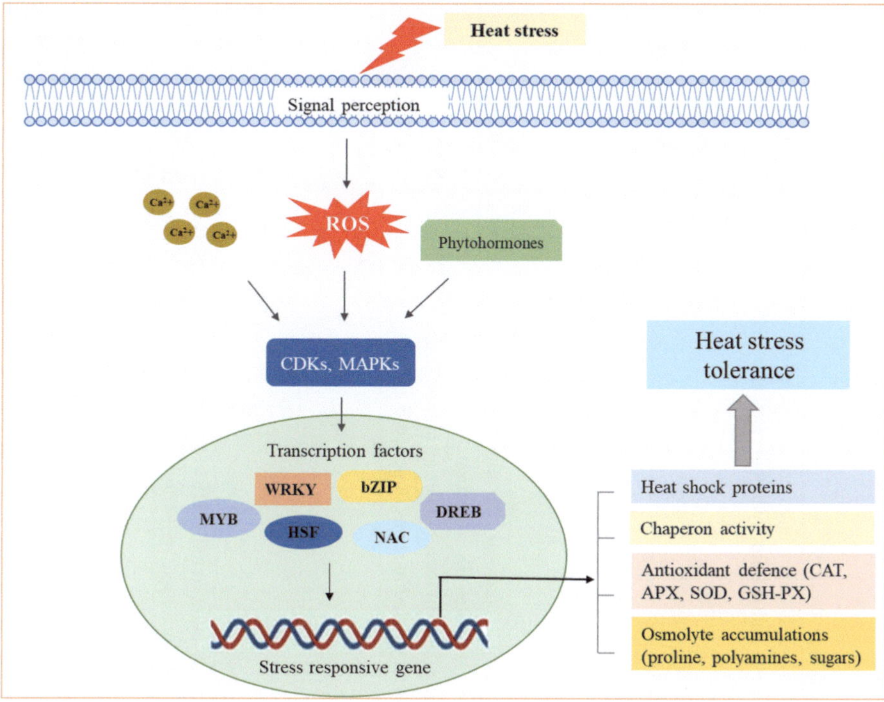

Fig. 4.2 Schematic representation of the signalling network underlying plant heat tolerance. Heat stress induces ROS production and calcium influx, which interact with hormonal signals to activate downstream cascades. These pathways converge on TFs, leading to antioxidant defence, HSP accumulation, and enhanced thermotolerance

managing and mitigating cellular damage caused by elevated temperatures (Table 4.1). The HSFs play a critical role in mediating plant responses to HS. In *Fagopyrum tataricum*, 29 *FtHSFs* were identified (Liu et al. 2019), while *Chenopodium quinoa* had 23 *CqHSFs* (Tashi et al. 2018), and *Amaranthus hypochondriacus* contains 13 *AhHSFs* (Goel et al. 2023). In *Amaranthus*, HS significantly reduced chlorophyll content and fluorescence (*Fv/Fm*) and increased malondialdehyde levels, indicating oxidative stress. *TCP* genes are important targets for responses to abiotic stress. Heat shock proteins in *Salvia hispanica* (chia) include *HSP1*, *HSP15.7*, *HSP17*, *HSP17.3*, *HSP17.4*, *HSP17.8*, *HSP18.2*, *HSP21.7*, *HSP21.9*, *HSP22*, *HSP26.5*, *HSP70*, *HSP80*, *HSP83*, and *HSP90*. These proteins act as molecular chaperones, helping to maintain proper protein folding, prevent aggregation, and stabilize cellular proteins under HS. By protecting cellular proteins and organelles, HSPs play a crucial role in enhancing thermotolerance and overall stress resilience in chia (Zare et al. 2025). Transcription factors are also key regulators of abiotic stress responses. In *Fagopyrum tataricum*, 26 *FtTCPs* (Yang et al. 2022a, b), 74 *FtWRKY* genes (He et al. 2019), and 10 *FtBES* TFs (Gao et al. 2022) have been reported. The *FtWRKY74* expression increased 12 h after HS, and

Table 4.1 Summary of major genes, TFs, signalling components, and proteins involved in HS responses in pseudocereals

Class	Genes/TFs/proteins	Plants/species	Physiological effect	References
HSFs	*AhHsfs* (13)	Amaranth	Chlorophyll content and fluorescence (Fv/Fm) reduced; malondialdehyde increased under HS	Goel et al. (2023)
	FtHsfs (1–29)	Buckwheat	–	Liu et al. (2019)
	ShHSFs	Chia		Zare et al. (2025)
	CqHsfs (1–23)	Quinoa	–	Tashi et al. (2018)
HSPs	*HSP1*, *HSP15.7*, *HSP17*, *HSP17.3*, *HSP17.4*, *HSP17.8*, *HSP18.2*, *HSP21.7*, *HSP21.9*, *HSP22*, *HSP26.5*, *HSP70*, *HSP80*, *HSP83*, *HSP90*	*Salvia hispanica* (Chia)		Zare et al. (2025)
TFs	*FtTCP* (26)	Buckwheat	–	Yang et al. (2022a, b)
	*FtWRKY*74	Buckwheat	Expression increased 12 h after HS	He et al. (2019)
	FtBES (1–10)	Buckwheat	*BES1* promotes HS protection	Gao et al. (2022)
	*CqTH*36	Quinoa	Upregulated in leaves under all abiotic stress	Li et al. (2022)
	CqBHLH 218	Quinoa	Downregulated in stems under cold and HS	Xue et al. (2023)
Ca^{2+} signalling	*CBPs*, *CLBs*, *CDPKs* (*CDPK1*, *CDPK4*, *SK5*, *SK10*, *SK11*, *SK13*, *SK28*, *SK32*), *ATPases* (*ATPase1*, *ATPase2*, *ATPase13*), *SCaMC*, *CSC1*, *CaSR*, *MCU2*, *CUP*, *CaMK*, *CCX1*, *CCX4*, *CMLs* (*CML1*, *CML13*, *CML36*, *CML48*)	*Salvia hispanica* (Chia)		Zare et al. (2025)
MAPKs	*AhERD*	Amaranth	Strongly induced under HS	Rosales et al. (2023)
	*FtMAPK*8	Buckwheat	Slightly increased under high temperature	Wang et al. (2022)

FADs	*AhHAB4-PAI*-1	*A. caudatus*, *A. cruentus*, *A. hypochondriacus*	–	Rosales et al. (2023)
	AhRIP		–	
	AhERD		–	
	AhTIL		–	
	AhDGR		–	
	AhBAMY		–	
	*Ah*2880		–	
	AhOEE		–	
	*ShFAD*2	*Salvia hispanica* (Chia)	Downregulated by HS	Xue et al. (2017)
Chaperones	*BAG1–BAG6*; *ClpB1*, *ClpB3*; *DnaJs* (*DnaJ1*, *DnaJ6*, *DnaJ8*, *DnaJ10*, *DnaJ11*, *DnaJ15*, *DnaJ16*, *DnaJ20*, *DnaJ49*, *DnaJ50*, *DnaJ76*); *CCS*; *HSCB*; *ASF1B*; *proteasome assembly chaperone 2*; *FKBP* (*12*, *16–20*, *43*, *62*); *Pin1*, *Pin4*; *CYPs* (*CYP21*, *CYP26*, *CYP28*, *CYP37*, *CYP38*, *CYP40*, *CYP59*, *CYP90*)	*Salvia hispanica* (Chia)		Zare et al. (2025)

(continued)

Table 4.1 (continued)

Class	Genes/TFs/proteins	Plants/species	Physiological effect	References
Proteins	Proteins including 26S proteasome non-ATPase regulatory subunit 4 homolog, 40S ribosomal protein S12-2, 6-phosphogluconate dehydrogenase, decarboxylating 1, carbonic anhydrase, chloroplastic, E3 ubiquitin-protein ligase UPL, eukaryotic translation initiation factor 3 subunit A, F-actin-capping protein subunit alpha-like, ferredoxin-dependent glutamatesynthase 2, guanosine nucleotide diphosphate dissociation inhibitor, heat shock 70 kDa protein, indole-3-glycerol phosphate synthase, chloroplastic-like isoform X2, kinesin-like protein KIN-12F, pentatricopeptide repeat-containing protein At3g02330, mitochondrial, protein SWEETIE, pyruvate, phosphate dikinase regulatory protein 2, structural maintenance of chromosomes protein 3, TF MYB98, trihelix transcription factor GTL2	Buckwheat	Tissue-specific expression in buds, open flowers, wilted flowers, and leaves under HS	Kopec et al. (2021)

Hormones	GH3.5, LAX (IAA)	Buckwheat	–	Liu et al. (2025)
	GID1B (GA)		–	
	AOS (JA)		–	
Antioxidant enzymes	FtCATs, FtSODs, FtPODs		–	Liu et al. (2025)

HS heat stress, HSFs, heat shock transcription factors, *HSPs* heat shock proteins, *TFs* transcription factors (including WRKY TCP, BES1, and bHLH families), *Ca^{2+}* calcium signalling components, *MAPKs* mitogen-activated protein kinases, *FADs* fatty acid desaturases, *BAG* Bcl-2-associated athanogene proteins, *ClpB* caseinolytic protease B, *DnaJs* DnaJ chaperone proteins, *CCS* copper chaperone for superoxide dismutase, *HSCB* heat shock cognate binding protein, *ASF1B* anti-silencing function 1B, *FKBP* FK506-binding proteins, *CYPs* cytochrome P450 proteins, *GH3* indole-3-acetic acid-amido synthetase, *LAX* auxin influx carrier, *GID1B* gibberellin receptor, *AOS* allene oxide synthase, *ABA* abscisic acid, *JA* jasmonic acid, *GA* gibberellic acid, *IAA* indole-3-acetic acid, *CAT* catalase, *SOD* superoxide dismutase, *POD* peroxidase

BES1 enhances heat tolerance. In *Chenopodium quinoa*, *CqTH36* was consistently upregulated in the leaves under abiotic stresses (Li et al. 2022), while *CqBHLH 202* was downregulated in the stems under cold and HS (Xue et al. 2023). Calcium (Ca^{2+}) signalling components in *Salvia hispanica* (chia) include CBPs, CLBs, CDPKs (CDPK1, CDPK4, SK5, SK10, SK11, SK13, SK28, SK32), ATPases (ATPase1, ATPase2, ATPase13), SCaMC, CSC1, CaSR, MCU2, CUP, CaMK, CCX1, CCX4, and CMLs (CML1, CML13, CML36, CML48). These proteins and enzymes mediate calcium signalling, which is a critical secondary messenger pathway in response to HS. Calcium signals activate downstream stress-responsive pathways, regulate TFs, and coordinate protective mechanisms such as antioxidant enzyme activity and HSP expression, thereby enhancing thermotolerance in chia (Zare et al. 2025). Mitogen-activated protein kinases (MAPKs) are crucial for stress signalling. *FtMAPK8* showed slight upregulation under high temperature in buckwheat (Wang et al. 2022), and *AhERD* was strongly induced in all grain amaranths under HS (Rosales et al. 2023). Fatty acid desaturases (FADs) and other stress-related genes also showed diverse responses. In *Salvia hispanica*, *ShFAD2* was downregulated under HS (Xue et al. 2017). Multiple genes in *Amaranthus* species (*AhHAB4-PAI-1*, *AhRIP*, *AhERD*, *AhTIL*, *AhDGR*, *AhBAMY*, *Ah2880*, and *AhOEE*) showed differential expression after 3–6 h of HS, with *AhOEE* strongly repressed in *A. cruentus* and *A. caudatus* (Rosales et al. 2023). Chaperones in *Salvia hispanica* (chia) include *BAG1–BAG6*, *ClpB1*, *ClpB3*, *DnaJs* (*DnaJ1*, *DnaJ6*, *DnaJ8*, *DnaJ10*, *DnaJ11*, *DnaJ15*, *DnaJ16*, *DnaJ20*, *DnaJ49*, *DnaJ50*, *DnaJ76*), *CCS*, *HSCB*, *ASF1B*, proteasome assembly chaperone 2, *FKBPs* (*12*, *16–20*, *43*, *62*), *Pin1*, *Pin4*, and *CYPs* (*CYP21*, *CYP26*, *CYP28*, *CYP37*, *CYP38*, *CYP40*, *CYP59*, *CYP90*). These proteins assist in proper protein folding, prevent aggregation, and help repair or degrade damaged proteins under HS, thereby maintaining cellular function and enhancing thermotolerance in chia (Zare et al. 2025). Proteins involved in stress responses are highly tissue-specific. For example, in buckwheat, Proteins including 26S proteasome non-ATPase regulatory subunit 4 homolog, 40S ribosomal protein S12–2 and TFs MYB98 and trihelix GTL2 are expressed. In open flowers, heat shock 70 kDa protein, E3 ubiquitin-protein ligase UPL1, and chloroplastic enzymes like indole-3-glycerol phosphate synthase are present. Wilted flowers express structural maintenance of chromosomes protein 3, heat shock 70 kDa protein, and guanosine nucleotide dissociation inhibitor. Leaf tissues contain carbonic anhydrase, pentatricopeptide repeat proteins, kinesin-like proteins, and ferredoxin-dependent glutamate synthase (Kopec et al. 2021).

4.4 Approaches to Enhance Heat Resilience in Pseudocereals

Addressing HS in pseudocereals requires targeted approaches that exploit natural genetic diversity and modern biotechnological tools for developing resilient cultivars (Fig. 4.3). Recent advances in omics technologies have opened new avenues for improving plant heat tolerance by unravelling transcriptional, translational, and

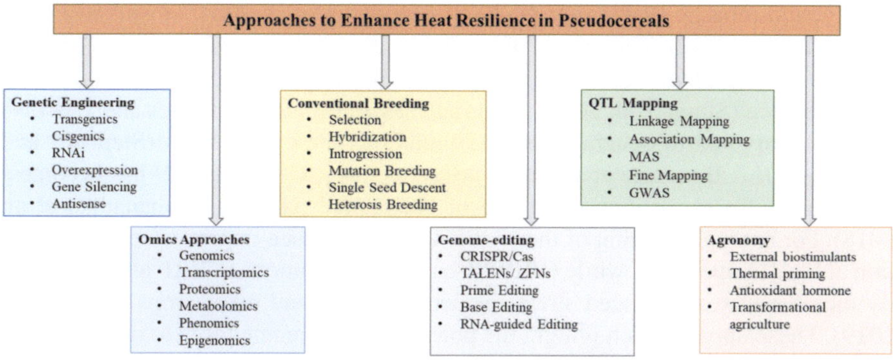

Fig. 4.3 Schematic representation of strategies to improve heat tolerance in pseudocereals, including genetic engineering, conventional breeding, exploration of genetic diversity, and modern molecular and omics-based approaches for identifying and deploying heat-resilient traits

post-translational mechanisms involved in stress responses (Zheng et al. 2022; Singh et al. 2024). Genomics and transcriptomics have enabled the identification of heat-responsive genes and regulatory elements, while proteomics links these transcripts to functional proteins that confer thermotolerance (Yang et al. 2022a, b; Li et al. 2023; Kamali and Singh 2023; Sinha et al. 2023; Liu et al. 2025). MicroRNAs play an additional role in post-transcriptional regulation, either repressing or enhancing stress-responsive genes, offering new opportunities for genetic engineering (Kushawaha et al. 2021). Metabolomics has revealed compatible solutes such as proline, glycine betaine, raffinose, and sucrose as key contributors to osmoprotection under HS (Raza 2022; Shunkao et al. 2023; Qian et al. 2023). Integrated omics approaches combining genome, transcriptome, proteome, and metabolome data provide a comprehensive framework to identify molecular targets, understand crosstalk between pathways, and accelerate the development of heat-tolerant crop varieties (Bhardwaj et al. 2022; Zheng et al. 2022).

Conventional breeding, although constrained by limited genetic diversity in modern cultivars, remains essential for improving crop resilience to HS by exploring new genetic resources and traits (Krishnamurthy et al. 2011). Heat stress tolerance is a complex quantitative trait controlled by multiple genes, and although traditional QTL mapping has helped identify genetic variability, its effectiveness was often limited by large genomic regions and low marker density (Wen et al. 2019; Sukumaran et al. 2018). Recent advances such as genotyping-by-sequencing and genome-wide association studies now enable high-resolution mapping and precise identification of SNPs, haplotypes, and candidate genes, enhancing the efficiency of breeding heat-tolerant cultivars (Rahaman et al. 2018; Maalouf et al. 2022; Li et al. 2023).

Genetic engineering offers promising strategies to enhance thermotolerance in crops by manipulating heat-responsive genes and proteins (Singh and Grover 2008; Grover et al. 2013). Transgenic studies have shown that overexpression of HSPs, small HSPs, and HSFs can improve tolerance in species such as *Arabidopsis*, rice, tomato, and tobacco (Katiyar et al. 2003; Murakami et al. 2004; Hoshikawa et al. 2021; Lu et al. 2022). The advent of complete crop genome sequences has enabled genome editing technologies to improve traits linked with abiotic stress, including heat tolerance (Singh et al. 2019). Tools such as zinc-finger nucleases and TALENs have been applied to generate targeted mutations in several crops (Stephens and Barakate 2017; Bonawitz et al. 2018). More recently, CRISPR/Cas9 has become a powerful platform for precise editing of HS-related pathways (Jaganathan et al. 2018). For example, editing of the NCED4 gene in lettuce improved seed germination at high temperature, while CRISPR-based modifications of ERF and HSF family members have enhanced stress resilience in flax and other crops (Saha et al. 2019). These approaches highlight the potential of genome editing to develop heat-tolerant cultivars with better productivity.

4.5 Conclusion and Future Perspective

Pseudocereals are important climate-resilient and nutrient-rich crops with growing global demand. However, HS poses a major challenge, affecting growth, yield, and grain quality. Plants respond to heat through complex signalling networks involving ROS, calcium, hormones, TFs (HSFs, WRKYs, TCPs), and protective proteins like HSPs and antioxidants. Recent advances in genomics, transcriptomics, proteomics, and metabolomics have helped identify key genes and pathways that contribute to thermotolerance. Future efforts should focus on using these molecular insights for breeding heat-resilient cultivars. Approaches such as genome editing, marker-assisted selection, and combining wild relatives and landraces can accelerate the development of stress-tolerant pseudocereals. Integrating molecular biology with conventional breeding and agronomic practices will help improve productivity and nutritional quality. These strategies will ensure pseudocereals remain reliable and sustainable crops under rising temperatures.

Acknowledgements The authors thank CSIR-IHBT (MLP-201) and ANRF (GAP-320, GAP-334) for funding. The CSIR-IHBT communication number is 5940.

References

Alenbrant R, Benetoli DST, Soares DVA, Mourao W, Corte EJ (2014) Chia cultivation in Brazil: future and perspectives. J Agric Sci 3:161–179

Anand SR, Rao AM, Jain JA (2024) Chia (*Salvia hispanica* L): a Nutri-rich potential crop, its present scenario and future perspectives in India. Mysore J Agric Sci 59(2):1–14

Angeli V, Miguel Silva P, Crispim Massuela D, Khan MW, Hamar A, Khajehei F, Graeff-Hönninger S, Piatti C (2020) Quinoa (*Chenopodium quinoa* Willd.): an overview of the potentials of the "Golden grain" and socio-economic and environmental aspects of its cultivation and marketization. Foods 9(2):216

Ayerza R, Coates W (2007) Effect of dietary alpha-linolenic fatty acid derived from chia when fed as ground seed, whole seed and oil on lipid content and fatty acid composition of rat plasma. Ann Nutr Metab 51(1):27–34

Ayerza R, Coates W (2009) Some quality components of four chia (*Salvia hispanica* L.) genotypes grown under tropical Coastal Desert ecosystem conditions. Asian J Plant Sci 8(4):301–307

Balfagon D, Sengupta S, Gomez-Cadenas A, Fritschi FB, Azad RK, Mittler R, Zandalinas SI (2019) Jasmonic acid is required for plant acclimation to a combination of high light and heat stress. Plant Physiol 181(4):1668–1682

Bazile D, Bertero HD, Nieto C (2013) State of the art report on Quinoa around the World in 2013. FAO, Rome

Bazile D, Jacobsen SE, Verniau A (2016) The global expansion of quinoa: trends and limits. Front Sci 7:622

Bender D, Schoenlechner R (2021) Recent developments and knowledge in pseudocereals including technological aspects. Acta Aliment 50(4):583–609

Bhardwaj A, Devi P, Chaudhary S, Rani A, Jha UC, Kumar S, Bindumadhava H, Prasad PV, Sharma KD, Siddique KH, Nayyar H (2022) Omics' approaches in developing combined drought and heat tolerance in food crops. Plant Cell Rep 41(3):699–739

Bhargava A, Shukla S, Ohri D (2006) *Chenopodium quinoa*—an Indian perspective. Ind Crop Prod 23:73–87

Bonawitz ND, Ainley WM, Itaya A (2018) Zinc finger nuclease-mediated targeting of multiple transgenes to an endogenous soybean genomic locus via non-homologous end joining. Plant Biotechnol J 17(4):750–761

Busilacchi H, Quiroga M, Bueno M, Di Sapio O, Flores V, Severin C (2013) Evaluation of *Salvia hispanica* L. cultivated in the south of Santa Fe (Argentine Republic). Tropical crops 34(4):55–59

Devireddy AR, Tschaplinski TJ, Tuskan GA, Muchero W, Chen JG (2021) Role of reactive oxygen species and hormones in plant responses to temperature changes. Int J Mol Sci 22(16):8843

Finka A, Cuendet AFH, Maathuis FJ, Saidi Y, Goloubinoff P (2012) Plasma membrane cyclic nucleotide gated calcium channels control land plant thermal sensing and acquired thermotolerance. Plant Cell 24:3333–3348

Gao A, Weng W, Yao X, Wu W, Bai Q, Xiong R, Ruan J (2022) Genome-wide identification, structural characterization, and gene expression analysis of BES1 transcription factor family in Tartary Buckwheat (*Fagopyrum tataricum*). Agronomy 12(11):2729

Goel K, Kundu P, Gahlaut V, Sharma P, Kumar A, Thakur S, Zinta G (2023) Functional divergence of Heat Shock Factors (Hsfs) during heat stress and recovery at the tissue and developmental scales in C4 grain amaranth (*Amaranthus hypochondriacus*). Front Plant Sci 14:1151057

Grover A, Mittal D, Negi M, Lavania D (2013) Generating high temperature tolerant transgenic plants: achievements and challenges. Plant Sci 205:38–47

Haider S, Iqbal J, Naseer S, Yaseen T, Shaukat M, Bibi H, Ahmad Y, Daud H, Abbasi NL, Mahmood T (2021) Molecular mechanisms of plant tolerance to heat stress: current landscape and future perspectives. Plant Cell Rep 40(12):2247–2271

Hasanuzzaman M, Nahar K, Alam MM, Roychowdhury R, Fujita M (2013) Physiological, biochemical, and molecular mechanisms of heat stress tolerance in plants. Int J Mol Sci 14(5):9643–9684

He M, He CQ, Ding NZ (2018) Abiotic stresses: general defenses of land plants and chances for engineering multistress tolerance. Front Plant Sci 9:1771

He X, Li JJ, Chen Y, Yang JQ, Chen XY (2019) Genome-wide analysis of the WRKY gene family and its response to abiotic stress in buckwheat (*Fagopyrum tataricum*). Open Life Sci 14(1):80–96

Hoshikawa K, Pham D, Ezura H, Schafleitner R, Nakashima K (2021) Genetic and molecular mechanisms conferring heat stress tolerance in tomato plants. Front Plant Sci 12:786688

Huang J, Zhao X, Bürger M, Wang Y, Chory J (2021) Two interacting ethylene response factors regulate heat stress response. Plant Cell 33(2):338–357

Iqbal N, Fatma M, Khan NA, Umar S (2019) Regulatory role of proline in heat stress tolerance: modulation by salicylic acid. In: Plant signaling molecules, 437, –448. https://doi.org/10.1016/B978-0-12-816451-8.00027-7

Islam MR, Feng B, Chen T, Tao L, Fu G (2018) Role of abscisic acid in thermal acclimation of plants. J Plant Biol 61(5):255–264

Jacobsen SE, Mujica A, Jensen CR (2003) The resistance of quinoa (*Chenopodium quinoa* Willd.) to adverse abiotic factors. Food Rev Int 19:99–109

Jaganathan D, Ramasamy K, Sellamuthu G, Jayabalan S, Venkataraman G (2018) CRISPR for crop improvement: an update review. Front Plant Sci 9:985

Janni M, Gulli M, Maestri E (2020) Molecular and genetic bases of heat stress responses in crop plants and breeding for increased resilience and productivity. J Exp Bot 71(33):3780–3802

Joshi DC, Sood S, Hosahatti R, Kant L, Pattanayak A, Kumar A, Yadav D, Stetter MG (2018) From zero to hero: the past, present and future of grain amaranth breeding. Theor Appl Genet 131(9):1807–1823

Joshi DC, Chaudhari GV, Sood S, Kant L, Pattanayak A, Zhang K, Fan Y, Janovska D, Meglic V, Zhou M (2019) Revisiting the versatile buckwheat: reinvigorating genetic gains through integrated breeding and genomics approach. Planta 250(3):783–801

Joshi DC, Sood S, Kudapa H, Zhou M, Santra D (2023) Trait mining and genetic enhancement of millets and potential crops: modern prospects for ancient grains. Front Plant Sci 14:1291893

Kamali S, Singh A (2023) Genomic and transcriptomic approaches to developing abiotic stress-resilient crops. Agronomy 13(12):2903

Katiyar AS, Agarwal M, Grover A (2003) Heat tolerant basmati rice engineered by over-expression of *hsp101*. Plant Mol Biol 51:677–686

Krishnamurthy L, Gaur PM, Basu PS, Chaturvedi SK, Tripathi S, Vadez V, Rathore A, Varshney RK, Gowda CL (2011) Large genetic variation for heat tolerance in the reference collection of chickpea (Cicer arietinum L.) germplasm. Plant Genet Resour 9(1):59–69

Kopec P, Hornyak M, Pastuszak J, Szczerba A, Rapacz M, Waga J, Płazek A (2021) Changes in the flower and leaf proteome of common buckwheat (*Fagopyrum esculentum* Moench) under high temperature. Int J Mol Sci 22(5):2678

Kumar S, Kaushal N, Nayyar H, Gaur P (2012) Abscisic acid induces heat tolerance in chickpea (*Cicer arietinum* L.) seedlings by facilitated accumulation of osmoprotectants. Acta Physiol Plant 34:1651–1658

Kushawaha AK, Khan A, Sopory SK, Sanan-Mishra N (2021) Priming by high temperature stress induces microRNA regulated heat shock modules indicating their involvement in thermopriming response in rice. Life 11(4):291

Lescano JL (1994) Genetica y mejoramiento de cultivos altoandinos: quinua, kanihua, tarwi, kiwicha, papa amarga, olluco, mashua y oca. Programa Interinstitucional de Waru Waru. Convenio inade/pelt-cotesu

Li K, Fan Y, Zhou G, Liu X, Chen S, Chang X, Zhang L (2022) Genome-wide identification, phylogenetic analysis, and expression profiles of trihelix transcription factor family genes in quinoa (*Chenopodium quinoa* Willd.) under abiotic stress conditions. BMC Genomics 23(1):499

Li P, Jiang J, Zhang G, Miao S, Lu J, Qian Y, Zhao X, Wang W, Qiu X, Zhang F, Xu J (2023) Integrating GWAS and transcriptomics to identify candidate genes conferring heat tolerance in rice. Front Plant Sci 13:1102938

Lin M, Han P, Li Y, Wang W, Lai D, Zhou L (2019) Quinoa secondary metabolites and their biological activities or functions. Molecules 24(13):2512

Liu M, Huang Q, Sun W, Ma Z, Huang L, Wu Q, Chen H (2019) Genome-wide investigation of the heat shock transcription factor (Hsf) gene family in Tartary buckwheat (*Fagopyrum tataricum*). BMC Genomics 20(1):871

Liu L, Li L, Feng Y, Wang T, Li C, Wu H, Zhao H (2025) Impact of heat stress on the development, physiological and biochemical characteristics of Tartary buckwheat flowers, and its transcriptomic analysis. Plant Physiol Biochem 220:109535

Lu XP, Shao KZ, Xu JY, Li JL, Ren W, Chen J, Zhao LY, Zhao Q, Zhang JL (2022) A heat shock transcription factor gene (*HaHSFA1*) from a desert shrub, *Haloxylon ammodendron*, elevates salt tolerance in *Arabidopsis thaliana*. Environ Exp Bot 201:104954

Luo J, Jiang J, Sun S, Wang X (2022) Brassinosteroids promote thermotolerance through releasing BIN2-mediated phosphorylation and suppression of HsfA1 transcription factors in Arabidopsis. Plant Commun 3(6):100419

Maalouf F, Abou-Khater L, Babiker Z, Jighly A, Alsamman AM, Hu J, Ma Y, Rispail N, Balech R, Hamweih A, Baum M (2022) Genetic dissection of heat stress tolerance in faba bean (*Vicia faba* L.) using GWAS. Plants 11(9):1108

Mankotia R, Singh H, Varsha N, Kumar M, Ghritlahare MS, Meena R, Murmu DK, Mani A (2025) Nutritional, functional, and environmental benefits of pseudocereals-A review. Int J Res Agron 8(4S):111–121

Martinez-Villaluenga C, Penas E, Hernandez-Ledesma B (2020) Pseudocereal grains: nutritional value, health benefits and current applications for the development of gluten-free foods. Food Chem Toxicol 137:111178

Medvedev SS (2005) Calcium signaling system in plants. Russ J Plant Physiol 52:249–270

Mir, Nisar Ahmad MN, Charanjit Singh Riar CS, Sukhcharn Singh S (2018) Nutritional constituents of pseudocereals and their potential use in food systems: a review. Trends Food Sci Technol 75:170–180

Mittler R, Blumwald E (2015) The roles of ROS and ABA in systemic acquired acclimation. Plant Cell 27:64–70

Murakami T, Matsuba S, Funatsuki H, Kawaguchi K, Saruyama H, Tanida M, Sato Y (2004) Overexpression of a small heat shock protein, sHSP17. 7, confers both heat tolerance and UV-B resistance to rice plants. Mol Breed 13(2):165–175

Niu Y, Xiang Y (2018) An overview of biomembrane functions in plant responses to high-temperature stress. Front Plant Sci 9:915

Prerostova S, Dobrev PI, Kramna B, Gaudinova A, Knirsch V, Spichal L, Zatloukal M, Vankova R (2020) Heat acclimation and inhibition of cytokinin degradation positively affect heat stress tolerance of Arabidopsis. Front Plant Sci 11:87

Pulvento C, Riccardi M, Lavini A, Iafelice G, Marconi E, d'Andria R (2012) Yield and quality characteristics of quinoa grown in open field under different saline and non-saline irrigation regimes. J Agron Crop Sci 198:254–263

Qian W, Zhu Y, Chen Q, Wang S, Chen L, Liu T, Tang H, Yao H (2023) Comprehensive metabolomic and lipidomic alterations in response to heat stress during seed germination and seedling growth of Arabidopsis. Front Plant Sci 14:1132881

Rahaman M, Mamidi S, Rahman M (2018) Genome-wide association study of heat stress-tolerance traits in spring-type *Brassica napus* L. under controlled conditions. Crop J 6(2):115–125

Raiger HL, Dua RP, Phogat BS, Sharma SK (2009) Improved varieties and culture practices of grain amaranth. Intensiv Agric 48:8–13

Raza A (2022) Metabolomics: a systems biology approach for enhancing heat stress tolerance in plants. Plant Cell Rep 41(3):741–763

Rodríguez JP, Rahman H, Thushar S, Singh RK (2020) Healthy and resilient cereals and pseudocereals for marginal agriculture: molecular advances for improving nutrient bioavailability. Front Genet 11:49

Rosales AR, Orona GC, Gallardo NAM, Segura LS, Escamilla JPP, Suarez PAP, Frier JPD (2023) Identification of genetic and biochemical mechanisms associated with heat shock and heat stress adaptation in grain amaranths. Front Plant Sci 14:1101375

Saha D, Mukherjee P, Dutta S (2019) Genomic insights into HSFs as candidate genes for high-temperature stress adaptation and gene editing with minimal off-target effects in flax. Sci Rep 9:5581

Schoenlechner R, Bender D (2020) Pseudocereals for global food production. Cereal Foods World 65(2). https://doi.org/10.1094/CFW-65-2-0014

Shunkao S, Theerakulpisut P, Wanichthanarak K, Pongdontri P, Thitisaksakul M (2023) Integrative physiological and metabolomics study reveals adaptive strategies of wheat seedlings to salt and heat stress combination. Plant Growth Regul 100(1):181–196

Singh A, Grover A (2008) Genetic engineering for heat tolerance in plants. Physiol Mol Biol Plants 14(1):155–166

Singh B, Salaria N, Thakur K (2019) Functional genomic approaches to improve crop plant heat stress tolerance. F1000 Res 8:1721
Singh S, Praveen A, Dudha N, Bhadrecha P (2024) Integrating physiological and multi-omics methods to elucidate heat stress tolerance for sustainable rice production. Physiol Mol Biol Plants 30(7):1185–1208
Sinha R, Induri SP, Pelaez-Vico MA, Tukuli A, Shostak B, Zandalinas SI, Joshi T, Fritschi FB, Mittler R (2023) The transcriptome of soybean reproductive tissues subjected to water deficit, heat stress, and a combination of water deficit and heat stress. Plant J 116(4):1064–1080
Stephens J, Barakate A (2017) Gene editing technologies–ZFNs, TALENs, and CRISPR/Cas9. In: Encyclopedia of applied plant sciences. https://doi.org/10.1016/B978-0-12-394807-6.00242-2
Sukumaran S, Reynolds MP, Sansaloni C (2018) Genome-wide association analyses identify QTL hotspots for yield and component traits in durum wheat grown under yield potential, drought, and heat stress environments. Front Plant Sci 9:81
Suvorova G, Zhou M (2018) Distribution of cultivated buckwheat resources in the world. In: Buckwheat germplasm in the world. Academic, pp 21–35
Suzuki N (2023) Fine tuning of ROS, redox and energy regulatory systems associated with the functions of chloroplasts and mitochondria in plants under heat stress. Int J Mol Sci 24(2):1356
Tashi G, Zhan H, Xing G, Chang X, Zhang H, Nie X, Ji W (2018) Genome-wide identification and expression analysis of heat shock transcription factor family in *Chenopodium quinoa*. Willd. Agronomy 8(7):103
Torres AC, Sepulveda G, Kahlaoui B (2017) Phytohormone interaction modulating fruit responses to photooxidative and heat stress on apple. Front Plant Sci 8:2129
Wang H, Wang H, Shao H, Tang X (2016) Recent advances in utilizing transcription factors to improve plant abiotic stress tolerance by transgenic technology. Front Plant Sci 9(7):67
Wang Z, Yan S, Ren W, Liu Y, Sun W, Liu M, Ma W (2022) Genome-wide identification of MAPK, MAPKK, and MAPKKK gene families in *Fagopyrum tataricum* and analysis of their expression patterns under abiotic stress. Front Genet 13:894048
Wanniarachchi PC, Pavalakumar D, Jayasinghe MA (2023) Pseudocereals: nutrition, health benefits, and potential applications in gluten-free food product developments. Curr Nutr Food Sci 19(4):377–385
Wen J, Jiang F, Weng Y, Sun M, Shi X, Zhou Y, Yu L, Wu Z (2019) Identification of heat-tolerance QTLs and high-temperature stress-responsive genes through conventional QTL mapping, QTL-seq and RNA-seq in tomato. BMC Plant Biol 19(1):398
Wolter F, Patrick S, Holger P (2019) Plant breeding at the speed of light: the power of CRISPR/Cas to generate directed genetic diversity at multiple sites. BMC Plant Biol 19(1):176
Xue Y, Yin N, Chen B, Liao F, Win AN, Jiang J, Chai Y (2017) Molecular cloning and expression analysis of two *FAD2* genes from chia (*Salvia hispanica*). Acta Physiol Plant 39(4):95
Xue G, Fan Y, Zheng C, Yang H, Feng L, Chen X, Ruan J (2023) bHLH transcription factor family identification, phylogeny, and its response to abiotic stress in *Chenopodium quinoa*. Front Plant Sci 14:1171518
Yabe S, Iwata H (2020) Genomics-assisted breeding in minor and pseudo-cereals. Breed Sci 70(1):19–31
Yan H, Sun M, Zhang Z, Jin Y, Zhang A, Lin C, Wu B, He M, Xu B, Wang J, Qin P (2023) Pangenomic analysis identifies structural variation associated with heat tolerance in pearl millet. Nat Genet 55(3):507–518
Yang M, He G, Hou Q, Fan Y, Duan L, Li K, He T (2022a) Systematic analysis and expression profiles of TCP gene family in Tartary buckwheat (*Fagopyrum tataricum* (L.) Gaertn.) revealed the potential function of FtTCP15 and FtTCP18 in response to abiotic stress. BMC Genomics 23(1):415
Yang Y, Zhang C, Zhu D, He H, Wei Z, Yuan Q, Li X, Gao X, Zhang B, Gao H, Wang B (2022b) Identifying candidate genes and patterns of heat-stress response in rice using a genome-wide association study and transcriptome analyses. Crop J 10(6):1633–1643
Yang X, Guan H, Yang Y, Zhang Y, Su W, Song S, Liu H, Chen R, Hao Y (2023) Extra-and intra-nuclear heat perception and triggering mechanisms in plants. Front Plant Sci 14:1276649

Zandalinas SI, Balfagon D, Arbona V, Gómez-Cadenas A, Inupakutika MA, Mittler R (2016) ABA is required for the accumulation of APX1 and MBF1c during a combination of water deficit and heat stress. J Exp Bot 67(18):5381–5390

Zare T, Kehelpannala C, Bhatnagar A, Rupasinghe TW, Ebert B, Fournier-Level A, Roessner U (2025) Thermotolerance in Chia (Salvia hispanica L.) is Mediated by Rapid Heat-Induced Transcriptomic Reprogramming and Lipid Remodelling in Leaves. BioRxiv 2:05

Zhao Y, Du H, Wang Y, Wang H, Yang S, Li C, Chen N, Yang H, Zhang Y, Zhu Y, Yang L (2021) The calcium-dependent protein kinase ZmCDPK7 functions in heat-stress tolerance in maize. J Integr Plant Biol 63(3):510–527

Zheng L, Wu W, Chen Q, Zhang G, Gao F, Zhou Y (2022) Integrated transcriptomics, proteomics, and metabolomics identified biological processes and metabolic pathways involved in heat stress response in jojoba. Ind Crops Prod 183:114946

Zhu JK (2016) Abiotic stress signaling and responses in plants. Cell 167(2):313–324

Proteomic Strategies to Uncover Insights into Abiotic Stress Responses

5

Vishal Saini, Monika Chouhan, Paramdeep Kumar, and Rajiv Kumar

Abstract

Proteomics has emerged as a key technique for identifying the molecular mechanisms of plant responses to abiotic stressors, offering insights beyond transcriptomics and genomics. The principles of plant proteomics are introduced at the beginning of this chapter, along with the importance of this field in comprehending stress adaptation. The dynamics of the proteome under significant abiotic stressors are then examined, emphasizing how protein expression, modifications, and interactions affect plant resilience. With an emphasis on crucial proteins that provide tolerance to oxidative stress, heavy metals, drought, salinity, and extreme temperatures are covered in detail. The chapter also looks at functional validation techniques, including transgenic, biochemical, and molecular approaches, which show causal links between stress-responsive proteins and plant performance. The application of proteomic findings to crop improvement is discussed, demonstrating how biomarkers and candidate proteins can be utilized in biotechnological and breeding methods to produce stress-tolerant crops. The chapter ends by outlining prospects for system-level approaches and high-throughput proteomic tools, while addressing contemporary concerns such as technical limitations, post-translational modification complexity, and integration with other omics platforms. By giving a comprehensive overview of the field in the context

V. Saini · M. Chouhan · R. Kumar (✉)
Department of Biotechnology, CSIR-Institute of Himalayan Bioresource Technology (IHBT), Palampur, Himachal Pradesh, India

Academy of Scientific and Innovative Research (AcSIR), Ghaziabad, Uttar Pradesh, India
e-mail: rajiv@ihbt.res.in; rajiv.kumar@csir.res.in

P. Kumar
Academy of Scientific and Innovative Research (AcSIR), Ghaziabad, Uttar Pradesh, India

V. Gahlaut et al. (eds.), *Smart Crop Development*,
https://doi.org/10.1007/978-981-95-7680-7_5

of plant abiotic stress, this chapter emphasizes the role proteomics in sustainable agriculture and global food security.

Keywords

Proteomics · Abiotic stress · Crop improvement · Stress-responsive proteins

5.1 Introduction

Plants, being sessile organisms, are continuously exposed to various abiotic stresses like drought, extreme temperatures, salinity, flooding, and heavy metal in their natural habitat. These abiotic stresses affect physiological and molecular processes that influence growth, development, and productivity or reproductive fitness (Fig. 5.1). However, in the era of global climate, the intensity and frequency of single, combined, and sequential abiotic stresses are increasing continuously, which is a serious threat to global food security (Uday et al. 2024). A recent report revealed that climate change reduced the global production of ten major crops (rice, wheat, maize, sugarcane, soybean, rapeseed, cassava, oil palm, sorghum, and barley) by 1% (Hultgren et al. 2025). How plants perceive and respond to stress conditions is the fundamental and common question of many plant biology researchers over the past few decades. Recent reports stated that plants have evolved sophisticated mechanisms to sense and respond to various abiotic stresses, which include an acclimation

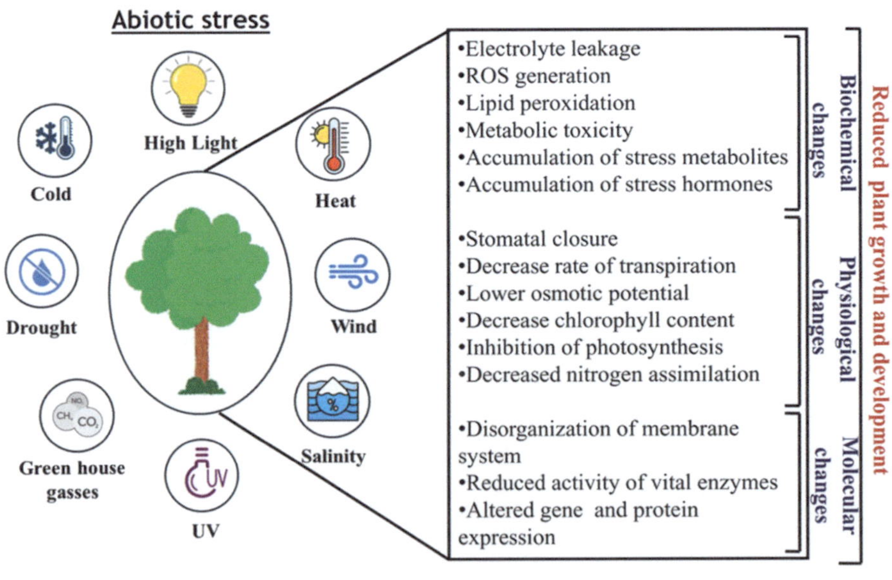

Fig. 5.1 Overview of abiotic stress effects on plant growth and development, highlighting key biochemical, physiological, and molecular responses, i.e., oxidative damage, stomatal closure, disrupted photosynthesis, and altered gene expression, which reduce growth and development

process at the cost of reduced growth (Borghi et al. 2019). For example, the plasma membrane is the site where plants sense various abiotic signals, control water and solute transport, and initiate downstream signals. Adapting to plant stress requires reorganization at the molecular level, spanning multiple biological layers: transcriptomic, proteomic, metabolomic, and epigenetic.

Several phenotypic and transcriptomic analyses have expanded our knowledge on the molecular mechanism of signal transduction, integration and the identification of stress-responsive genes involved in abiotic stress; however, they lack the functional proteomic aspects. Proteins are the direct players of biological processes as they participate as enzymes, scaffolds, regulators, and cell signaling. It is vital to note that proteomics offers the study of post-translational modifications (PTMs), protein interactions, and protein function. Advancements in proteomic technologies have transformed large-scale protein analysis of cells. We observe that classical gel-based methods like two-dimensional polyacrylamide gel electrophoresis (2D-PAGE) and differential gel electrophoresis (DIGE) have been the most popular proteomic techniques in the last three decades. Later on, second-generation techniques like label-based and label-free quantification techniques, which overcome the limitations of the classical method, became more popular. Moreover, next-generation proteomic techniques have been developed, which are supported by high-throughput mass spectrometry (MS)-based platforms, and have become a powerful tool nowadays for protein quantification, PTMs, subcellular localization, and protein–protein interactions (PPIs).

Proteomics provides a powerful lens to understand how plants respond to abiotic stresses—such as drought, salinity, extreme temperatures, heavy metals, and flooding—by profiling stress-responsive proteins and interrogating PTMs and subcellular dynamics (Kumari et al. 2021). Advanced quantitative approaches (e.g., iTRAQ, TMT, label-free MS/MS) have enabled high-throughput identification of differentially expressed proteins involved in signaling cascades, redox regulation, carbohydrate and energy metabolism, heat shock response, antioxidant defense, and secondary metabolism (Chouhan et al. 2025; Kosová et al. 2015). Such analyses uncover proteins that function as molecular chaperones, transporters, or transcriptional regulators, ultimately contributing to cellular homeostasis and stress adaptation. Crop-specific proteomic investigations provide valuable insights into tolerance mechanisms. For example, in wheat, proteomic profiling has identified drought-responsive proteins linked to root plasticity, photosynthesis, and energy balance (Azarkina et al. 2025). Similarly, stress-responsive proteomes have revealed proteins mediating salt tolerance, osmolyte biosynthesis, and reactive oxygen species (ROS) detoxification in rice and maize. The integration of proteomics with transcriptomics, metabolomics, and allows the construction of complex regulatory networks and has expanded its utility in precision breeding (Kumari et al. 2025). Emerging studies also highlight the role of advanced nanomaterials in modulating plant proteomes under salt stress, suggesting new strategies for resilience enhancement (Cherati and Khodakovskaya 2025). Despite these advances, significant challenges limit the full application of proteomics in plant stress biology. Plant tissues often pose difficulties for protein extraction due to rigid cell walls, abundant secondary metabolites, and proteolytic enzymes, leading to biased representation. Additionally, current technologies less efficient to identify low-abundance proteins such as transcription factors (TFs). The lack of comprehensive and curated proteomic databases for polyploid crops, coupled with inter-platform variability, reduces reproducibility and hampers accurate functional annotation (Uday et al. 2024). Proteomics is indispensable for deciphering plant stress tolerance mechanisms and identifying candidate proteins for crop improvement, but its potential remains constrained by technical, biological, and database-related limitations.

5.2 Proteomic Techniques

Proteomics entails the study of the entire protein complement of a cell, tissue, or organism at a particular developmental stage or conditions (Jorrin-Novo 2020). In addition, proteomics portrays a cell's protein levels, activity, interactions, subcellular localization, and modifications (Hossain and Komatsu 2013). The proteome of a cell is highly dynamic; it varies from time to time and cell to cell and with different types of stimuli. Unlike genomics and transcriptomics, which offer valuable insights into plants' genetic makeup and transcriptional activity, they do not always result in functional protein outcomes. Several studies underscore the limited correlation between mRNA and protein abundance in plants under abiotic stress. In rice, transcriptome–proteome comparisons under drought stress revealed partial overlap between upregulated transcripts and protein accumulation, indicating strong post-transcriptional control (Chintakovid et al. 2017). Several post-transcriptional, translational efficiency and various PTMs contribute to this disparity (Fan et al. 2025). Proteins are the real molecular effectors that fills the knowlegde gap between signaling to phenotype. Proteomics incorporates diverse approaches such as expression proteomics (quantitative profiling of proteins), structural proteomics (characterizing protein architecture and complexes), functional proteomics (uncovering protein activity and networks), and subcellular proteomics (compartment-specific proteomes). The major application of proteomics is protein profiling, which is involved in biological processes like signal perception, signal initiation, and stress response, to understand the physiology of stress and develop biomarkers for crop improvement. For example, proteome analyses in rice under salinity stress have identified enhanced antioxidant enzymes in tolerant genotypes (Frukh et al. 2020), while drought proteomics in wheat highlighted maintenance of photosynthetic proteins in tolerant cultivars (Michaletti et al. 2018). Apart from stress physiology, proteomics contributes to developmental biology, nutrient quality research, plant–pathogen interactions, and precision agriculture applications (Punia et al. 2024; Iacobucci et al. 2021). With advancements in MS and computational biology, proteomics has become a cornerstone of plant systems biology and crop improvement.

Sample preparation is the proteomic study's first and most critical step to address a biological question. We must consider tissue type (leaf, stem, root, rhizome, flower, inflorescence, and seed), tissue age (young and old tissue), and harvesting time; daytime (morning, afternoon, and evening) or nighttime has a significant impact on protein composition. Protein isolation from plant tissue is more challenging than animal tissue, as plant tissue is rich in polyphenols and various secondary metabolites that interfere with further proteomic analysis. In addition, abundant proteins, such as ribulose bisphosphate carboxylase/oxygenase (Rubisco) in the leaves and albumin in the seeds often mask the low abundant protein during protein identification. It is advisable to perform all the protein work at 4 °C to inactivate the metabolism through enzymatic or oxidative degradation of protein. We should consider a higher number of biological replicates to minimize sample variations. Further, due to differences in the protein properties like charge, molecular weight, hydrophobicity, and PTM, a single protein isolation method would be less effective; rather, a combination of two methods is preferred. For example, the combination of the TCA and acetone method is more effective than the TCA and acetone method alone (Fang et al. 2024). However, the choice of protein extraction methods largely depends on plant material and the specific biological question.

The 2D-PAGE coupled with MS, is one of the versatile classical tools in proteomics for separation (according to their pI and molecular weight) and identification of proteins. This enables the resolution of hundreds to thousands of proteins within a single gel, making it valuable for comparative studies. This technique's primary benefits include its ease of use, repeatability, ability to separate moderately hydrophilic proteins, and analysis of protein complexes and PTM (Ha et al. 2021). The 2D-PAGE has been used to understand plant response toward drought (Yahoueian et al. 2021), salinity (Şekerci et al. 2024), cold (Khan et al. 2022), heat (Chunduri et al. 2021), and UV radiation (Sah et al. 2022; Jan et al. 2023). However, the main disadvantage of this technique is that it is tedious, time-consuming, and unable to resolve highly acidic and basic proteins and membrane proteins, has gel-to-gel variation, and is unable to detect low-abundant proteins. To separate hydrophobic and membrane proteins and analysis of protein complexes later on, blue-native polyacrylamide gel electrophoresis has been used in tomato and plastid (Sharma et al. 2024). Moreover, the implementation of fluorescent dye for DIGE improves this method by allowing multiple samples to be resolved within the same gel, and thus minimizing gel-to-gel variability, increasing reproducibility and dynamic range of protein identification. The DIGE has been used to study a variety of abiotic stresses (Liu et al. 2014). However, the requirement of fluorescent imaging instruments and associated software limits their application for proteomic studies. To overcome the limitation of gel-based proteomics, the so-called "second-generation" proteomic techniques like multidimensional protein identification, isotope-coded affinity tags, isobaric tags for relative and absolute quantification (iTRAQ), and stable isotope labeling by amino acids in cell culture (SILAC) have been successfully used for quantitative proteomic studies (Jorrin et al. 2007). These techniques have a broader dynamic range, higher sensitivity, and the ability to identify low-abundance proteins, phosphorylation and its site, and protein complexes. Further, recent advancements in triple quadrupole MS and orbitraps and next-generation proteomic tools like multiple reaction monitoring, selected reaction monitoring, proteogenomics, MALDI imaging, and protein microarrays were used for quantitative proteomic analysis (Thikekar et al. 2023). Multiple reaction monitoring is a highly reproducible, absolute, and targeted quantification method in which a precursor or fragment ions are selected for quantification. More recently, data-independent acquisition approaches such as Sequential Window Acquisition of All Theoretical Mass Spectra (SWATH-MS) have gained precedence. SWATH-MS represents all fragment ions across a mass range in a single run, providing reproducible and quantitative proteome coverage (Guzman et al. 2024).

The advancement of proteomic technologies has resulted in the generation of complex datasets that require robust bioinformatic pipelines for a meaningful interpretation of biological significance. These techniques support protein identification, functional annotation, quantification, and system-level network analysis. Core databases such as UniProtKB and NCBI RefSeq provide comprehensive protein sequence information, while plant-focused resources like The Arabidopsis Information Resource and the Plant Proteome Database offer curated annotations and experimentally validated datasets for model plants (Subba et al. 2019). To decipher PPIs, resources such as STRING and visualization platforms like Cytoscape are widely employed (Szklarczyk et al. 2019). Furthermore, Gene Ontology annotation and Kyoto Encyclopedia of Genes and Genomes pathway mapping enable researchers to place stress-responsive proteins into functional categories and metabolic pathways.

5.3 Proteome Dynamics Under Abiotic Stresses

Abiotic stresses such as salinity, cold, heat, and heavy metal exposure often disrupt protein functionality, causing misfolding or aggregation. In response, plants mobilize a robust defense system: reprogramming gene expression and reshaping their proteome both qualitatively and quantitatively. This involves the induction of stress-responsive proteins that adjust metabolic pathways to shift the plant's priority from growth to survival (Table 5.1). Proteomic and transcriptomic profiling has revealed that selenium application under cadmium stress enhances levels of antioxidant proteins—such as glutathione S-transferases, G6PDH, and chlorophyll-binding proteins—effectively strengthening photosynthesis and ROS detoxification in rice seedlings (Mishra et al. 2024). Similarly, quantitative proteomics using iTRAQ in rice exposed to cadmium, with or without nitric oxide, demonstrated changes in plasma membrane proteins including PLDs, ATPases, kinases, and glutathione accumulation, highlighting a coordinated antioxidant defense and signaling response (Yang et al. 2016). Beyond detoxification, plants redirect metabolism toward cell wall reinforcement and energy balance. Selenium-mediated cadmium detoxification in rice roots was shown to upregulate glycosyl hydrolases, O-methyltransferases, polygalacturonases, and carbohydrate-processing enzymes—underscoring the importance of cell wall remodeling for stress tolerance (Zhu et al. 2025). Such metabolic adjustments are critical in maintaining osmotic balance and protecting cellular integrity under adverse conditions. The role of PTMs in stress signaling is increasingly recognized. Recent studies demonstrate that phosphorylation, ubiquitination, SUMOylation, and acetylation modulate key regulatory

Table 5.1 List of proteins, their functions in abiotic stress, and techniques used for identification

S. no.	Protein family	Functions	Technique used	Stress type	References
1.	Late embryogenesis abundant proteins	Protect cellular structures by stabilizing proteins and membranes	2-DE, MALDI-TOF MS	Salt	Saleethong et al. (2016)
2.	Heat shock proteins	Prevent protein aggregation and assist protein folding	iTRAQ-based proteomics	Heat, salt	Iacobucci et al. (2021)
3.	Metallothioneins	Metal chelation and detoxification	2-DE, LC-MS/MS	Heavy metal	Ahsan et al. (2009)
4.	Glutathione S-transferases	Detoxification of ROS and xenobiotics under cd toxicity	2-DE, MALDI-TOF MS	Heavy metal	Ahsan et al. (2009)
5.	Aquaporins (PIP, TIP, NIP family)	Facilitate water transport	Western blot	Drought	Alexandersson et al. (2005)
6.	RuBisCO-binding protein	Maintains photosynthetic machinery	2-DE, MS	Drought	Ashoub et al. (2013)
7.	Plasma membrane H^+-ATPase	Maintains ion balance	2-DE, MS	Salt	
8.	Peroxiredoxins	ROS detoxification	2-DE, MS/MS	Cold	Byun et al. (2014)
9.	Stress-related kinases	Phosphorylation signaling during abiotic stress	2-DE, LC-MS	General abiotic	Carpentier et al. (2005)
10.	Pathogenesis-related proteins	Contribute to osmoprotection and defense	2-DE, MS	Salt	Yan et al. (2005)
11.	Antioxidant enzymes (SOD, CAT, APX)	Detoxification of ROS	2-DE, LC-MS/MS	Drought	Fan et al. (2025)
12.	Phytochelatin synthases	Detoxification of Cd via chelation	iTRAQ-based proteomics	Heavy metal	Zhong et al. (2017)
13.	Drought-inducible proteins (LEAs, chaperones)	Adaptation under stress	LC-MS	Drought	
14.	Small heat shock proteins (sHSPs)	Protect membranes and proteins	2-DE, MS/MS	Heat, drought	Hu et al. (2010)

(continued)

Table 5.1 (continued)

S. no.	Protein family	Functions	Technique used	Stress type	References
15.	Cold-responsive proteins (CORs, dehydrins)	Membrane stabilization under freezing stress	Label-free LC-MS/MS	Cold	Jiang et al. (2021)
16.	Proteome-wide markers	Broad stress tolerance functions	MS-based proteomics	General abiotic	Jorrín-Novo et al. (2019)
17.	Drought stress proteome (LEAs, ROS enzymes)	Multiple protective functions	SWATH-MS, LC-MS	Drought	Kausar et al. (2022)
18.	Aquaporin RWC3	Improves water uptake in drought	Western blot	Drought	Lian et al. (2004)
19.	WRKY/NAC/MYB TFs	Regulate stress response networks	MS	General abiotic	Muthuramalingam et al. (2018)
20.	MAPKs	Mediate signal transduction under abiotic stress	Phosphoproteomics, MS	General abiotic	Nakagami et al. (2010)
21.	Conserved phosphoproteins	Regulatory role under stress	LC-MS/MS	General abiotic	Nakagami et al. (2010)
22.	Zn–nicotianamine transporters	Metal ion transport under stress	LC-MS	Heavy metal	Nishiyama et al. (2012); Hossain and Komatsu (2013)
23.	Stress-responsive leaf proteins	Drought adaptation	2-DE, LC-MS	Drought	Michaletti et al. (2018)
24.	*Quercus ilex* stress proteins	Potential drought tolerance markers	LC-MS/MS	Drought	San-Eufrasio et al. (2021)
25.	Antifreeze proteins	Prevent ice formation in cold stress	Protein purification	Cold	Sidebottom et al. (2000)
26.	Heat stress proteins (RuBisCO activase, HSPs)	Protect photosynthesis under heat	2-DE, LC-MS	Heat	Wang et al. (2015)
27.	Drought-responsive proteins	Root adaptation in soybean	SWATH-MS, LC-MS/MS	Drought	Wang et al. (2020)
28.	Salt-responsive proteins (transporters, chaperones)	Ion balance and osmoprotection	LC-MS/MS	Salt	Zhang et al. (2025)
29.	Salt-responsive maize proteins	Mitigate salt symptoms	Proteomics, LC-MS	Salt	Zörb et al. (2004)

proteins and TFs in horticultural crops under drought, salinity, cold, and heat stress (Zhang et al. 2025). Moreover, site-specific SUMOylation mapping in *Arabidopsis* revealed new regulatory layers in plant stress responses, expanding our understanding of proteome plasticity under abiotic stress (Sang et al. 2024).

Heat shock proteins (HSPs) are highly conserved molecular chaperones that safeguard cellular proteome integrity by preventing protein misfolding, aggregation and facilitating refolding through ATP-dependent cycles. They are rapidly induced under adverse environmental conditions such as heat, drought, salinity, and oxidative stress, which often destabilize proteins and disrupt proteostasis (Waters and Vierling 2020). In plants, HSPs are classified into five major families based on molecular weight: HSP100, HSP90, HSP70, HSP60, and small HSPs (sHSPs). Large HSPs such as HSP70 and HSP90 function as central regulators, stabilizing client proteins including kinases, hormone receptors, and TFs, thereby linking stress perception with signaling and developmental pathways (Haider et al. 2021). Small HSPs, in contrast, act as the first line of defense by transiently binding denatured proteins during stress events, preventing irreversible aggregation until refolding can be completed by ATP-dependent chaperones (Waters and Vierling 2020). Functional studies have highlighted the critical role of HSPs in abiotic stress tolerance. For example, HSP101 in *Arabidopsis thaliana* contributes to acquired thermotolerance and recovery after heat stress, while sHSPs are consistently upregulated during drought and salinity, contributing to ROS detoxification, stabilization of membranes, and metabolic reprogramming (Tiwari et al. 2021; Kumar et al. 2024). Proteomic analyses further indicate that HSP70/HSP90 complexes not only stabilize the proteome but also regulate stress-responsive TFs and signaling proteins, making them crucial for cross-talk between stress adaptation and growth.

Abiotic stresses such as drought, salinity, heat, and heavy metals lead to excessive accumulation of ROS such as superoxide ($O_2^{\bullet -}$), hydrogen peroxide (H_2O_2), and hydroxyl radicals (•OH). While controlled ROS levels function as secondary messengers in stress signaling and gene regulation, uncontrolled accumulation causes oxidative injury to lipids, proteins, nucleic acids, and organelles, ultimately impairing photosynthesis and growth (Hasanuzzaman et al. 2020; Mhamdi and Van Breusegem 2022). Plants rely on a highly coordinated enzymatic antioxidant defense system to maintain ROS homeostasis. Superoxide dismutase (SOD), located in chloroplasts, mitochondria, cytosol, and peroxisomes, catalyzes the dismutation of $O_2^{\bullet -}$ into H_2O_2, thus initiating the detoxification cascade (Devireddy et al. 2021). Hydrogen peroxide is subsequently removed by catalase (CAT) in peroxisomes through a rapid, high-capacity reaction that converts H_2O_2 into water and oxygen. However, CAT lacks substrate specificity in different compartments, making ascorbate peroxidase (APX) a key complementary enzyme (Esnarriaga et al. 2020). The APX uses ascorbate as an electron donor to reduce H_2O_2 and operates within multiple cellular compartments, including chloroplasts, mitochondria, and cytosol. Its action is tightly coupled with the ascorbate–glutathione (AsA–GSH) cycle, in which monodehydroascorbate reductase and dehydroascorbate reductase regenerate ascorbate, while glutathione reductase sustains reduced glutathione (GSH) pools using NADPH (Gao et al. 2021). This cycle ensures continuous scavenging of ROS, maintaining the redox state required for metabolic homeostasis. Beyond scavenging, these enzymes regulate signaling pathways by modulating H_2O_2 levels, which act as second messengers in activating stress-responsive TFs (Magno Massuia de Almeida et al. 2022). For example, elevated SOD and APX activities in drought-tolerant rice and wheat genotypes limit ROS accumulation, preserving photosynthetic efficiency and protecting photosystem II. Similarly, enhanced glutathione reductase activity in maize under combined drought and heat stress ensures faster recovery by restoring redox balance (Magno Massuia de Almeida et al. 2022). Thus, the antioxidant defense system functions both as a detoxification network and a regulatory hub, integrating redox signals with stress-responsive metabolic reprogramming to enhance plant resilience under adverse environments.

Signal perception and downstream reprogramming during abiotic stress are tightly regulated by reversible protein phosphorylation, orchestrated by protein kinases and phosphatases. These enzymes act as molecular switches, controlling TFs, ion channels, and metabolic enzymes, thereby enabling plants to balance stress tolerance with growth and development (Singh et al. 2017). Mitogen-activated protein kinases (MAPKs) constitute one of the most conserved signaling cascades, consisting of MAPKKKs, MAPKKs, and MAPKs. Stress perception at the plasma membrane activates MAPK cascades, which transmit signals to the nucleus. In crops such as rice and wheat, MAPK cascades modulate TFs involved in ABA-dependent and ABA-independent pathways, linking environmental stress perception with transcriptional reprogramming (Singh et al. 2017). Calcium-dependent protein kinases (CDPKs/CPKs) act as critical sensors of intracellular Ca^{2+} signatures generated during stress. By directly binding Ca^{2+}, they undergo conformational changes that activate kinase activity, enabling phosphorylation of ion channels, antioxidant enzymes, and TFs. For example, OsCPK21 in rice enhances drought tolerance by modulating ABA signaling and ROS detoxification (Asano et al. 2012). Protein phosphatases, particularly PP2Cs, serve as negative regulators of stress signaling. In ABA signaling, PP2Cs inhibit SNF1-related kinases (SnRK2s), thereby preventing constitutive activation of stress responses. Under drought or salinity stress, elevated ABA levels inhibit PP2Cs, releasing SnRK2s to phosphorylate downstream TFs (AREBs/ABFs), ion transporters, and metabolic regulators (Hayford et al. 2022). This reversible regulation ensures fine-tuned activation of stress responses without excessive metabolic penalties.

In summary, kinases and phosphatases constitute central hubs in stress signaling, translating environmental cues into transcriptional and metabolic reprogramming. Their tight regulation ensures efficient adaptation, and recent studies highlight them as promising targets for developing stress-resilient crops. Aiotic stresses such as drought, salinity, heat, cold, and oxidative stress severely compromise cellular homeostasis by inducing osmotic imbalance, ion toxicity, oxidative damage, and reduced photosynthetic efficiency. Plants counter these stresses through a multilayered defense system involving TFs, metabolic reprogramming, and membrane-associated transport proteins. Together, these mechanisms reprogram gene expression, remodel metabolism, and stabilize cellular structures to improve survival under adverse conditions.

Transcription factors act as master regulators of stress signaling pathways. The DREB/CBF proteins are rapidly activated by cold and drought signals and bind to dehydration-responsive element (DRE/CRT) motifs, inducing genes encoding late embryogenesis abundant (LEA) proteins, osmolytes, and antioxidant enzymes, thereby mediating ABA-independent pathways. The NAC and MYB TFs contribute to remodeling of the cell wall, lignin deposition, and regulation of flavonoid and phenylpropanoid biosynthesis, enhancing structural resilience and ROS detoxification. The bZIP family members, including OsbZIP46 in rice, undergo phosphorylation under drought stress, which enhances their ability to induce ABA-dependent stress-responsive genes such as dehydrins and proline biosynthetic enzymes. Importantly, TF activity is tightly regulated by PTMs like phosphorylation, ubiquitination, and SUMOylation, ensuring a dynamic and context-specific response (Wang et al. 2024).

Abiotic stress triggers profound shifts in primary and secondary metabolism. Carbohydrate metabolism is redirected toward the synthesis of osmoprotectants such as trehalose and sucrose, which not only stabilize proteins and membranes but also act as signaling molecules to activate stress defense cascades. Nitrogen metabolism is modulated through increased glutamine synthetase and glutamate dehydrogenase activity, contributing to the biosynthesis of amino acids like proline and arginine, which function as osmolytes and ROS scavengers. Lipid metabolism undergoes adaptive remodeling: cold stress induces fatty acid desaturases, enhancing unsaturated fatty acid content and membrane fluidity, whereas heat stress triggers restructuring of membrane phospholipids to reduce leakage and oxidative damage. Energy metabolism also shifts under drought, with glycolytic enzymes upregulated and Calvin cycle proteins downregulated, reflecting a strategy to conserve ATP while maintaining essential cellular functions (Ma et al. 2024).

Membrane-associated proteins play frontline roles in sensing stress cues and regulating ionic and osmotic balance. Aquaporins (AQPs) such as *Arabidopsis* PIP2;1 regulate cellular hydraulic conductivity, ensuring water uptake under drought but closing under excess stress to prevent uncontrolled water loss. Na^+/H^+ exchangers (NHXs) sequester excess sodium into vacuoles, maintaining cytosolic ion homeostasis under salinity. Similarly, HKT transporters retrieve sodium from the xylem, preventing toxic accumulation in photosynthetic tissues. The ABC transporters participate in the transport of secondary metabolites, phytohormones, and xenobiotics, playing key roles in detoxification and signaling under oxidative or heavy metal stress (Mishra et al. 2024; Zhu et al. 2024).

These mechanisms function in a highly interconnected manner. Stress-induced TFs activate genes encoding AQPs, ion transporters, and osmoprotectant enzymes. Metabolic reprogramming provides the substrates and cofactors needed for these defense processes, while transport proteins ensure ionic stability and signaling continuity. Collectively, these multilayered mechanisms establish a robust adaptive network, allowing plants to balance growth with survival under abiotic stress. With climate change intensifying the frequency of environmental stresses, recent studies emphasize that integrating proteomic, transcriptomic, and metabolomic data will be essential to fully unravel these mechanisms and to design crops with improved resilience.

5.4 Integration of Proteomic Changes

The stress-responsive proteome works as a highly connected network instead of a set of separate proteins. For example, HSPs collaborate with ROS-scavenging enzymes to stop protein denaturation and oxidative damage. Combining proteomics with transcriptomics and metabolomics through integrated omics approaches has strengthened this system-level view. In rice under drought stress, multi-omics analyses showed that ABA signaling, ROS detoxification, and carbohydrate metabolism are closely linked together, they help improve tolerance (Zargar et al. 2022). This complex communication makes sure that protective mechanisms, energy metabolism, and signaling pathways stay well-coordinated during stress. Overall, proteome dynamics under abiotic stress reflect a multifaceted defense strategy, encompassing molecular chaperones, antioxidants, signaling proteins, TFs, metabolic enzymes, and membrane transporters. These proteins act together to stabilize macromolecules, detoxify ROS, reprogram primary and secondary metabolism, and maintain water and ion homeostasis. The advent of high-resolution quantitative proteomics, coupled with powerful bioinformatics and integrative omics, continues to transform plant stress biology from descriptive cataloging toward predictive, mechanistic frameworks.

5.5 Proteomics in Abiotic Stress Conditions

Abiotic stresses such as salinity, drought, heat, heavy metals, and oxidative stress severely impair the growth, development, and productivity of the plant. Proteomics provides the understanding of how plants reprogram their proteomes under stress, which helps in revealing the stress-specific protein signatures crucial for their survival. This section elaborates on stress-specific proteomic signatures with examples and mechanistic insights.

Proteome dynamics are apparent across diverse conditions. Under drought and saline conditions, photosynthetic proteins including Rubisco and chlorophyll a/b--binding proteins get downregulated, while enzymes related to osmolyte biosynthesis (e.g., pyrroline-5-carboxylate synthetase for proline accumulation) are upregulated (Gupta and Huang 2014). Heat stress induces the synthesis of HSPs (HSP70, HSP90, small HSPs) and proteases that protect protein folding and eliminate damaged proteins, whereas cold stress favors the accumulation of lipid desaturases and antifreeze proteins (AFPs), which help in maintaining the membrane fluidity and ice recrystallization inhibition (Jiang et al. 2021). Heavy metal exposure enhances the accumulation of metallothioneins, phytochelatins, and glutathione S-transferases, which are crucial for chelation and detoxification of toxic ions (Hossain and Komatsu 2013). In addition to these expression changes, PTMs act as important regulators of stress signaling and adaptation. Phosphorylation cascades involving MAP kinases mediate stress-responsive signaling pathways, ubiquitination directs the selective degradation of misfolded proteins, and oxidative carbonylation often marks proteins for degradation under severe stress (Nakagami et al.

2010; Yu et al. 2019). Likely, phosphorylation of AQPs has been shown to regulate water transport efficiency during drought, while ubiquitination of key TFs ensures rapid turnover and fine-tuning of stress responses (Alexandersson et al. 2005).

5.5.1 Proteomics Under Drought Stress

Drought stress induces biochemical and physiological changes in plants, which include stomatal closure, activation of respiration, and impairment of photosynthesis, nutrient uptake, and growth. Plants respond and adapt to drought stress at both cellular and molecular levels. Proteomic studies reveal reprogramming of protein expression involved in osmotic adjustment, antioxidant defense, water transport, and metabolic regulation. These changes maintain cellular homeostasis and ensure survival under low water availability (Kosová et al. 2018). Proteins linked to the synthesis of osmolytes such as proline, glycine betaine, and trehalose are accumulated under drought stress. In *Triticum aestivum*, proteomic analysis revealed increased expression of enzymes for proline biosynthesis (Δ^1-pyrroline-5-carboxylate synthetase (P5CS)), supporting osmoprotection and protein stability (Hassan et al. 2015). The LEA proteins act as molecular shields by stabilizing membranes and preventing protein aggregation. In rice, drought stress induced group 3 LEA proteins, protecting chloroplast proteins from denaturation. Water channel proteins regulate cellular and subcellular water fluxes. Proteomics in maize showed that plasma membrane intrinsic proteins were strongly upregulated under water deficit, optimizing water-use efficiency (Song et al. 2022). Drought stress redefines energy pathways, reducing photosynthetic proteins while increasing glycolysis and TCA cycle enzymes. In barley, RuBisCO levels decreased, whereas ATP synthase and glycolytic enzymes increased, indicating a shift toward energy conservation (Ashoub et al. 2013). Shotgun proteomics identified accumulation of seed storage proteins and stress-related enzymes in roots under drought stress, suggesting enhanced metabolic resilience (Mohammadi et al. 2012). Moreover, LC-MS/MS revealed drought-induced phosphorylation of AQPs and kinases, highlighting post-translational regulation (Xiong et al. 2019).

5.5.2 Proteomics Under Salinity Stress

Salinity stress leads to significant changes in the proteome of plants by affecting processes such as ion homeostasis, biosynthesis of compatible solutes, antioxidant defense mechanisms, and stress-related signalling pathways. "Na^+/H^+" antiporters (NHXs) play an important role in salinity tolerance by maintaining cytosolic Na^+/K^+ homeostasis. These antiporters actively sequester the excess Na^+ into vacuoles, preventing the ionic toxicity and preserving essential cellular functions. In *Arabidopsis thaliana*, vacuolar NHX1 protein accumulation was confirmed via proteomics under salinity stress (Apse et al. 1999). H^+-ATPases and H^+-PPases generate proton gradients, which drive the ion transport. Under salinity stress, enzymes including proline (P5CS), glycine betaine (choline monooxygenase), and mannitol are accumulated. For example, in sugar beet (*Beta vulgaris*), the proteomic study revealed induction of glycine betaine biosynthetic enzymes during saline irrigation (Yolcu et al. 2024). ROS-scavenging enzymes like CAT, SOD, and APX are consistently upregulated under salinity stress. Signal transduction proteins like MAPKs and CDPKs orchestrate downstream transcriptional reprogramming. In rice, proteomics showed phosphorylation of salt overly sensitive kinases, directly linked to ion homeostasis (Zhu, 2016). Proteomic approaches such as iTRAQ identified over 370 salinity-responsive proteins, including transporters, antioxidant enzymes, and stress-associated proteins (Hussain et al. 2019). Similarly, DIGE-based analyses detected significant accumulation of LEA proteins and molecular chaperones in salt-tolerant cultivars, which were largely absent in sensitive genotypes (Zuther et al. 2007).

5.5.3 Proteomic Insights into Temperature Stress

5.5.3.1 Heat Stress

Heat stress disrupts membrane integrity, protein stability, and photosynthetic efficiency, which leads to impaired growth and productivity. Proteomic studies have identified several proteins (HSPs) that play crucial roles in thermotolerance. These proteins act as molecular chaperones, which help in the stabilization of denatured proteins and facilitate refolding. In wheat, HSP70 and HSP90 were strongly induced in flag leaves under high temperature, contributing to sustained grain filling and thermotolerance (Wang et al. 2015). High temperatures accelerate ROS accumulation, necessitating redox homeostasis. In maize, APX and peroxiredoxins were significantly upregulated under heat stress, protecting cellular integrity (Hu et al. 2010). Heat stress leads to RuBisCO inactivation and destabilization of PSII complexes, but certain HSPs counteract these effects. In rice, proteomics revealed enhanced abundance of chloroplast-localized OsHSP26.7, which preserved photosystem integrity and improved seedling survival under heat stress (Mahmood et al. 2010).

5.5.3.2 Cold Stress

Low-temperature stress reduces membrane fluidity, slows down the enzymatic activity, and perturbs metabolic balance. During cold stress, proteomic alteration involves stabilization of cellular structures and osmoprotection. For example, AFPs suppress the ice crystal nucleation and growth of plants. In ryegrass, AFPs were shown to accumulate in crown tissues during freezing, preventing cellular injury (Sidebottom et al. 2000). Membrane remodeling proteins such as desaturases increase unsaturated fatty acids, preserving membrane flexibility. In *Arabidopsis*, cold-induced desaturases enhanced lipid unsaturation, enabling cellular homeostasis under freezing stress (Romero et al. 2014). Sugars act as cryoprotectants by stabilizing proteins and membranes. Proteomic profiling of tea revealed upregulation of β-amylase and sugar transporters during chilling stress, supporting raffinose and sucrose accumulation for osmoprotection (Li et al. 2019).

5.5.4 Proteomics Under Heavy Metal Stress

Heavy metals such as Cd, Pb, Hg, and As pose severe challenges by inducing oxidative stress, which causes protein misfolding and ion imbalance. Proteomic investigations have identified major protein classes that are involved in providing tolerance. Metallothioneins and phytochelatins help in sequestering the toxic metals and prevent the oxidative damage. For example, in *Oryza sativa* roots, multiple metallothionein isoforms were strongly upregulated under Cd exposure, enhancing detoxification capacity (Ahsan et al. 2009). ROS-scavenging enzymes like glutathione-S-transferases, peroxidases, and thioredoxins neutralize ROS, generated by heavy metal stress. In *Brassica juncea*, proteomic profiling showed marked induction of GSTs under arsenic stress, facilitating redox balance (Yadav et al. 2005). The ABC transporters and ZIP family proteins mediate sequestration and compartmentalization of metals to mitigate toxicity. The iTRAQ-based proteomic study in rice revealed upregulation of antioxidative enzymes and metal transporters under Cd stress, highlighting root detoxification strategies. The DIGE analysis further identified enhanced expression of pathogenesis-related proteins under Hg stress, indicating cross-talk between heavy metal and biotic stress responses.

5.5.5 Proteomics Under Oxidative Stress

Abiotic factors like UV radiation, ozone, and secondary effects of different kinds of stress lead to oxidative stress. During stress, excessive production of ROS takes place, which not only damage the cellular components but also act as signaling molecule. Key proteins involved in the ROS detoxification are SOD (converts superoxide to hydrogen peroxide), CAT and APX (breakdown of hydrogen peroxide into water and oxygen), and peroxiredoxins and glutathione peroxidases (ROS regulation). The proteomic study of *Arabidopsis* under oxidative stress revealed the upregulation of Cu/Zn-SOD and peroxiredoxin (Mittler et al. 2004). Enzymes involved in the secondary metabolism like phenylpropanoid and flavonoid biosynthetic proteins get accumulated and produce antioxidants. Chaperones and proteases help in eliminating the oxidized proteins, maintaining proteostasis. Aquaporin proteins in drought, AFPs in cold, and metallothioneins in heavy metal stress collectively safeguard the cellular integrity and metabolic balance. Advanced MS combined with bioinformatics has enabled the identification of stress-specific signatures and common molecular players, offering new targets for developing stress-resilient crops.

5.6 Techniques for Functional Validation of Stress-Responsive Proteins

The identification of stress-responsive proteins through proteomics is only the first step in understanding plant abiotic stress tolerance. To establish the functional roles of these proteins, it is essential to perform validation experiments that directly link protein expression to plant phenotypes under stress. Functional validation employs genetic, biochemical, and integrative systems biology approaches. One of the most widely employed strategies to validate the role of a protein in stress tolerance is to manipulate its expression level in transgenic plants. Overexpression of stress-responsive proteins often enhances plant resilience, while suppression can reduce tolerance, thus confirming functional involvement. For example, overexpression of OsHSP17.0 in rice enhanced tolerance to heat and drought by stabilizing cellular proteins and membranes (Do et al. 2023). Similarly, AtP5CS1, an enzyme involved in proline biosynthesis, when overexpressed in *Arabidopsis*, led to increased proline accumulation and improved drought tolerance (Mahmud et al. 2023). Overexpression of wheat LEA proteins in maize and rice has been reported to enhance tolerance against water deficit by maintaining osmotic balance and membrane integrity (Guo et al. 2023). Similarly, transgenic tobacco plants expressing a sorghum AQP (SbPIP1) showed improved water-use efficiency and drought resistance (Fan et al. 2023). Instead of constitutive expression, stress-inducible promoters are often used to avoid metabolic burden under normal growth. For instance, the rd29A promoter in *Arabidopsis* has been used to drive stress-responsive genes under drought and salinity, ensuring targeted expression only under stress conditions (Cheng et al. 2023).

Stress responses rarely rely on individual proteins acting in isolation. Instead, they depend on dynamic protein networks, signaling cascades, and molecular complexes. The PPI mapping provides insight into how stress-responsive proteins interact with others to execute cellular functions. The Y2H system is a classical method used to screen binary interactions. Y2H revealed that the *Arabidopsis* kinase SnRK2 interacts with ABA-responsive TFs, linking stress signaling with downstream gene expression (Zhang et al. 2021). Further, co-immunoprecipitation and pull-down assays were used to validate in vivo interactions between two proteins. For example, co-immunoprecipitation showed that DREB2A, a drought-responsive TF, physically interacts with co-activators that regulate its stability and transcriptional activity (Huang et al. 2024). Proteomics has advanced PPI mapping through affinity purification coupled with MS. This approach was used in *Arabidopsis* to identify interactors of HSP90, revealing its involvement in multiple signaling complexes under stress (Trinh et al. 2025). In plants, PPIs can be visualized using bimolecular fluorescence complementation or Förster resonance energy transfer. For instance, bimolecular fluorescence complementation confirmed the interaction between CDPKs and ROS-scavenging enzymes under oxidative stress (Melicher et al. 2022). Mapping interaction networks not only validates a protein's role but also helps identify upstream regulators and downstream effectors in stress pathways. Loss-of-function studies are critical for validating the importance of stress-responsive proteins. By silencing or disrupting genes encoding candidate proteins, researchers can evaluate their contribution to stress tolerance. RNAi-based knockdown of OsHKT1;5, a sodium transporter in rice, resulted in increased sensitivity to salt stress, confirming its role in maintaining ion homeostasis (Wangsawang et al. 2021). Similarly, suppression of antioxidant enzymes (e.g., APX) leads to higher ROS accumulation and stress susceptibility. Large collections of *Arabidopsis* T-DNA mutants have facilitated high-throughput validation. For instance, T-DNA mutants of AtNHX1, a vacuolar Na^+/H^+ antiporter, exhibited impaired salt tolerance, confirming its critical role in ion sequestration (Pabuayon et al. 2021). The CRISPR/Cas system has revolutionized functional validation due to its precision and efficiency. For example, CRISPR-mediated knockout of OsERF922, a TF, increased drought and blast resistance in rice by modulating stress-related pathways (Huang et al. 2016). Similarly, CRISPR disruption of SlMAPK3 in tomato reduced stress tolerance, linking it to ROS signaling and defense responses (Yu et al. 2019). Since stress responses often involve redundant gene families, multiplex CRISPR allows simultaneous targeting of multiple genes. For example, editing of TaDREB2 homologs in wheat has provided insights into redundancy and specialization among TF family members (Abdallah et al. 2025). These approaches provide robust evidence by directly correlating gene/protein function with phenotypic outcomes under stress.

5.7 Integration with Transcriptomics and Metabolomics

Proteins act in a larger context of gene regulation and metabolic reprogramming. Integrating proteomics with transcriptomic and metabolomic data allows a holistic understanding of stress-responsive proteins. For example, combined transcriptome–proteome analysis of rice under drought stress showed partial correlation, highlighting proteins regulated post-transcriptionally, such as AQPs and ROS-detoxifying enzymes (Muthuramalingam et al. 2018). Proteins often regulate metabolite levels, and validating proteins through their metabolic effects strengthens functional evidence. In *Arabidopsis*, proteomic–metabolomic integration revealed that drought-induced accumulation of raffinose and proline was tightly associated with upregulation of biosynthetic enzymes, validating their protective role (Lozano-Elena et al. 2022). Similarly, multi-omics integration allows the construction of stress response networks. For example, in maize, combined proteomics, transcriptomics, and metabolomics under salinity stress identified ZmNHX transporters as central regulators of ion homeostasis and osmolyte balance. Functional validation of stress-responsive proteins is crucial to bridge the gap between proteomic discovery and practical applications in crop improvement.

5.8 Proteomics in Crop Improvement

Traditional breeding has achieved significant progress, but it remains time-consuming and is often limited by the complexity of stress tolerance traits, which are typically quantitative, multigenic, and environment-dependent. Proteomics, by capturing the dynamic protein expression and modifications that underpin plant stress responses, has emerged as a powerful approach to bridge fundamental discovery with applied crop improvement.

Molecular markers have long been used in plant breeding, but proteomics adds dimension by identifying protein-based markers that reflect not only the genetic potential but also the physiological state of plants under stress. Proteomic markers are often linked to phenotypically relevant traits, making them more reliable for screening stress tolerance. For example, proteomic studies in wheat under drought stress revealed increased accumulation of dehydrins and LEA proteins, which helps to stabilize membranes and proteins during dehydration. These proteins have been proposed as molecular markers for selecting drought-tolerant genotypes. Similarly, in rice, the expression of salt overly sensitive pathway proteins has been linked with salinity tolerance, providing potential markers for marker-assisted selection (Haque et al. 2021). In maize, proteomic profiling under water deficit identified upregulation of antioxidant enzymes such as SOD and CAT, which are now considered promising markers for oxidative stress tolerance. These protein markers complement conventional DNA-based QTLs, enabling integrated molecular breeding strategies that combine genomics and proteomics.

Translational proteomics aims to convert basic proteomic discoveries into practical tools for crop improvement. By integrating protein discovery, functional validation, and biotechnological applications, proteomics paves the way for engineering stress-resilient crops. Candidate proteins identified from proteomics, such as AQPs, have been overexpressed in rice, leading to enhanced water-use efficiency and improved drought tolerance. Similarly, overexpression of phytochelatin synthase, identified through proteomics, improved cadmium and arsenic tolerance in rice (Li et al. 2004). CRISPR-based editing, combined with proteomic data, allows precise manipulation of stress-responsive proteins. For example, editing of ROS-detoxifying enzymes and ion transporters is being explored for improving salinity and oxidative stress tolerance in crops. Proteomics is most powerful when integrated with transcriptomics and metabolomics, enabling a holistic understanding of stress responses. Such system-level integration allows identification of master regulators of stress tolerance, which are prime targets for translational applications. Thus, translational proteomics represents a bridge from bench to field, ensuring that laboratory discoveries are effectively harnessed for crop breeding programs.

5.9 Challenges and Future Prospects

Despite significant progress, several challenges limit the full potential of proteomics in plant abiotic stress research. One major issue is protein extraction and solubilization, especially in tough tissues that are rich in secondary metabolites and cell wall components. These factors make it hard to recover low-abundance proteins (Carpentier et al. 2005). Additionally, the broad range of protein levels in the proteome creates difficulty. Highly abundant proteins, like Rubisco, often overshadow stress-inducible low abundant proteins. This can lead to an underrepresentation of important regulatory, signaling proteins (Jorrín-Novo et al. 2019). Another key limitation is the detection of PTMs such as phosphorylation, ubiquitination, and glycosylation. These modifications play vital roles in stress signaling but are still technically difficult to fully characterize (Beltrao et al. 2013). To tackle these challenges, we need high-throughput and more precise proteomic platforms. Improvements in MS, such as SWATH-MS and DIA-MS, along with better protein fractionation and labeling methods, are allowing for deeper coverage of the proteome with greater reproducibility (Kausar et al. 2022). Moreover, bioinformatics and machine learning tools play a crucial role in combining large proteomic datasets, enhancing protein annotation, and predicting stress-responsive networks. Future directions focus on integrating multi-omics approaches. This means combining proteomics with genomics, transcriptomics, metabolomics, and phenomics to create a complete understanding of stress responses. Systems biology can model complex regulatory networks (Varadharajan et al. 2025). At the same time, synthetic biology provides new chances to redesign crops for specific stress responses. Additionally, proteomics will be vital in precision agriculture. Protein biomarkers can be used in field diagnostics for real-time monitoring of crop stress status (Pandey et al. 2018). By addressing current challenges and adopting new technologies and ideas, proteomics is ready to greatly speed up the development of climate-resilient crops that are crucial for sustainable agriculture.

5.10 Conclusions

Proteomics has become a valuable tool for understanding how plants detect, respond to, and adjust to different environmental stresses. By providing a direct view of functional proteins and their post-translational changes, proteomic methods reveal important details about stress-responsive pathways that genomics or transcriptomics alone cannot show. Research has identified important groups of proteins, such as chaperones, ROS-detoxifying enzymes, signaling kinases, and metabolic regulators. Together, these proteins help plants stay resilient during challenges like drought, salinity, extreme temperatures, and heavy metal stress. These findings highlight proteomics' special ability to connect genotype to phenotype under stress. Proteomics has the potential to change sustainable agriculture and global food security. Protein markers found through stress proteome profiling can help speed up marker-assisted selection. Translational proteomics connects laboratory findings with crop improvement programs in the field. Moreover, combining proteomics with multi-omics platforms allows us to model stress responses at a system level. This approach can help develop climate-resilient, high-yielding cultivars. In the future, improvements in high-throughput MS, bioinformatics, and synthetic biology will broaden the role of proteomics in crop improvement strategies. This makes it a key part of precision agriculture and sustainable food production as we face climate change.

Acknowledgments We acknowledge the director of CSIR-IHBT for providing research infrastructure. The CSIR-IHBT publication number is 5971.

References

Abdallah NA, Elsharawy H, Abulela HA, Thilmony R, Abdelhadi AA, Elarabi NI (2025) Multiplex CRISPR/Cas9-mediated genome editing to address drought tolerance in wheat. GM Crops & Food 16(1):1–17

Ahsan N, Renaut J, Komatsu S (2009) Recent developments in the application of proteomics to the analysis of plant responses to heavy metals. Proteomics 9(10):2602–2621

Alexandersson E, Fraysse L, Sjövall-Larsen S, Gustavsson S, Fellert M, Karlsson M, Johanson U, Kjellbom P (2005) Whole gene family expression and drought stress regulation of aquaporins. Plant Mol Biol 59(3):469–484. https://doi.org/10.1007/s11103-005-0352-1

Apse MP, Aharon GS, Snedden WA, Blumwald E (1999) Salt tolerance conferred by overexpression of a vacuolar Na+/H+ antiport in Arabidopsis. Science 285(5431):1256–1258

Asano T, Hayashi N, Kikuchi S, Ohsugi R (2012) CDPK-mediated abiotic stress signaling. Plant Signal Behav 7(7):817–821

Ashoub A, Beckhaus T, Berberich T, Karas M, Brüggemann W (2013) Comparative analysis of barley leaf proteome as affected by drought stress. Planta 237(3):771–781

Azarkina R, Makeeva A, Mamaeva A, Kovalchuk S, Ganaeva D, Tikhonovich I, Fesenko I (2025) The proteomic and peptidomic response of wheat (*Triticum aestivum* L.) to drought stress. Plants 14(14):2168. https://doi.org/10.3390/plants14142168

Beltrao P, Bork P, Krogan NJ, van Noort V (2013) Evolution and functional cross-talk of protein post-translational modifications. Mole Syst Biol 9(1):MSB134521

Borghi M, Perez de Souza L, Yoshida T, Fernie AR (2019) Flowers and climate change: a metabolic perspective. New Phytol 224(4):1425–1441

Byun YJ, Koo MY, Joo HJ, Ha-Lee YM, Lee DH (2014) Comparative analysis of gene expression under cold acclimation, deacclimation and reacclimation in Arabidopsis. Physiol Plant 152(2):256–274

Carpentier SC, Witters E, Laukens K, Deckers P, Swennen R, Panis B (2005) Preparation of protein extracts from recalcitrant plant tissues: an evaluation of different methods for two-dimensional gel electrophoresis analysis. Proteomics 5(10):2497–2507

Cheng L, Wei S, Liu K, Zhao X, Zhang J, Zhao Y (2023) Identification of the inducible activity in the promoter of the soybean BBI-DII gene exposed to abiotic stress or abscisic acid. Physiol Mol Biol Plants 29(7):947–957

Cherati SR, Khodakovskaya MV (2025) Integration of transcriptomics and proteomics data for understanding the mechanisms of positive effects of carbon-based nanomaterials on plant tolerance to salt stress. Environ Sci Nano 12:3772

Chintakovid N, Maipoka M, Phaonakrop N, Mickelbart MV, Roytrakul S, Chadchawan S (2017) Proteomic analysis of drought-responsive proteins in rice reveals photosynthesis-related adaptations to drought stress. Acta Physiol Plant 39(10):240

Chunduri V, Kaur S, Kaur S, Kumar S, Sharma P, Sharma R, Singh R, Kapoor P, Kaur N, Kumari P, Roy S, Kaur S, Garg R (2021) Gene expression and proteomics studies suggest an early and late response to heat stress in wheat. Front Plant Sci 12:660446. https://doi.org/10.3389/fpls.2021.660446

Chouhan M, Kumar P, and Kumar R (2025) iTRAQ-based proteomic approach to study stress memory responses in plants. In: Plant stress memory: methods and protocols. New York, NY: Springer US, pp 93–101

Devireddy AR, Zandalinas SI, Fichman Y, Mittler R (2021) Integration of reactive oxygen species and hormone signaling during abiotic stress. Plant J 105(2):459–476

Do J-M, Kim H-J, Shin S-Y, Park S-I, Kim J-J, Yoon H-S (2023) *OsHSP* 17.9, a small heat shock protein, confers improved productivity and tolerance to high temperature and salinity in a natural Paddy field in transgenic Rice plants. Agriculture 13(5):931. https://doi.org/10.3390/agriculture13050931

Esnarriaga DN, Mariotti M, Cardelli R, Arduini I (2020) The importance of root interactions in field bean/triticale intercrops. Plants 9(11):1474

Fan S, Chen J, Yang R (2023) Candidate genes for salt tolerance in forage sorghum under saline conditions from germination to harvest maturity. Genes 14(2):293

Fan X, Zhang Y, Gu P, Naz M (2025) Epitranscriptomic control of drought tolerance in Rice: the role of RNA methylation. Plants 14(13):2002

Fang F, Gao G, Wang Q, Wang Q, Sun L (2024) Combining SDS-PAGE to capillary zone electrophoresis-tandem mass spectrometry for high-resolution top-down proteomics analysis of intact histone proteoforms. Proteomics 24(17):2300650

Frukh A, Siddiqi TO, Khan MIR, Ahmad A (2020) Modulation in growth, biochemical attributes and proteome profile of rice cultivars under salt stress. Plant Physiol Biochem 146:55–70

Gao Z, Guo L, Chen M, Yu F, Wei Q (2021) Characterization of the development dynamics within the linear growth bamboo leaf. Physiol Plant 172(3):1518–1534

Guo B, Zhang J, Yang C, Dong L, Ye H, Valliyodan B, Nguyen HT, Song L (2023) The late embryogenesis abundant proteins in soybean: identification, expression analysis, and the roles of GmLEA4_19 in drought stress. Int J Mol Sci 24(19):14834. https://doi.org/10.3390/ijms241914834

Gupta B, Huang B (2014) Mechanism of salinity tolerance in plants: physiological, biochemical, and molecular characterization. Int J Genomics 2020:1–18

Guzman UH, Østergaard O, Rydbirk R, Aznar S, Stewart H, Xuan Y, Hermanson D, Horning S, Hock C, Makarov A, Zabrouskov V, Olsen JV (2024) Ultra-fast label-free quantification and comprehensive proteome coverage using narrow-window data-independent acquisition. Nat Biotechnol 42(12):1855–1866. https://doi.org/10.1038/s41587-023-02099-7

Ha J, Park J-Y, Choi Y, Chang P-S, Park K-M (2021) Comparative analysis of universal protein extraction methodologies for screening of lipase activity from agricultural products. Catalysts 11(7):816. https://doi.org/10.3390/catal11070816

Haider S, Iqbal J, Naseer S, Yaseen T, Shaukat M, Bibi H, Ahmad Y, Daud H, Abbasi NL, Mahmood T (2021) Molecular mechanisms of plant tolerance to heat stress: current landscape and future perspectives. Plant Cell Rep 40(12):2247–2271

Haque MA, Rafii MY, Yusoff MM, Ali NS, Yusuff O, Datta DR, Anisuzzaman M, Ikbal MF (2021) Advanced breeding strategies and future perspectives of salinity tolerance in Rice. Agronomy 11(8):1631. https://doi.org/10.3390/agronomy11081631

Hasanuzzaman M, Bhuyan MHMB, Zulfiqar F, Raza A, Mohsin SM, Mahmud JA, Fujita M (2020) Reactive oxygen species and antioxidant defense in plants under abiotic stress: revisiting the crucial role of ROS. Front Plant Sci 11:601495. https://doi.org/10.3389/fpls.2020.601495

Hayford RK, Serba DD, Xie S, Ayyappan V, Thimmapuram J, Saha MC, Wu CH, Kalavacharla VK (2022) Global analysis of switchgrass (Panicum virgatum L.) transcriptomes in response to interactive effects of drought and heat stresses. BMC Plant Biol 22(1):107

Hossain Z, Komatsu S (2013) Contribution of proteomic studies toward understanding plant heavy metal stress response. Front Plant Sci 3:310

Hassan N, El-Bastawisy Z, Ebeed H, Nemat Alla M (2015) Role of defense enzymes, proteins, solutes and Δ1-pyrroline-5-carboxylate synthase in wheat tolerance to drought. Rendiconti Lincei 26(3):281–291

Hu X, Li Y, Li C, Yang H, Wang W, Lu M (2010) Characterization of small heat shock proteins associated with maize tolerance to combined drought and heat stress. J Plant Growth Regul 29(4):455–464

Huang Y, Chen Y, Zeng B, Wang Y, James AA, Gurr GM, Yang G, Lin X, Huang Y, You M (2016) CRISPR/Cas9 mediated knockout of the abdominal–A homeotic gene in the global pest, diamondback moth (Plutella xylostella). Insect Biochem Mol Biol 75:98–106

Huang Y, Du B, Yu M, Cao Y, Liang K, Zhang L (2024) *Picea wilsonii* NAC31 and DREB2A cooperatively activate *ERD1* to modulate drought resistance in transgenic *Arabidopsis*. Int J Mol Sci 25(4):2037. https://doi.org/10.3390/ijms25042037

Hultgren A, Chatterjee S, Gerber JS (2025) Impacts of climate change on global agriculture accounting for adaptation. Nature 617(7957):1–6. https://doi.org/10.1038/s41586-025-09085-w

Hussain S, Zhu C, Bai Z, Huang J, Zhu L, Cao X, Nanda S, Hussain S, Riaz A, Liang Q, Wang L (2019) iTRAQ-based protein profiling and biochemical analysis of two contrasting rice genotypes revealed their differential responses to salt stress. Int J Mol Sci 20(3):547

Iacobucci I, Monaco V, Cozzolino F, Monti M (2021) From classical to new generation approaches: an excursus of-omics methods for investigation of protein-protein interaction networks. J Proteomics 230:103990

Jan N, Majeed U, Wani MA, Wani ZA, Chakraborty N, John R (2023) Analysis of physiological and proteomic changes in marigold (Calendula officinalis) in response to short term cold stress. S Afr J Bot 158:31–48

Jiang Y, Yang B, Harris NS, Deyholos MK (2021) Comparative proteomics analysis of cold stress responses in Arabidopsis. Front Plant Sci 12:637733

Jorrín JV, Maldonado AM, and Castillejo MA (2007) Plant proteome analysis: a 2006 update. Proteomics 7(16):2947–2962

Jorrin-Novo JV (2020) What is new in (plant) proteomics methods and protocols: the 2015–2019 Quinquennium. In: Plant proteomics: methods and protocols. Springer US, New York, pp 1–10

Jorrín-Novo JV, Komatsu S, Sánchez-Lucas R, Rodríguez de Francisco LE (2019) Gel electrophoresis-based plant proteomics: past, present, and future. J Proteome 198:1–10. https://doi.org/10.1016/j.jprot.2018.08.016

Kausar R, Wang X, and Komatsu S (2022) Crop proteomics under abiotic stress: from data to insights. Plants 11(21):2877

Khan AH, Wu Y, Luo L, Ma Y, Li Y, Ma H, Luo A, Zhang R, Zhu L, Lin Y, Zhang X (2022) Proteomic analysis reveals that the heat shock proteins 70-17 and BiP5 enhance cotton male fertility under high-temperature stress by reducing the accumulation of ROS in anthers. Ind Crop Prod 188:115693

Kosová K, Vítámvás P, Urban MO, Klíma M, Roy A, Prášil IT (2015) Biological networks underlying abiotic stress tolerance in temperate crops—a proteomic perspective. Int J Mol Sci 16(9):20913–20942

Kumari M, Pradhan UK, Joshi R, Punia A, Shankar R, Kumar R (2021) In-depth assembly of organ and development dissected Picrorhiza kurroa proteome map using mass spectrometry. BMC Plant Biol 21(1):604

Kumari M, Kumar P, Saini V, Joshi R, Shankar R, Kumar R (2025) Transcriptional landscape illustrates the diversified adaptation of medicinal plants to multifactorial stress combinations linked with high altitude. Planta 261(5):111

Kosová K, Vítámvás P, Urban MO, Prášil IT, Renaut J (2018) Plant abiotic stress proteomics: the major factors determining alterations in cellular proteome. Front Plant Sci 9:122

Kumar P, Paul D, Jhajhriya S, Kumar R, Dutta S, Siwach P, Das S (2024) Understanding heat-shock proteins' abundance and pivotal function under multiple abiotic stresses. J Plant Biochem Biotechnol 33(4):492–513

Li Y, Dhankher OP, Carreira L, Lee D, Chen A, Schroeder JI, Balish RS, Meagher, RB (2004) Overexpression of phytochelatin synthase in Arabidopsis leads to enhanced arsenic tolerance and cadmium hypersensitivity. Plant Cell Physiol 45(12):1787–1797

Li Y, Wang X, Ban Q, Zhu X, Jiang C, Wei C, Bennetzen JL (2019) Comparative transcriptomic analysis reveals gene expression associated with cold adaptation in the tea plant Camellia sinensis. BMC Genomics 20(1):624

Lian HL, Yu X, Ye Q, Ding XS, Kitagawa Y, Kwak SS, Su WA, Tang ZC (2004) The role of aquaporin RWC3 in drought avoidance in rice. Plant Cell Physiol 45(4):481–489

Liu CW, Chang TS, Hsu YK, Wang AZ, Yen HC, Wu YP, Wang CS, Lai CC (2014) Comparative proteomic analysis of early salt stress responsive proteins in roots and leaves of rice. Proteomics 14(15):1759–1775

Lozano-Elena F, Fàbregas N, Coleto-Alcudia V et al (2022) Analysis of metabolic dynamics during drought stress in Arabidopsis plants. Sci Data 9:90. https://doi.org/10.1038/s41597-022-01161-4

Ma L, Zhang X, Wang Y (2024) Metabolic responses to multi-stress: an update. Trends Plant Sci 29(5):123–135. https://doi.org/10.1016/j.tplants.2024.01.005

Magno Massuia de Almeida L, Coulon M, Avice JC, Morvan-Bertrand A, Bessoule JJ, Le Guédard M, Kim TH, Niare A, Mollier A, Bertin N, Brunel-Muguet S (2022) Effects of two-generational heat stress exposure at the onset of seed maturation on seed yield and quality in Brassica napus L. Environ Exp Bot 195:104788

Mahmood T, Safdar W, Abbasi BH, Naqvi SS (2010) An overview on the small heat shock proteins. Afr J Biotechnol 9(7):927–939

Mahmud S, Kamruzzaman M, Bhattacharyya S, Alharbi K, Abd El Moneim D, Mostofa MG (2023) Acetic acid positively modulates proline metabolism for mitigating PEG-mediated drought stress in maize and Arabidopsis. Front Plant Sci 14:1167238. https://doi.org/10.3389/fpls.2023.1167238

Melicher P, Dvořák P, Šamaj J, Takáč T (2022) Protein–protein interactions in plant antioxidant defense. Front Plant Sci 13:1035573. https://doi.org/10.3389/fpls.2022.1035573

Mhamdi A, Van Breusegem F (2022) Reactive oxygen species in plant development, metabolism, and stress responses. Annu Rev Plant Biol 73:209–236. https://doi.org/10.1146/annurev-arplant-070221-030014

Michaletti A, Naghavi MR, Toorchi M, Zolla L, Rinalducci S (2018) Metabolomics and proteomics reveal drought-stress responses of leaf tissues from spring-wheat. Sci Rep 8(1):5710

Mishra G, Mohapatra SK, Rout GR (2024) Plant membrane transporters function under abiotic stresses: a review. Planta 260(6):125

Mittler R, Vanderauwera S, Gollery M, Van Breusegem F (2004) Questions and future challenges. Trends Plant Sci 9(10):490–498

Muthuramalingam P, Krishnan SR, Saravanan K et al (2018) Genome-wide identification of major transcription factor superfamilies in rice identifies key candidates involved in abiotic stress dynamism. J Plant Biochem Biotechnol 27:300–317

Nakagami H, Sugiyama N, Mochida K, Daudi A, Yoshida Y, Toyoda T, Tomita M, Ishihama Y, Shirasu K (2010) Large-scale comparative phosphoproteomics identifies conserved phosphorylation sites in plants. Plant Physiol 153(3):1161–1174

Nishiyama R, Kato M, Nagata S, Yanagisawa S, Yoneyama T (2012) Identification of Zn–nicotianamine and Fe–2′-deoxymugineic acid in the phloem sap from rice plants (Oryza sativa L.). Plant Cell Physiol 53(2):381–390

Pabuayon ICM, Jiang J, Qian H, Chung JS, Shi H (2021) Gain-of-function mutations of AtNHX1 suppress sos1 salt sensitivity and improve salt tolerance in Arabidopsis. Stress Biol 1(1):14. https://doi.org/10.1007/s44154-021-00014-1. PMID: 37676545; PMCID: PMC10441915

Pandey R, Teig-Sussholz O, Schuster S, Avni A, Shacham-Diamand Y (2018) Integrated electrochemical Chip-on-Plant functional sensor for monitoring gene expression under stress. Biosens Bioelectron 117:493–500

Pour Mohammadi P, Nouri MZ, Komatsu S (2012) Proteome analysis of drought-stressed plants. Curr Proteom 9(4):232–244

Punia A, Kumari M, Chouhan M, Saini V, Joshi R, Kumar A, Kumar R (2024) Proteomic and metabolomic insights into seed germination of Ferula assafoetida. J Proteomics 300:105176

Romero P, Lafuente MT, Alférez F (2014) A transcriptional approach to unravel the connection between phospholipases A2 and D and ABA signal in citrus under water stress. Plant Physiol Biochem 80:23–32

Sah SK, Jumaa S, Li J, Reddy KR (2022) Proteomic analysis response of rice (Oryza sativa) leaves to ultraviolet-B radiation stress. Front Plant Sci 13:871331

Saleethong P, Roytrakul S, Kong-Ngern K, Theerakulpisut P (2016) Differential proteins expressed in rice leaves and grains in response to salinity and exogenous spermidine treatments. Rice Sci 23(1):9–21

San-Eufrasio B, Bigatton ED, Guerrero-Sánchez VM, Chaturvedi P, Jorrín-Novo JV, Rey MD, Castillejo MÁ (2021) Proteomics data analysis for the identification of proteins and derived proteotypic peptides of potential use as putative drought tolerance markers for Quercus ilex. Int J Mol Sci 22(6):3191

Sang T, Xu Y, Qin G, Zhao S, Hsu CC, Wang P (2024) Highly sensitive site-specific SUMOylation proteomics in Arabidopsis. Nat Plants 10(9):1330–1342

Şekerci K, Higashitani N, Ozgur R, Uzilday B, Higashitani A, Turkan I (2024) Proteomic profiling of an extreme halophyte Schrenkiella parvula with accelerated root elongation under mild salt stress. Plant Stress 11:100357

Sharma N, Pradhan S, Majumder S (2024) Proteomics-based strategies to develop food crops with enhanced nutritional quality. In: Plant proteomics, pp 226–242. https://doi.org/10.1201/b23255-15

Sidebottom C, Buckley S, Pudney P, Twigg S, Jarman C, Holt C et al (2000) Heat-stable antifreeze protein from grass. Nature 406(6793):256. https://doi.org/10.1038/35018640

Singh A, Sagar S, Biswas DK (2017) Calcium dependent protein kinase, a versatile player in plant stress management and development. Crit Rev Plant Sci 36(5–6):336–352

Song J, Lu D, Niu Y, Sun H, Zhang P, Dong W, Li Y, Zhang Y, Lu L, Men Q, Zhang X (2022) Label-free quantitative proteomics of maize roots from different root zones provides insight into proteins associated with enhance water uptake. BMC Genom 23(1):184

Szklarczyk D, Gable AL, Lyon D, Junge A, Wyder S, Huerta-Cepas J, Simonovic M, Doncheva NT, Morris JH, Bork P, Jensen LJ (2019) STRING v11: protein–protein association networks with increased coverage, supporting functional discovery in genome-wide experimental datasets. Nucleic Acids Res 47(D1):D607–D613

Subba P, Narayana Kotimoole C, Prasad TSK (2019) Plant proteome databases and bioinformatic tools: an expert review and comparative insights. Omics: A Journal of Integrative Biology 23(4):190–206

Thikekar AK, Rathod VS, Panchal VP, Raut SA, Raut RS, Jain KS (2023) A review on-analytical tools in proteomics. Journal of Proteins and Proteomics 14(3):201–221

Trinh CS, Shrestha R, Conner WC, Reyes AV, Karunadasa SS, Liu G, Hu K, Xu SL (2025) Mapping the architecture of protein complexes in *Arabidopsis* using cross-linking mass spectrometry. bioRxiv [Preprint]. https://doi.org/10.1101/2025.04.28.651104. PMID: 40777349; PMCID: PMC12330729

Tiwari LD, Kumar R, Sharma V, Sahu AK, Sahu B, Naithani SC, Grover A (2021) Stress and development phenotyping of Hsp101 and diverse other Hsp mutants of Arabidopsis thaliana. J Plant Biochem Biotechnol 30(4):889–905

Uday G, Shardha HB, Priyanka K, Jagirdhar S (2024) Advances in plant proteomics toward improvement of crop productivity and stress resistance. In: Plant proteomics. CRC Press, pp 11–29

Varadharajan V, Rajendran R, Muthuramalingam P, Runthala A, Madhesh V, Swaminathan G, Murugan P, Srinivasan H, Park Y, Shin H, Ramesh M (2025) Multi-omics approaches against abiotic and biotic stress—a review. Plants 14(6):865. https://doi.org/10.3390/plants14060865

Wang X, Dinler BS, Vignjevic M, Jacobsen S, Wollenweber B (2015) Physiological and proteome studies of responses to heat stress during grain filling in contrasting wheat cultivars. Plant Sci 230:33–50

Wang T, Ma G, Ang CS, Korhonen PK, Stroehlein AJ, Young ND, Hofmann A, Chang BC, Williamson NA, Gasser RB (2020) The developmental phosphoproteome of Haemonchus contortus. J Proteome 213:103615

Wang Z, Liu J, Zhang H (2024) Genome-wide identification of bZIP transcription factors and their role in abiotic stress tolerance. Front Plant Sci 15:1403220. https://doi.org/10.3389/fpls.2024.1403220

Wangsawang T, Chuamnakthong S, Ueda A (2021) Na+ exclusion mechanism in the roots through the function of OsHKT1;5 confers improved tolerance to salt stress in the salt-tolerant developed rice lines. ScienceAsia 47(6):717–726. https://doi.org/10.2306/scienceasia1513-1874.2021.106

Waters ER, Vierling E (2020) Plant small heat shock proteins – evolutionary and functional diversity. J Exp Bot 71(2):361–372. https://doi.org/10.1093/jxb/erz425

Xiong Q, Cao C, Shen T, Zhong L, He H, Chen X (2019) Comprehensive metabolomic and proteomic analysis in biochemical metabolic pathways of rice spikes under drought and submergence stress. Biochimica et Biophysica Acta (BBA)-Proteins and Proteomics 1867(3):237–247

Yadav SK, Singla-Pareek SL, Ray M, Reddy MK, Sopory SK (2005) Methylglyoxal levels in plants under salinity stress are dependent on glyoxalase I and glutathione. Biochem Biophys Res Commun 337(1):61–67

Yahoueian SH, Bihamta MR, Babaei HR, Bazargani MM (2021) Proteomic analysis of drought stress response mechanism in soybean (Glycine max L.) leaves. Food Sci Nutr 9(4):2010–2020

Yan S, Tang Z, Su W, Sun W (2005) Proteomic analysis of salt stress-responsive proteins in rice root. Proteomics 5(1):235–244

Yang L, Ji J, Harris-Shultz KR, Wang H, Wang H, Abd-Allah EF, Luo Y, Hu X (2016) The dynamic changes of the plasma membrane proteins and the protective roles of nitric oxide in rice subjected to heavy metal cadmium stress. Front Plant Sci 7:190

Yu W, Wang L, Zhao R, Sheng J, Zhang S, Li R, Shen L (2019) Knockout of SlMAPK3 enhances tolerance to heat stress involving ROS homeostasis in tomato plants. BMC Plant Biol 19(1):354

Yolcu S, Skorupa M, Uras ME, Mazur J, and Ozyiğit II (2024) Genome-wide identification, phylogenetic classification of histone acetyltransferase genes, and their expression analysis in sugar beet (Beta vulgaris L.) under salt stress. Planta 259(4):85

Zargar SM, Mir RA, Ebinezer LB, Masi A, Hami A, Manzoor M, Salgotra RK, Sofi NR, Mushtaq R, Rohila JS, Rakwal R (2022) Physiological and multi-omics approaches for explaining drought stress tolerance and supporting sustainable production of rice. Front Plant Sci 12:80360

Zhang X, Huang J, Zhang S, Yang C, Wang W (2021) The Arabidopsis sucrose non-fermenting-1-related protein kinase AtSnRK2.4 interacts with a transcription factor AtMYB21 that is involved in salt tolerance. Plant Sci 303:110779. https://doi.org/10.1016/j.plantsci.2020.110779

Zhang Z, Dong Y, Gao Y, Xian X, Li J (2025) Protein post-translational modifications (PTMs) unlocking resilience to abiotic stress in horticultural crops: a review. Int J Biol Macromol 306:141772

Zhu J-K (2016) Abiotic stress signaling and responses in plants. Cell 167(2):313–324

Zhong M, Li S, Huang F, Qiu J, Zhang J, Sheng Z, Tang S, Wei X, Hu P (2017) The phosphoproteomic response of rice seedlings to cadmium stress. Int J Mol Sci 18(10):2055

Zhu S, Sun S, Zhao W, Yang X, Mao H, Sheng L, Chen Z (2024) Utilizing transcriptomics and proteomics to unravel key genes and proteins of Oryza sativa seedlings mediated by selenium in response to cadmium stress. BMC Plant Biol 24(1):360

Zhu S, Du X, Zhao W, Yang X, Sheng L, Mao H, Su S (2025) Integrated proteomic and transcriptomic analysis reveals the mechanism of selenium-mediated Cell Wall polysaccharide in Rice (Oryza sativa L.) cadmium detoxification. Toxics 13(8):642

Zörb C, Schmitt S, Neeb A, Karl S, Linder M, Schubert S (2004) The biochemical reaction of maize (Zea mays L.) to salt stress is characterized by a mitigation of symptoms and not by a specific adaptation. Plant Sci 167(1):91–100

Zuther E, Koehl K, Kopka J (2007) Comparative metabolome analysis of the salt response in breeding cultivars of rice. In: Advances in molecular breeding toward drought and salt tolerant crops. Dordrecht: Springer Netherlands, pp 285–315

Biomacromolecular Pattern–Guided Breeding for Nutritional Quality and Safety Under Heat Stress in Soybean and Peanut

6

Jyoti Prasad Kakati, Hrishikesh Ingole, and Sachin Rustgi

Abstract

Soybean and peanut are major oilseed crops and important global sources of protein. They also provide significant amounts of carbohydrates and micronutrients. These crops are cultivated worldwide for both human consumption and animal feed. In addition, because of their ability to fix atmospheric nitrogen, soybean and peanut are often used in crop rotations as green manures. However, changing climatic conditions, particularly high temperatures and irregular rainfall during the growing season, adversely affect both yields and seed quality. While numerous studies have been conducted globally to understand and mitigate the effects of heat stress on yield (as reviewed by many research groups), relatively little attention has been given to its impact on nutritional quality. Given the importance of soybean and peanut as major global sources of oil and protein, it is essential to study the effects of heat stress on both harvest quality and final yield. This article examines how heat stress at different developmental stages influences the nutritional quality of soybean and peanut. Additionally, it explores the development and application of lipid, protein, and carbohydrate biomarkers to study and map genes regulating heat stress tolerance in these two crops.

Keywords

Peanut · Soybean · Heat stress · Nutritional quality · Biomarkers

J. P. Kakati · H. Ingole · S. Rustgi (✉)
Department of Plant and Environmental Sciences, Clemson University, Pee Dee Research and Education Center, Florence, SC, USA
e-mail: srustgi@clemson.edu

V. Gahlaut et al. (eds.), *Smart Crop Development*,
https://doi.org/10.1007/978-981-95-7680-7_6

6.1 Introduction

Population growth and unfavorable climatic conditions have increasingly become a concern for ensuring food security. Over the past 50 years, the world population has doubled and is projected to reach 9.8 billion by 2050, which will require 60% more food to meet the global demand (United Nations 2012; United Nations 2022). Intensifying existing farming systems through higher-yielding varieties and sustainable management practices is one approach to achieving this goal. However, climate change, which causes yield losses, remains a major challenge in realizing this objective.

Biotic and abiotic factors affect crop yields. Among the abiotic factors, heat and drought invariably affect all major crops (Lamaoui et al. 2018; Rustgi et al. 2021a; Kakati et al. 2022a, 2024). Heat stress is typically defined as a condition during which temperatures remain high for an extended period of time, causing irreversible damage to a plant's functional processes or development (Bhattacharya 2019). Unfortunately, heat waves are expected to be more prevalent (frequent and prolonged) during cropping season in the foreseen future, impacting crop productivity. In line with these assumptions, 2024 was the warmest year on record since 1850, with temperatures averaging 1.18 °C above the twentieth-century average and 1.35 °C above the pre-industrial era (1850–1900) average. Additionally, it is projected that global temperatures may rise by around 1.5 °C by 2050, creating further challenges for feeding the predicted population of 9.8 billion (NOAA 2025).

The increased global temperature has impacted the Global North more intensely than Global South, imparting a negative impact on agriculture in countries in the Global North including the USA. According to a prediction, global temperatures are expected to rise between 2 and 4 °C by the end of this century, further magnifying the impacts (Hsiang et al. 2017). Indeed Hsiang et al. (2017) estimated that the temperature rise in this magnitude will result in up to 90% crop yield losses for major crops including corn, wheat, soybean (*Glycine max*), and cotton (Fig. 6.1). Much like soybean (*Glycine max*), the yields of other major US oilseed crops, such as peanut (*Arachis hypogaea*) (Zoong Lwe et al. 2020), rapeseed (*Brassica napus*) (Chen et al. 2021), sunflower (*Helianthus annuus*) (Agüera and De La Haba 2021), and cotton (*Gossypium hirsutum*) (El Sabagh et al. 2020), are also impacted by the rise in temperature (Rustgi et al. 2021a).

The severity of heat stress on plants depends on the intensity, duration, and timing of the stress during plant development. Though these stresses can harm the plant at any developmental stage, they majorly impact it when imposed during the germination and reproductive stages (Egli et al. 2005; Onemli 2012; Jagadish et al. 2021). Reports show that the reproductive stage is crucial for heat stress in corn (Ray and Sinclair 1998), rice (Krishnan et al. 2011), canola (Aksouh-Harradj et al. 2006), barley (Ghotbi-Ravandi et al. 2021), wheat (Ullah et al. 2022), soybean (Kakati et al. 2024), and other legumes, including peanuts (Cox 1979; Zoong Lwe et al. 2020), due to their direct impact on the final yield. At the molecular level, this occurs because stress hinders the major enzymatic and biochemical processes required for the biosynthesis of key macromolecules—proteins, lipids, and

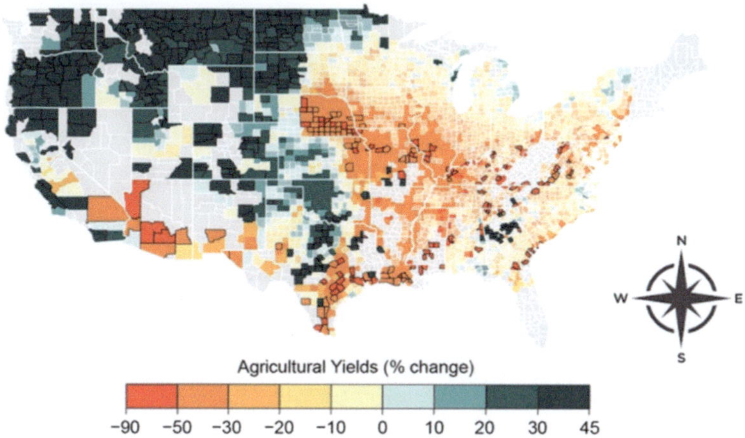

Fig. 6.1 Map of the United States showing changes in agricultural yields by county, with a color gradient from red to green. Red indicates a decrease in yields, while green indicates an increase. The western regions show more green, suggesting yield increases, while the central and eastern regions show more red, indicating decreases. A compass rose is present in the bottom right corner. Data were unavailable for the US Caribbean, Alaska, Hawai'i, and US-affiliated Pacific Islands regions (Modified from Hsiang et al. (2017) and Lipton et al. (2018))

carbohydrates—impacting a myriad of physiological processes and ultimately affecting yield (Sehgal et al. 2018) (Fig. 6.2).

Proteins, oils (lipids), and carbohydrates are the major macronutrients found in seeds; however, their content and composition vary by crop (Miransari 2016; Wang et al. 2023). The demand for dietary plant proteins has risen higher than ever, as they are increasingly viewed as sustainable alternatives to animal protein (Fig. 6.3).

Leguminous crops play a key role in meeting this demand, as they are rich sources of balanced proteins, do not require nitrogen fertilization, and are affordable. Among leguminous crops, soybean and peanut are particularly important due to their high protein content. An additional advantage of soybean and peanut is that they also produce oil, an economically valuable seed commodity (Table 6.1).

In the USA, soybean and peanut are two major oilseed crops grown primarily for oil and protein, both of which have high consumer demand. However, heat stress poses a significant threat to these crops, affecting peanut protein content and composition, which in turn influences overall yield, seed quality, safety, and various biochemical and physiological traits (Carrera et al. 2009; Chakraborty et al. 2018; Katam et al. 2020; Rustgi et al. 2021a). Like many other crops, soybean and peanut experience negative effects on growth and development under heat stress, with the reproductive stages being the most severely affected (Fig. 6.4). Recent studies have largely focused on the impact of heat stress on yield-related traits, and this topic has been reviewed extensively (Egli et al. 2005; Onemli 2012; Miransari 2016; Veas et al. 2022). Similarly, the effects of heat stress on lipid content and composition, as well as the role of fatty acid desaturase genes in conferring heat stress tolerance in soybean and peanut, have also been reviewed (Rustgi et al. 2021a). However,

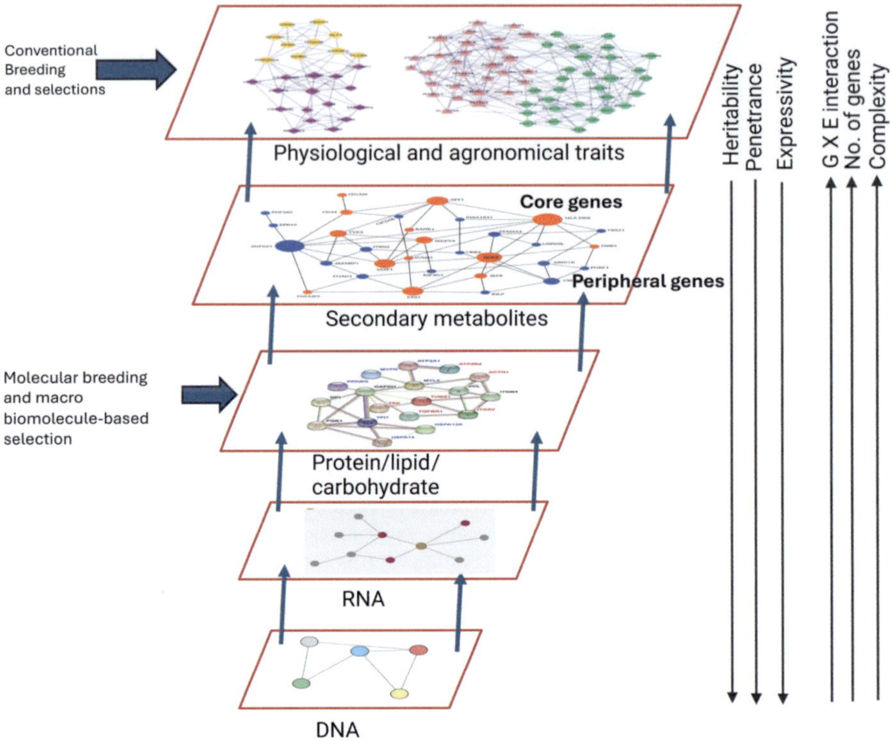

Fig. 6.2 This diagram illustrates the increasing complexity of interactions, including gene × gene and gene × environment interactions, as well as epistatic interactions that influence the penetrance and expressivity of a trait. It also indicates the levels at which traits are typically measured, highlighting where macro-biomolecular traits—the focus of this chapter—are recorded, along with the expected interactions and detectability of these traits. Some traits are quantitative, influenced by many small-effect genes with both positive and negative contributions, showing additive and non-additive effects. Others are qualitative, controlled by one to a few large-effect genes (oligogenic or ordinal traits), mostly through cis effects and with less epistasis. A third category is omnigenic, where two-thirds of the genes in the genome contribute to most phenotypic traits. Genes within these interaction networks are categorized into core and peripheral genes. Core genes are relatively few and directly affect trait expression, often producing large, detectable effects on heritability. Peripheral genes, on the other hand, indirectly influence core genes. Despite their small individual effects, peripheral genes collectively explain more heritability than core genes due to their greater numbers

relatively little attention has been given to how heat stress at different developmental stages affects other macromolecules—such as proteins (in terms of content and composition)—and, consequently, the nutritional quality and safety of soybean and peanut.

Heat stress affects the plant proteome through two main response types, such as acclimatization (general) response and specialized response. In acclimatization response, most plants respond to heat stress by overaccumulating heat shock proteins (HSPs), catalase, superoxide dismutase (SOD), and other antioxidant enzymes.

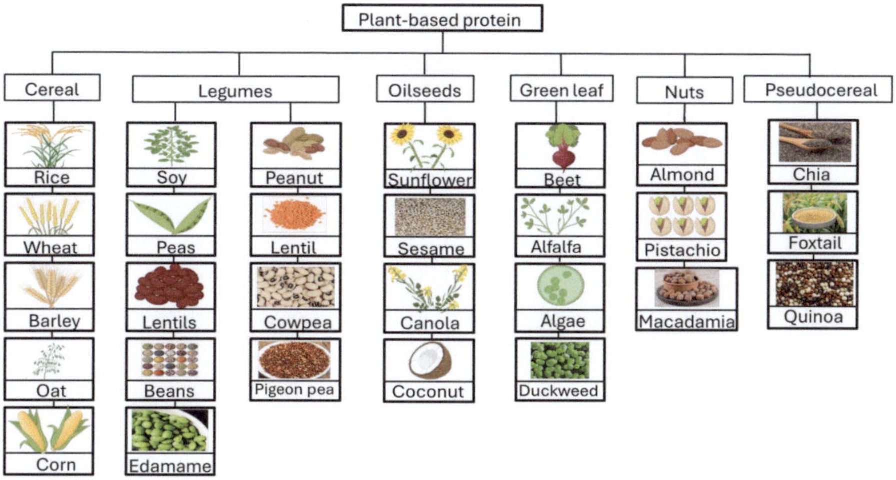

Fig. 6.3 Diagram of plant-based protein sources organized into categories: Cereal, Legumes, Oilseeds, Green leaf, Nuts, and Pseudocereal. Cereal includes rice, wheat, barley, oat, and corn. Legumes feature soy, pea, lentil, cowpea, beans, and edamame. Oilseeds list peanut, sunflower, sesame, canola, and coconut. Green leaf includes beet, alfalfa, algae, and duckweed. Nuts feature almond, pistachio, and macadamia. Pseudocereal includes chia, foxtail, and quinoa. Each category is visually represented with images of the respective items (Modified from Wang et al (2023))

Table 6.1 Protein, lipid, and carbohydrate content in seeds of various legumes, highlighting soybean and peanut as major sources of protein and oil

Legume	Carbohydrate (%)	Lipid (%)	Protein (%)
Soybean (*Glycine max*)	14	21	43
Broad bean (*Vicia faba*)	55	1	32
Peanut (*Arachis hypogaea*)	14	50	30
Pea (*Pisum sativa*)	60	3	28
Lentil (*Lens esculenta*)	62	3	30
Cowpea (*Vigna unguiculata*)	62	5	27
Pigeon pea (*Cajanus cajan*)	62	5	27
Bean (*Phaseolus vulgaris*)	68	1	27

Adopted from Miransari et al. (2016)

Changes in free amino acid profiles also occur—proline commonly accumulates as both an osmoprotectant and reactive oxygen species (ROS) scavenger, whereas quantities of amino acids such as lysine often decline due to reduced synthesis or increased degradation.

On the other hand, in specialized response, heat-tolerant genotypes uniquely express high levels of transcription factors (TFs) and stress-signaling and regulatory proteins. These proteins are rarely observed in susceptible lines and contribute to enhanced stress resilience. In addition to protein accumulation, severe heat stress can damage cellular and organellar membranes, leading to protein misfolding, loss of structural integrity, and the need for proteasome-mediated degradation to recycle

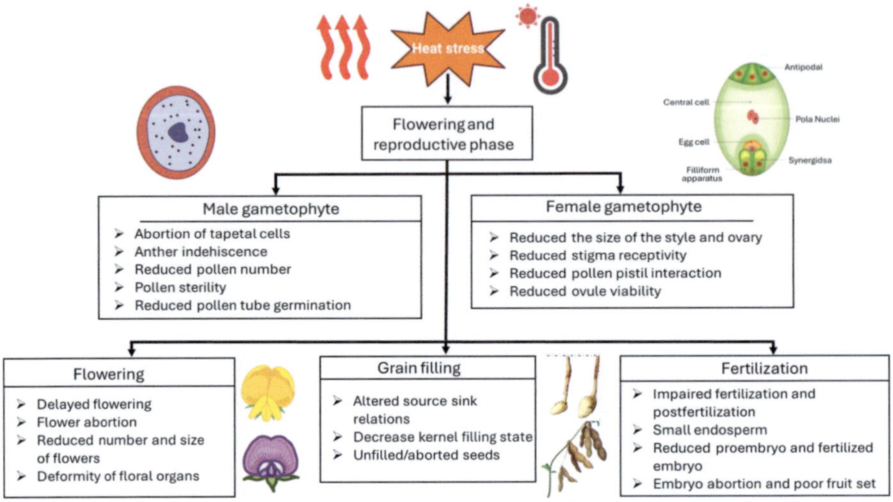

Fig. 6.4 Effect of heat stress on soybean and peanut during the reproductive stage (Modified from Bhattacharya (2019), illustrating the consequences of heat stress on the reproductive and grain-filling stages)

damaged proteins into amino acids. Proteomic study in cotton (*Gossypium hirsutum*) showed that under high heat (45 °C), tolerant genotypes maintained higher levels of proteins involved in photosynthesis, oxidoreductase activity, translation, transporter function, and carbohydrate/protein metabolism, whereas susceptible genotypes exhibited pronounced downregulation (Perveen et al. 2025). The changes in amino acid composition under heat and drought stress arise from protein degradation, altered synthesis, and overaccumulation of free amino acids serving as osmolytes and ROS scavengers. High-lysine mutants in barley exemplify this effect, where reduced gluten protein accumulation coincides with increased albumin/globulin proteins (Munck 1992). Such shifts indirectly indicate broader alterations in seed protein quality and content. While heat stress effects on wheat immunogenicity are well documented (Rustgi et al. 2021b; Juhász et al. 2018), studies in dicots like soybean and peanut remain limited. For example, soybean and peanut research using gel-based protein separation and Western blotting provided low resolution (Krishnan et al. 2020); proteomics-based approaches could yield deeper insights into heat stress impacts on seed development and composition, as was done in faba bean (Yu et al. 2025; Machado et al. 2025). Overall, integrative proteomics studies across multiple crops are needed to fully understand how heat stress reshapes protein profiles, amino acid composition, and seed quality.

The impact of heat stress during the reproductive stage of soybean and peanut is well documented (Bellaloui et al. 2015; Miransari 2016; Kakati et al. 2022b; Kakati et al. 2024). However, little attention has been given to its impact on nutritional quality, though soybean and peanut have been global sources of protein and oil. Hence, we place particular emphasis on the effects of heat stress during the reproductive stage in soybean and peanut, highlighting its influence on nutritional quality and safety. In addition, we explore the identification of biomacromolecular patterns (oil, protein, and carbohydrate) associated with heat tolerance and the discovery of molecular markers linked to these patterns for developing heat stress–tolerant soybean and peanut varieties.

6.2 Agronomical Importance of Soybean

Soybean is an important legume crop that is widely cultivated as a source of vegetable oil, protein, carbohydrates, macronutrients, and minerals for human consumption and animal feed (Chen et al. 2013; Govindasamy et al. 2017). Soybean cultivation initiated in East Asia and gradually spread to other countries during the twentieth century (Stark 2006). Recently, soy-based foods and their bioactive compounds have received significant attention because of their health benefits and are recently designated as the "functional food of the century" (Hasler 2002; Shea et al. 2020). Due to population growth, changes in diet, and increased biofuel consumption, the global demand for soybean has increased and is expected to excel in the future.

Protein and oil are the two major constituents of soybean seeds, typically accounting for 35–45% proteins, 15–25% oil, and about 33% carbohydrates (Hou et al. 2009; Clemente and Cahoon 2009). Due to their high nutritional quality and protein content, soybeans are often used as raw materials to manufacture processed foods or used as food additives for human consumption and in poultry feed. Soybean protein contains all the essential amino acids, making it a complete protein and highly bioavailable. Like eggs and cow's milk, soybean protein has a Protein Digestibility Corrected Amino Acid Score (PDCAAS) of 1.00 (1 is the highest, and 0 is the lowest), making it a superior alternative to many other protein sources, such as beef (PDCAAS, 0.92) and chicken (PDCAAS, 0.95) (Hughes et al. 2011). Additionally, soybean seeds are a rich source of several macro- and micronutrients, such as N, P, K, Ca, Mg, Fe, Cu, Mn, Zn, Co, and vitamins B1, B2, and B6, as well as bioactive compounds like isoflavones (Wijewardana et al. 2019). Soybean-based food products such as tofu, soymilk, and yuba are a source of high-quality vegetable protein and are free of lactose and cholesterol, which makes soybean a preferable protein source for people with lactose intolerance and high blood cholesterol (Rizzo and Baroni 2018). Soymilk is frequently used as a dairy substitute, whereas the hydrolyzed vegetable protein derived from processed soybeans can be used as a meat substitute (Qin et al. 2022). Additionally, soybean flour is used in numerous processed foods as a stabilizer. Soy oil is used as a cooking oil and as biofuel. After oil extraction, soy meal was used to make fibers, textiles, adhesives, and livestock feed (Brentin 2014).

Soybean protein is also reported to have many health benefits, such as lowering LDL (low-density lipoprotein) cholesterol levels, which reduces the risk of coronary heart disease (Erdman 2000). Studies showed that consuming soybean protein may reduce the risk of breast and prostate cancer (Wu et al. 2008; Korde et al. 2009; Shu et al. 2001; Zhang et al. 2017a, b). It also benefits people with diabetes due to its low starch content (Fan et al. 2022). As stated before, with the shift in consumer preference from animal-based to plant-based proteins, soybeans and other legumes have caught global attention (Akharume et al. 2021).

Soybean accounts for 60% of global oilseed production for 2023–2024 (SOYSTATS 2025). In the USA, soybean constitutes 90% of oilseed production, while other oilseeds (peanut, sunflower seed, canola, and flax) constitute the remainder. The major fatty acids in soybean oils are palmitic (C16:0), stearic (C18:0), oleic (C18:1), linoleic (C18:2), and linolenic (C18:3), which account for more than 98% of total oil composition (Table 6.2). The saturated fatty acids palmitic and stearic account for 10–12% and 3–5% of the total oil, respectively. The predominant unsaturated fatty acid is linoleic acid, which accounts for 54% of the oil, whereas oleic and linolenic acids account for 24% and 8% of the oil composition (Huth et al. 2015).

From a consumer point of view, oil with higher unsaturated and lower saturated fatty acids is desirable as it has numerous health benefits (Galão et al. 2014; Bellaloui et al. 2015). In contrast to the industrial perspective, soybean oil with higher oleic acid is preferred as it improves the storability and shelf-life of food products. Since polyunsaturated fatty acids (PUFA) are prone to oxidation and reduce storability, soybean oil with higher linoleic and linolenic acids often undergoes an additional process called hydrogenation that increases saturation and improves oil stability (Bellaloui et al. 2015). However, *trans*-fatty acids produced through hydrogenation have negative consequences on human health. They are reported to increase the risk of coronary heart disease and are associated with higher LDL and lower high-density lipoprotein (HDL) (Ference et al. 2017). Use of products with *trans*-fat is already banned in California. Since soybean oil accounts for 29% of the world's vegetable oil production, developing soybean genotypes with high oleic and low linolenic acids has been a breeding target, and many researchers have reported the successful development of such genotypes, besides increased vitamin E α-tocopherol (Fehr et al. 1992; Hagely et al. 2021).

Table 6.2 Fatty acid composition of peanut and soybean seeds

Fatty acids	Numerical abbreviation	Soybean (%)[a]	Peanut (%)[b]
Palmitic	C16:0	10.98–13.65	10.83–11.99
Palmitoleic	C16:1	–	–
Stearic	C18:0	3.12–3.67	1.94–2.34
Oleic	C18:1	12.12–21.84	43.07–49.94
Linoleic	C18:2	52.66–60.82	30.07–37.77
Linolenic	C18:3	5.93–9.39	0.9–1.07
Arachidic	C20:0	–	0.95–1.65
Eicosenoic	C20:1	–	–
Behenic	C22:0	–	1.69–2.77
Lignoceric	C24:0	–	0.72–1.28

[a]For soybeans, fatty acids C16:0, C18:0, C18:1, C18:2, and C18:3, which together account for approximately 98% of the total oil content, are presented in the table above (Adopted from Kakati et al. 2024)

[b]For peanuts, the C18:3 and C22:1 fatty acids are omitted, as they each account for less than 0.5% of the total oil content (Adopted from Kakati et al. 2022b)

6.3 Agronomical Importance of Peanut

Peanuts, commonly known as groundnut, is grown worldwide, and it plays a crucial role in US agriculture. Asia stands out as the leading peanut producer, accounting for approximately 60% of the global production. Nutrient-rich and affordable peanut kernels comprise oil (50%), protein (25%), and other macro- and micronutrients, such as dietary fibers, vitamins, and carbohydrates (Bonku and Yu 2020), making it a popular choice. Peanuts are generally consumed as peanut oil in cooking and baking, snacks (roasted and boiled), peanut butter, or a confectionery ingredient (Arya et al. 2016).

Runner, Virginia, Spanish, and Valencia are commonly grown peanut market types in the USA; are unique in size, shape, and flavor; and have specific end uses. Runner-type peanuts are generally used for peanut butter production and accumulate less oil in seeds than Virginia-type peanuts, which are famous for fresh consumption and export and are known for large seeds and their high oleic acid content. Valencia-type peanuts produce high-quality forage, which increases economic returns. Spanish-type peanuts produce small kernels with unique reddish-brown skin. These peanuts are used for producing peanut candies and snacks (Gong et al. 2018; Kunta et al. 2021; Kunta et al. 2022). The defatted peanut meal, a by-product of the oil extraction industry, contains around 50% protein. It is also high in dietary fibers, antioxidants, vitamins, and minerals and is either used as animal feed or further processed for human foods. Peanuts are naturally low in indigestible carbohydrates such as raffinose and stachyose. The oil composition of peanuts is enriched for health-promoting fatty acids. Additionally, the presence of flavonoids and isoflavones contributes to the health benefits of peanuts (Ros 2010; Morgillo et al. 2019; Abubakar 2022). Epidemiological studies have linked peanut consumption with health benefits, including reduced incidences of coronary heart disease, gallstones (cholelithiasis), and type 2 diabetes—particularly in women—while also suggesting possible protective effects against obesity, hypertension, cancer, and inflammation, though evidence for these is more limited (Jiang 2002). Peanut consumption (42 g per day) provided significant benefits for participants with elevated serum lipids and blood pressure in a controlled study (Jones et al. 2014).

The primary proteins in peanut, arachin and conarachin, contain all essential amino acids as per the FAO's standards, except for methionine (Chiou 1990). Due to its rich nutritional profile, higher digestibility, and sensory properties, peanut protein holds great potential as an alternative to animal proteins (Cui et al. 2023). Besides that, peanuts are recognized for their role in sustainable agriculture. Their nitrogen-fixing ability increases soil quality, thus supporting eco-friendly farming practices (Akram et al. 2018). Due to its high nitrogen solubility index, it can also be incorporated into both plant- and animal-based products to improve nutritional quality and functionality. Additionally, peanut proteins can also be a substitute for lactose-intolerant individuals.

On the molecular level, peanuts are known to have narrow genetic diversity. However, sufficient variation exists among the quality traits such as oil content and fatty acid composition, which is beneficial for breeding programs aiming to improve the health benefits. The cultivation and breeding for nutritionally enriched peanut varieties can contribute to an overall increase in peanut consumption (Akram et al. 2018).

Overall, peanuts are an essential crop that provides significant nutritional, economic, health, and environmental benefits. They support the economy, and their high protein and oil content, along with abundant bioactive compounds, make them a vital component of global food systems.

6.4 Heat Stress and Its Impact on Soybean and Peanut Production in the USA

Until recently, the USA was the top soybean producer; however, in 2019, Brazil took over the USA. Soybean accounts for about 90% of oilseed production in the USA and 67% of global soybean production in 2025. In the USA, soybeans are widely cultivated in the Midwest. Illinois, Iowa, and Minnesota are the top producers, and these three states accounted for over 38% of total US production in 2022. Similarly, the USA is the fifth-largest producer of peanuts globally. In the USA, peanut is grown in the South, primarily in Georgia, Florida, Alabama, North Carolina, South Carolina, and Texas, where the climate is warmer and growing seasons are longer than in northern zones. In 2022, Georgia accounted for over 50% of the US peanut production (2.9 billion pounds).

Climate change is expected to become more severe by the middle of this century. The impact of heat on vegetation is particularly pronounced in the Northern Hemisphere. Therefore, like other countries in the Global North, soybean and peanut producers in the USA will also face climate-related challenges that result in heat stress for their crops (Fig. 6.1). For example, peanut growers in Texas and Oklahoma often experience above-average temperatures and below-average rainfall in June, July, and August—a critical period for crop growth and development—which negatively affects both yield and harvested area. Additionally, as of 2017, only 10% of the soybean acreage in the USA is irrigated, posing a challenge for soybean production. Although management practices can help improve yields for both soybean and peanut, understanding the impact of heat stress is crucial for developing climate-resilient crops.

Recent studies have shown the impact of heat stress on various developmental stages of soybean and peanut, from seed germination to physiological maturity. Heat stress has been reported to affect root growth, alter morphology, and hinder root elongation in both crops. It also reduces carbon assimilation, causes stomatal closure and membrane injury, decreases enzymatic activities, disrupts metabolic pathways, and leads to abnormal plant growth (Sehgal et al. 2018; Bhattacharya 2019). As heat stress intensifies each year, it poses a significant threat to global food security, making the development of heat-tolerant soybean and peanut genotypes a critical priority.

6.4.1 Impact of Heat Stress on Soybean

A day/night temperature of 30/22 °C (an average temperature of 26 °C) is optimal for soybean growth (Jumrani and Bhatia 2018). However, an increase in average temperature by one degree during the growing season is predicted to cause a 17% loss in soybean yield (Lobell and Asner 2003). Further, heat stress was reported to cause up to 6% daily suppression in soybean yield under rainfed conditions if the growing season temperature exceeds 30 °C (Schauberger et al. 2017). With soybean being mostly rainfed crop and climate change-associated high temperatures being a regular occurrence during the growing season, it is expected that heat stress will become a serious threat to soybean production in the USA and the world (Schauberger et al. 2017). As temperatures continue to rise, soybean yield in the USA is anticipated to decline by nearly 20% by the year 2050 (Mathur et al. 2023).

Recently, a novel management strategy named the Early Soybean Production System (ESPS) has been adopted to mitigate the effects of high-temperature stress in the mid-southern USA that capitalizes on beneficial interactions between local weather patterns and soybean development, including planting, growth, and maturation. Under the ESPS, the short-season soybean cultivars are planted early to avoid heat stress during flowering. Besides that, it also helps germination and canopy development before most summer annual weeds emerge and become competitive. For example, Bateman et al. (2020) demonstrated the importance of early planting dates for soybean producers in the southern USA (Mississippi) to ensure profitability by maximizing yield potential. The finding showed that soybean yield decreased 26.7 kg/ha per day when soybean (Maturity group IV and V) was planted after April 20 (Bateman et al. 2020). Soybean genotypes with excellent seed traits were also identified by Smith et al. (2009) for use in high-temperature production environments (Smith et al. 2009). Research conducted across the southern USA, particularly in Kentucky, Arkansas, Mississippi, Texas, and Florida, generally shows ESPS to be as good as or better than standard production systems. However, the adoption of this system introduced a unique problem. Under the ESPS, early sown plants generally mature during the growing season's highly humid and warm spell (between mid-August and mid-September) (Bellaloui et al. 2009). The consistently elevated temperatures during this period coincide with seed maturation and impair the seed physiologically in its ability to germinate and emerge (Bellaloui et al. 2009). For example, Logan et al. (1998) reported that ESPS led to lower yields than the recommended system for Tennessee. No significant difference in oil content between ESPS and plants grown according to the recommended system was observed. Though the percentage of protein content in the seed was higher in ESPS, the seeds exhibited poor quality. This led to the conclusion that ESPS may not hold much potential as an alternative production system in Tennessee (Logan et al. 1998).

Seed growth rate or filling rate (SGR) and seed-filling duration (SFD) determine the seed weight. Heat stress during seed filling negatively affects these parameters, resulting in lower yield (Fig. 6.4). Between SGR and SFD, the SFD is highly affected; short SFD results in reduced seed size and weight (Veas et al. 2022). Elevated temperatures exceeding 30 °C during seed development can diminish seed vigor, leading to lower soybean germination. However, prolonged exposure to temperatures above this threshold is common in many soybean production regions, particularly in the Southern USA. Heat stress during seed development also affects the seed composition, mainly the oil and protein content and quality (Sehgal et al. 2018). Interestingly, the seed-filling rate and seed weight may be considered as selection criteria for heat stress tolerance (Dias and Lidon 2009).

More recently, to better understand plant responses to heat stress under semi-realistic conditions, Van der Laan et al. (2025) examined the impact of long-term heat stress on four soybean genotypes, (1) Williams82 (control), (2) IAS 19C3 (elite variety), (3) PI 89008 (heat susceptible), and (4) PI 639693 (heat tolerant), both with and without the microbial community. Plants were exposed to 38/28 °C for 33 days under controlled conditions. The study also investigated how soil microbial communities influence plant responses to heat stress by growing plants in sterile and non-sterile soils under both heat-stress and optimal conditions, followed by RNA-seq analysis. Heat shock factors (HSFs) were highly expressed under short-term stress and remained overexpressed during prolonged heat stress in reproductive tissues, with multiple HsfA2 genes identified. Additionally, MIKC-type MADS-box TF family (MICK-MADS TFF) was reported as a key component of the soybean heat stress response—previously unreported in any plant species.

6.4.2 Impact of Heat Stress on Peanut

In the USA and globally, peanut production is challenged by reduced yield due to heat stress. Roughly 90% of peanuts worldwide are grown in tropical and semi-arid regions, while approximately 65% of peanuts in the USA are cultivated under dryland and rainfed conditions (Hamidou et al. 2013). The temperature in the range of 25–30 °C promotes peanut growth and productivity during the vegetative stage. However, the optimum temperature for the reproductive stage is from 22 to 24 °C, and temperatures above this optimum range can cause significant yield loss in peanut production. Wood (1968) and Cox (1979) reported that daily average temperatures above 30 °C significantly retard vegetative and reproductive growth and affect peanut pod yield (Wood 1968; Cox 1979). In line with these early observations, Prasad et al. (1999a, b) found that a daytime temperature of 38 °C causes severe yield reduction in peanuts, specifically when the heat stress occurs between 6 days before and 15 days after flowering. The reduced number of flowers producing pegs (budding ovaries) was identified as the primary cause for the lower yield. The same authors also reported a reduction in fruit set upon exposure to heat stress (≥34/28 °C, day/night) around the flowering time, the most sensitive developmental stage (Prasad et al. 1999a, b). This was primarily due to reduced pollen viability. Decreases in seed yield by 14, 59, and 90% were reported when day/night temperatures increased from 32/22 °C to 36/26 °C, 40/30 °C, and 44/34 °C, respectively (Prasad et al. 2003). The fatty acid composition of the oil is a major seed-quality trait of peanuts and is highly affected by heat stress (Kumar et al. 2023). Since, as in soybeans, peanuts' storage life and nutritional quality are largely influenced by the relative proportion of saturated and unsaturated fatty acids in the oil, heat stress often leads to lower-quality seeds. Under moderate temperatures, drought alone can reduce peanut pod yield by up to 55%, where combined heat and drought stress are more severe, resulting in a yield reduction of up to 72% (Hamidou et al. 2013).

Heat stress in peanut plants has previously been known to cause many disabilities, majorly impairing pollen germination and pollen tube growth, which acts as a hindrance in fertilization and, eventually, leads to a lower seed set. Major effects of heat stress conditions also included deformed pollen structures and a significant reduction in pollen count. Key research by Prasad et al. (1999a, b) reported that pollen germination declines at temperatures exceeding 30 °C. The study also reported crucial findings, which included the adverse effects of high temperatures on floral development stages and noted genotypic variations in heat tolerance in some cases. A similar study by Kakani et al. (2002) also suggests that pollen viability is negatively impacted because of reproductive dysfunction, which is caused by elevated temperature. Thus, suggesting pollen viability serves as a marker to identify heat-tolerant genotypes. A comprehensive list of the effects of heat stress during reproductive stages is highlighted in Table 6.3.

A recent study by Ingole et al. (2025) identified a major QTL on chromosome 20 by screening 72 diverse peanut genotypes for pollen viability under heat stress. In this study, peanut genotypes were exposed to heat stress at the five-leaf stage for 2 weeks (Ingole et al. 2025). This work suggested that pollen viability is genotype specific. Other than pollen viability, Selvaraj et al. (2011) used chlorophyll accumulation upon light exposure in high-temperature acclimatized plants to indicate heat tolerance. This experiment includes detaching and exposing the leaf to high temperatures to trigger the acquired thermotolerance (Selvaraj et al. 2011), and this method can be used for screening multiple genotypes simultaneously and identifying QTLs controlling acquired thermotolerance. In a separate study, a set of RILs derived from a cross between a heat-tolerant (ICGS-76) and a heat-susceptible (Tamrun OL02) peanut genotype was screened to study the impact of heat stress (38/28 °C for 14 days) on lipid remodeling (Spivey et al. 2023). This study identified a genomic region associated with lipid remodeling, which could potentially be used to select for this trait. Sharma et al. (2023) assessed the impact of heat stress on physiological traits and reported 39 QTLs from a screen of a biparental population derived from genotypes with contrasting levels of heat stress tolerance, JL 24 (susceptible) and 55–437 (tolerant). In a separate study investigating the regulatory mechanisms underlying peanut responses to heat stress during flowering and pod development, two contrasting peanut genotypes—heat-tolerant ICGV 16553 (resistance derived from 55–437, one of the genotypes studied by Sharma et al. 2023) and heat-sensitive ICGV 16516—were exposed to heat stress (40 °C for 2 h) under controlled conditions.

The tolerant genotype exhibited significantly higher antioxidant activity and phenol content than the sensitive genotype during flowering. Additionally, nutrient translocation studies revealed efficient sucrose loading into the phloem and increased glucose levels in pods, indicating active sucrose metabolism in the tolerant genotype under heat stress. Transcriptome profiling identified differentially

Table 6.3 Effect of heat stress on soybean and peanut during reproductive stages

Temp. (D/N) in °C	Duration (Days)	Growth stages	Effects					References
			Pollen viability	Pollen germination	Pod number	Seed size	Seed weight	
Soybean								
39/20	10	Flowering	Decrease	Decrease	Decrease	–	Decrease	Djanaguiraman et al. (2013a)
30	5	Early flowering and full bloom	Less effect	Less effect	Less effect	Less effect	Less effect	Puteh et al. (2013)
30	14	Early flowering to pod initiation	Less effect	Less effect	Less effect	Less effect	Less effect	
35	5	Early flowering and full bloom	High effect	High effect	High effect	High effect	High effect	
35	14	Early flowering to pod initiation	High effect	High effect	High effect	High effect	High effect	
38/28	14	Flowering	–	–	Decrease	–	Decrease	Djanaguiraman et al. (2011)
38/28	14	Flowering stage	Decrease	Decrease	Decrease	–	–	Djanaguiraman et al. (2013b)
38/30	50	Starting from the seedling	Decrease	Decrease	–	–	–	Koti (2005)
15–50 °C at 5 °C intervals	7	Flowering	Decrease	Decrease	–	–	–	Salem et al. (2007)
Peanuts								
42 and 48	6	Flowering	–	-	Reduced	Reduced	Reduced	Prasad et al. (2000)
34, 42 and 48	6	Flowering	Reduced	-	Reduced	–	–	Prasad (1999)
36/26, 40/30, 44/34	–	Starting from the seedling until maturity	Reduced	-	–	Reduced	Reduced	Prasad et al. (2003)
40	1	6 days before flowering	-	–	Reduction	–	–	Craufurd et al. (2003)
38/26 °C	14	Five-leaf stage	–	Reduced	–	–	–	Ingole et al. (2025)
41/27 and 38/26	17–18	Beginning of flowering	–	Reduced	Reduced	Reduced	–	Zoong Lwe et al. (2020) and Kakati et al. (2022b)

expressed genes (DEGs) encoding heat shock proteins, sucrose metabolism enzymes, transcriptional regulators, and antioxidants, corroborating the physiological observations (Bagudam et al. 2025). More recently, the effects of heat, drought, and combined stresses were evaluated in controlled environments, while combined stress (heat and drought) was also studied under field conditions using rainout shelters. Stress responses were assessed using physiological parameters such as stomatal conductance, transpiration rate, and leaf temperature. Varieties like NC 20 and Sullivan consistently showed higher stomatal conductance and transpiration under stressful environments, indicating greater physiological resilience. The study also reported similarities in plant responses between field and controlled conditions, highlighting the need for testing in both settings (Vennam et al. 2025).

Studies conducted on different populations, where plants were exposed to heat stress (38–40 °C during the day and 28–30 °C at night) at the flowering stage under controlled environments or field conditions (using heat tents or rainout shelters), revealed multiple mechanisms contributing to heat tolerance in various genotypes. These mechanisms were regulated differently, suggesting that the genes involved in heat tolerance could be combined through crossbreeding to develop peanut lines with durable heat tolerance (Fig. 6.2). In addition to these studies, research on peanuts demonstrated that the topical application of 24-epibrassinolide improved the antioxidant capacity, photosynthesis, and kernel weight of heat-stressed plants (Lai et al. 2024). This highlights the importance of identifying mechanisms that either enable plants to tolerate heat stress or reduce its negative impacts. Similarly, a study conducted on *Arabidopsis*, soybean, and wheat suggested that applying ABA receptor agonists could help plants avoid drought stress (Vaidya et al. 2019). Together, these findings point to new opportunities for identifying diverse mechanisms of heat and drought tolerance and, in practical terms, provide chemical options for managing these abiotic stresses.

6.5 Impact of Heat Stress on Soybean and Peanut Seed Protein Composition

Heat stress negatively impacts seed protein synthesis and influences its composition in soybean and peanut. Soybean plants exposed to heat stress show yield loss and produce seeds with wrinkles, exhibiting discoloration, poor germination, and increased susceptibility to disease, which reduces the seed quality. DS-25-1 and DT-97-4290 are heat-tolerant and heat-susceptible soybean lines, respectively, used in soybean heat stress studies (Krishnan et al. 2020; Narayanan et al. 2020). However, in a recent study, we evaluated these lines along with high-protein soybean lines for the impact of heat stress on protein and oil content at seed-filling stage (Kakati et al. 2024). The heat-tolerant line, DS-25-1, showed a significant increase in protein content under heat stress (38/28 °C) compared to optimal conditions (30/20 °C), whereas the high protein lines (NLM09–77, N14-7017, and Benning HP) used in the study maintained their protein content. However, the change in protein composition was not analyzed in this study; it can be found in a study conducted by Krishnan et al. (2020). The study showed that extreme heat stress (42/26 °C) adversely impacts the accumulation of lipoxygenase (an enzyme involved in fatty acid oxidation), the β-subunit of β-conglycinin (a storage protein, ~94 kDa), sucrose binding protein (~52 kDa), and Bowman-Birk protease inhibitor (an anti-nutritional protein, ~14 kDa), compared to the plants grown under control conditions (28/22 °C) or under moderate heat stress (36/24 °C) imposed at the seed-filling stage (Krishnan et al. 2020). It is well proven that heat shock proteins play an important role in conferring heat stress tolerance in plants, and soybean is no exception. In this study, Krishnan et al. (2020) also reported changes in the accumulation of heat shock proteins under elevated temperatures. The accumulation of HSP70 and HSP17.6 was observed in both heat-susceptible and heat-tolerant lines with rise in temperature. Interestingly, the immunoglobulin heavy chain-binding protein (BiP), a member of the HSP70 family and a resident endoplasmic reticulum (ER) protein, also showed increased accumulation under heat stress. Since BiP plays an important role in chaperoning newly synthesized proteins and helps degrade misfolded proteins, increasing BiP may have a negative impact on the total protein content of soybean seeds.

Over 16 allergenic proteins have been identified in soybeans, with the 34 kDa oil body–associated protein Gly m Bd 30 K (P34) recognized as the immunodominant allergen, affecting over 65% of soy-sensitive individuals. P34 accounts for less than 1% of total seed protein and shares sequence similarity with papain-like thiol proteases. It retains its allergenicity even after autoclaving (Watanabe et al. 2017). Screening over 13,000 accessions of the USDA *Glycine max* collection revealed only two genotypes expressing lower levels of P34 (Joseph et al. 2006). Interestingly, Krishnan et al. (2020) reported that extreme stress also reduces P34 levels. This suggests a potentially positive effect of heat stress on soybean seed protein composition by lowering P34 content. However, because heat stress negatively impacts yield, further research is needed to identify the balance between reducing allergen content and maintaining yield, ensuring significant allergen reduction without compromising productivity. Efforts are also underway to transfer the low-P34 trait into adapted soybean germplasm (Watanabe et al. 2017). Additionally, Nakagawa et al. (2020) reported a negative impact of heat stress on overall seed protein content. This reduction was attributed to the downregulation of genes associated with seed storage proteins (GmGy1, GmGy2, GmGy4, GmGy5, and Gmβ-conglycinin) and the upregulation of genes encoding cysteine and aspartate proteinases. Altered protein composition was associated with lower seed weight and yield (Nakagawa et al. 2020). On the other hand, Wei et al. (2020) demonstrated that short-term heat stress—lasting a week or less during physiological maturity (R7)—may not significantly impact seed vigor. These findings indicate that the effects of heat stress depend on both its duration and timing during plant development.

Similar studies related to the reduction in protein content also existed in peanuts. For instance, a RIL population derived from a cross between 'JL-24' (heat susceptible) and '55–437' (heat tolerant) was screened to examine the effect of heat stress (45 °C) during the peak reproductive stage (Aravind et al. 2024). The study showed stronger antioxidant defense, higher catalase and peroxidase activity, lower membrane damage (reduced malondialdehyde), and greater expression of HSP17 in the heat-tolerant genotype. The authors proposed a correspondence between these biochemical and molecular traits and enhanced heat stress tolerance during the reproductive stage in peanut.

Furthermore, given the known influence of heat stress on seed protein composition and allergen content in soybean, an experiment was conducted to examine the impact of heat stress on the allergen content of popular peanut varieties (Ingole, Jones, and Rustgi, unpublished data). Four commonly grown varieties were selected for the experiment: two runner types [Florunner (released in 1969) and Sunoleic 95R (released in 1994)] and two Virginia types [NC-7 (released in 1979) and Champs (released in 2006)]. When these genotypes developed their first peg, they were exposed to heat stress (38/26 °C) for 14 days in a growth chamber. Seeds were harvested at maturity, and proteins were profiled using SDS-PAGE to study the impact of heat stress on major allergenic proteins: Ara h1, Iso-Ara h3, Ara h3, Ara h2, and Ara h6 (Fig. 6.5). Results showed a significant decrease in Ara h1, Ara h3, and Iso-Ara h3 content in Sunoleic 95R under heat stress. Similarly, NC-7 showed significant reductions in Iso-Ara h3, Ara h2, and Ara h6 content under heat stress compared to the control (seeds collected from plants grown under optimal conditions). The Virginia-type variety Champs also showed a significant reduction in Ara h3 and Iso-Ara h3, although no significant decrease was observed for Florunner. Although a reduction in allergen proteins seems promising, as mentioned earlier,

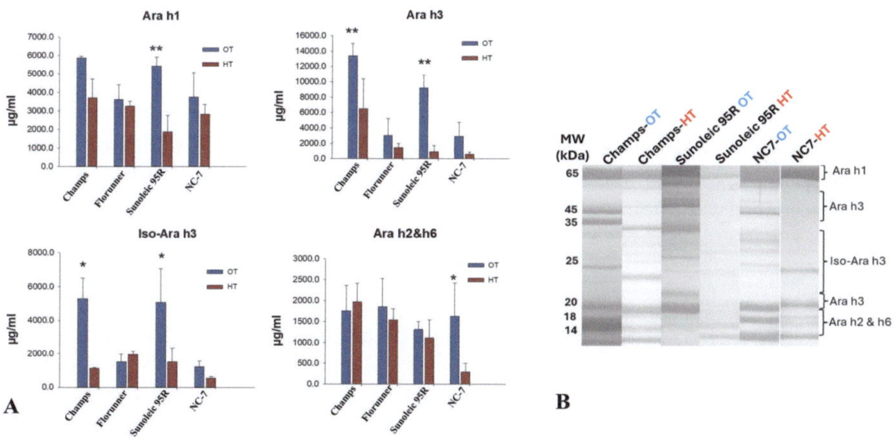

Fig. 6.5 Effect of heat stress (HT, 38/26 °C) on major peanut seed allergens when imposed at the first peg formation (R2) stage for 14 days. (**a**) Comparative bar graph showing major allergens (Ara h1, Ara h2, Iso-Ara h3, Ara h2, and Ara h6) in four peanut varieties (Champs, Florunner, Sunoleic 95R, and NC-7). Variety names are on the X-axis; the Y-axis shows densitometric values (μg/ml) derived from SDS-PAGE gel analysis and normalized to seed weight using ImageJ software. Data represent mean ± SD ($n = 3$). Asterisks indicate significant differences (one-tailed t-test) from the optimum temperature (OT) at $\alpha = 0.05$ ($*p < 0.05$, $**p < 0.01$). (**b**) SDS-PAGE images showing the major allergens and their reduction under heat stress in the peanut varieties (except Florunner)

the impact of heat stress on yield still needs to be determined to find a balance that avoids yield penalties.

Unlike soybean and peanut, the impact of environmental stressors—specifically drought, high temperatures, and elevated CO_2—has been extensively studied in legumes, including chickpea, common bean, faba bean, lentil, lupin, mung bean, and pea. These studies established that primary metabolites, including proteins, and secondary metabolites can serve as biomarkers for heat and drought stress (Machado et al. 2025). For example, Abid et al. (2024) conducted a comprehensive physiological and proteomic analysis of drought-stressed faba bean leaves. Drought affected all physiological parameters, and the faba bean mitigated oxidative stress via proline accumulation and involved the synergistic action of antioxidant enzymes (CAT, SOD, APX, and GPOX). Proteomic analysis identified 2000 proteins, 81 of which were differentially expressed. KEGG and GO enrichment revealed proteins associated with photosynthesis, ROS detoxification, amino acid and secondary metabolite biosynthesis, molecular chaperones, signal transduction, energy and carbohydrate metabolism, and other metabolic processes, indicating a complex network in which ROS detoxification and photoprotection are central to water-deficit tolerance. In a separate study, Li et al. (2018) exposed 2-week-old seedlings of the cultivar 'Ga da dou' to drought for 1 week. Proteins from drought-stressed and well-watered seedlings were analyzed via 2D electrophoresis and MALDI-TOF/TOF. Over 300 proteins were detected, 50 were differentially expressed, and 30 were successfully identified. Of these, 25 were downregulated and 5 upregulated, including chitinase, which was upregulated, and 50S ribosomal protein, which was downregulated, indicating their roles in stress response. Abdellatif and Zakaria (2012) assessed drought effects on eight Egyptian faba bean varieties using 1D SDS-PAGE of seed proteins. Drought-tolerant 'Giza 843' and susceptible 'Giza 3' showed distinct protein patterns, with unique bands in each variety, highlighting varietal differences in stress response. Collectively, these studies emphasize the importance of using diverse protein-resolving techniques—from gel-based to mass spectrometry-based methods—to identify protein biomarkers associated with environmental stress in faba bean.

6.6 Impact of Heat Stress on Soybean and Peanut Seed Oil Content

Different fatty acids, including C14:0, C16:0, C18:0, C18:1, C18:2, C18:3, C20:0, C20:1, C22:0, and C24:0, are present in soybean and peanut oil (Kakati et al. 2022b, 2024). In soybeans, however, C16:0 (palmitic acid), C18:0 (stearic acid), C18:1 (oleic acid), C18:2 (linoleic acid), and C18:3 (linolenic acid) account for approximately 98% of the total oil content, with C18:2 being a particularly abundant constituent (Table 6.2). In peanuts, C18:1 is the predominant fatty acid, whereas C18:3 and C22:1 together contribute less than 0.5% of the total composition. Oleic acid (C18:1) is especially important in cooking oils, as higher oleic acid content enhances oxidative stability. Conventional soybeans contain around 20% oleic acid, though increases of up to 80% have been achieved through conventional breeding and gene-editing approaches.

During soybean embryogenesis, carbon is primarily allocated to protein and oil, which constitute the major carbon reserves at maturity, representing roughly 40% and 20% of seed dry matter, respectively. The remaining seed dry matter is predominantly carbohydrate, which has relatively lower economic value. Historically, the inverse relationship between protein and oil content in soybeans has complicated efforts to enhance seed protein levels. Typically, a 2% increase in total protein content results in a 1% decrease in total oil content (Clemente and Cahoon 2009). However, our previous research identified soybean lines NLM09–77, N14–7017, and N16–9924 that do not exhibit this negative correlation. In these lines, the concentrations of saturated fatty acids (palmitic and stearic acids), the monounsaturated fatty acid (oleic acid), and polyunsaturated fatty acids (linoleic and linolenic acids) remain stable even under heat stress (Kakati et al. 2024).

Heat stress has also become a focus in peanut studies due to contrasting findings in soybean. Our previous research on select peanut genotypes—including Bailey (heat-susceptible), SPT-06-07 (heat-tolerant), Tifguard, and Phillips (moderately heat-tolerant)—indicates that a short period (2.5 weeks) of heat stress at the onset of flowering does not affect the major fatty acids (Kakati et al. 2022b). This suggests that seed fatty acid composition remains stable if heat stress subsides before the seed-filling stage. However, heat stress during flowering can reduce pod and seed numbers, ultimately decreasing yield.

6.7 Impact of Heat Stress on Soybean and Peanut Seed Starch Content

Carbohydrates are the primary seed constituents that play a vital role in seed quality and are involved in various crucial plant functions. Starch is the primary carbohydrate in seeds, and its monomer, glucose, serves as both a signaling and metabolic molecule, regulating almost every aspect of plant growth and development. Glucose helps store energy besides being a signaling molecule. Furthermore, glucose is known for supplementing energy for various metabolic processes when there are drops in photosynthetic efficiency, particularly at night or during low light periods (Wu et al. 2021). The conversion of starch to sugars also plays an essential role in plant responses to environmental stimuli, linking energy storage with developmental and stress-related responses.

Starch is predominantly present in species such as cereals (e.g., maize and rice), whereas soybean and peanut are known as low-starch legumes (Table 6.1). Various breeding strategies, which involve selective breeding, are being used to increase the storage starch content, targeting an increase in the grain starch content by 60–80%. Elevated starch content generally enhances the value and utility of a crop as a staple food source, contributing significantly to the human diet. On the other hand, legumes with lower starch content tend to have higher levels of resistant starch and dietary fiber, due to their unique cellular structures and higher amylose content (Ghoorchi et al. 2013). This characterization highlights the need for different strategies when studying and improving starch traits in legumes compared to cereals.

Citing the above reasons, understanding the importance of starch as a biomarker can improve the existing knowledge of the subject. Starch can indicate stress response, developmental stages, and overall seed quality in soybean and peanut. Ren et al. (2019) suggested using starch levels in assessing germination efficiency, seed vigor, and plant responses to abiotic stresses. Exploring the importance of starch as a biomarker can be helpful in studies, especially for increasing the nutritional value of legume crops and improving resilience.

Information generated in model organism, like *Arabidopsis*, can guide the research for identifying starch biomarkers in soybean and peanut. The outcome of these studies can lead to improving the crops currently in a race to meet nutritional needs while being challenged by changing climate.

Starch metabolism is known to occur during seed development in both peanut and soybean. In soybeans, β-amylase plays an important role in starch degradation and mobilization during seed development and germination (Hildebrand and Hymowitz 1981). Due to the decrease in starch content, a fluctuating level of enzymatic activity was observed in maturing soybean seeds, which results in the ability of the plant to use the stored carbohydrates effectively (Hildebrand and Hymowitz 1981; Jeong et al. 2010). The above trend hints that starch profiling observed during different developmental stages could be used as a biomarker to examine seeds' physiological status and capacity for successful germination. Previous studies have already identified a positive correlation between seed starch concentration and yield-related traits; hence, these traits can be considered as indicators of cultivar performance under various stress conditions (Zhang et al. 2019; Patel et al. 2022). A recent study by Chen et al. (2016) demonstrates that a breeding pipeline can be developed where various QTLs can be identified related to starch content, which can assist in breeding peanut varieties optimized for favorable growth conditions while also being resilient to stress.

Notably, there is a significant effect of environmental factors on starch accumulation in soybean and peanut. A common phenomenon of photoperiodicity significantly affecting the partitioning of photosynthate was successfully demonstrated in both peanut and soybean, which indicates a complex interaction between environmental signals and starch biosynthesis (Chatterton and Silvius 1979; Dwivedi et al. 1993). Studying these mechanisms allows breeders to predict better how different cultivars might respond to climate variability and changing agricultural conditions. In peanut, kernel size positively correlated with crude protein and seed weight. In contrast, the carbohydrate content negatively correlated with moisture, fat, ash, crude fiber, and total polyphenol content (Nayak et al. 2020). Given the above evidence, analyzing starch content and its related traits in soybean and peanut is important for diagnosing stress responses, understanding developmental biology, and improving seed quality.

In soybean, starch accumulation and degradation are tightly regulated processes influenced by the expression of specific enzymes. The *BAM5* (beta-amylase 5) gene, which is proactively involved in starch metabolism, demonstrates higher expression during the early developmental stage, leading to a decrease in the starch content, which is evident in soybean seeds in relation to other legumes (Cheng et al. 2021). The transient nature of starch is crucial, as it allows stored energy to be mobilized efficiently, particularly when seeds encounter varying environmental conditions.

Soybean deals with various stresses during its growing cycle, among which abiotic stresses like drought and heat have an oversized impact on the starch metabolism. The symptoms of drought conditions include drop in leaf starch content; furthermore, a shift in the metabolism is observed, leading to the degradation of starch to release sugars that are used for osmotic adjustment and energy (Coutinho et al. 2024). A study by Dong et al. (2025) has identified several genes that are upregulated in response to drought, including transcription factors that modulate stress responses and metabolic pathways. The changes highlighted above in the metabolism highlight how carbohydrates can serve as biomarkers to reflect a plant's physiological status under heat stress.

A potential and effective breeding strategy to improve soybean resilience and productivity should include identifying QTLs for starch content under heat stress and optimal growth conditions. The identification of trait-associated molecular markers in genetic studies could uncover some of the important genes responsible for variations in starch content and composition (Vidal et al. 2012). In fact, molecular marker-assisted selection (MAS) was used to breed soybean varieties with enhanced stress tolerance and optimized starch content (Rabara et al. 2017).

Starch plays a vital role in peanut breeding for enhanced nutritional properties through its influence on seed storage, germination, and morphological features. Starch in peanut seeds acts as the primary energy source, which is critical during germination. Starch is hydrolyzed into sugars that fuel the metabolic activities necessary for seedling growth. The interaction between temperature and moisture significantly affects germination rates and seed vigor in peanuts (Jeammuangpuk et al. 2020; Weaver et al. 2021). The ability to store the seeds under optimum conditions directly correlates with its starch content. This further signifies the importance of examining its role as a marker for seed viability post-harvest.

Regarding morphological features, starch accumulation is closely related to various morphological traits. Chavarro et al. (2020) studied genetic variation in pod and seed traits, which are associated with differences in starch storage capacity, which in turn affects overall seed quality. Furthermore, seeds with larger sizes often exhibit higher starch content, which is critical for sustaining seedling growth and improving yield potential (Zhang et al. 2017a, b). In addition, a few other traits, such as shelling percentage and pod weight, have been found to be associated with starch-related traits (Huang et al. 2019).

In recent years, peanut breeding has increasingly focused on enhancing starch content and associated nutritional traits. Associations with key traits, including starch metabolism, oil content, and protein quality, have enabled the identification of molecular markers that can be used in marker-assisted breeding programs (Huang et al. 2019; Zhang et al. 2023). QTL mapping has already facilitated the discovery of molecular markers linked to traits such as high oleic acid content and protein levels in peanuts, which are crucial for both nutrition and flavor (Sun et al. 2022;

Wang et al. 2025). Moreover, studies have shown that the accumulation of oligosaccharides such as raffinose under heat stress helps stabilize cellular structures, including membranes and proteins, while scavenging reactive oxygen species (ROS) to reduce oxidative damage to lipids, proteins, and nucleic acids. Similarly, the accumulation of trehalose under heat stress acts as a protective osmolyte, preventing protein denaturation, stabilizing cellular structures, and exhibiting antioxidant effects (Eh et al. 2024; Raza et al. 2024). Collectively, these findings highlight the importance of studying the impact of heat stress on peanut starch content to better understand its effects on key traits such as seed storage compounds, germination, and morphological characteristics, as well as to develop breeding strategies for enhanced nutritional qualities.

6.8 Starch Metabolism and Seed Components

Starch metabolism is an essential process that mainly occurs in plastids, primarily in amyloplasts and non-photosynthetic tissues such as roots and seeds. Furthermore, these plastids play a key role in storing carbohydrates and producing starch from glucose through a complicated enzyme-driven process. Starch metabolism is regulated and controlled by various environmental factors, signaling molecules, and circadian rhythms. Starch biosynthesis is governed by enzymes such as granule-bound starch synthases (GBSS) and soluble starch synthases (SSS), which are in the plastid stroma. GBSS plays a key role in amylose synthesis, while SSS assists in amylopectin production (Wang et al. 2016). The localization of these enzymes determines the structure of starch granules and the efficiency of starch metabolism, indicating the significance of plastids for storing and mobilizing energy within plant cells (Goren et al. 2018).

Seed filling relies on two primary carbon sources: current assimilates from photosynthesis and stored carbohydrates translocated from vegetative tissues like leaves and stems. Most legume crops contain over 50% of starch in their seed composition, though soybean and peanut are exceptions to this (Table 6.1). Irrespective of the carbohydrate content, heat stress impacts photosynthesis in plants, reducing the available assimilates to the seed. Heat stress during the seed-filling phase can trigger premature leaf senescence, hindering photosynthetic capacity and impairing seed development and yield. Senescence is marked by disrupted cellular structures, significant chlorophyll loss, and cessation of photosynthesis, limiting the production and transport of assimilates to growing seeds. Moreover, heat stress during this phase suppresses sucrose metabolism in leaves and diminishes sucrose transport to developing seeds, leading to the production of shriveled seeds in chickpea and mung bean. The rate of chlorophyll degradation correlates closely with levels of non-structural carbohydrates, nitrogen, and their remobilization efficiency. High temperatures markedly reduce carbon fixation by inhibiting key enzymes like PEP carboxylase and RuBP carboxylase, as shown in wheat during grain filling (Tahir et al. 2005; Sehgal et al. 2018).

In soybean seed, developmental rates vary with genotype and environmental conditions. Dornbos and Mullen (1992) characterized the seed mass and composition changes of 'Williams 79' soybean at different reproductive growth stages (R4 to R8) under field conditions. They found protein, oil, and sugar levels were low at R5 (immature seed), while moisture content was high. However, the protein content increased gradually from the R5 to R7, reaching 380 g per kg dry weight at the end of R8. There were also changes in the fatty acid compositions, from R5 to R7 (beginning maturity). A significant increase in fatty acids occurred between R4 and R6. At R4, the palmitic, stearic, and linolenic acids comprised 68% of fatty acids, though it dropped to 21% at R6. The oleic and linoleic acids accumulated rapidly at R6 (79%) and maintained a higher level until R8. Similarly, a significant carbohydrate composition shift was observed from R5 to R6. For example, the monosaccharides that accounted for 82% of soluble sugars at R5 decrease to 8% by R6. However, the sucrose levels doubled between R5 and R6 and remained high until R8 (full maturity). The raffinose and stachyose were absent until R7, though they reached 11% to 34% at the end of R7. As expected, the seed moisture decreased gradually toward maturity (Dornbos and Mullen 1992). Dias et al. (2009) showed that high temperatures during grain filling reduced wheat grain weight by shortening the filling period, but genotypes with a higher grain filling rate and greater potential grain weight showed better heat tolerance. Therefore, these traits may need to be explored in both soybean and peanut.

However, under heat stress, the cotyledon cell number and cell expansion may decrease, reducing the rate of seed filling and adversely affecting seed size. Krishnan et al. (2020) reported that heat-stress-tolerant soybean maintained cotyledonary cellular integrity better than heat-sensitive genotypes, regardless of heat stress. Similar studies on peanut are limited and represent a potential area for exploration in developing climate-resilient peanut varieties.

6.9 Conclusions

As affordable sources of oil and protein, soybean and peanut have received global attention. With climate change threatening global crop production, collaborations beyond geographic boundaries are essential to address the forecasted challenges to global food security. Achieving this goal—feeding the world—requires a multidisciplinary approach and a comprehensive understanding of plant responses to environmental stressors. Importantly, while increasing yield remains a priority, maintaining or improving crop quality is equally critical. Therefore, studying the effects of heat and drought stress on different plant developmental stages, from planting to harvest, is essential to producing crops with higher yields and improved nutritional quality.

Research reviewed here suggests that the macromolecules—protein, oil, and starch—function as versatile biomarkers in soybean, peanut, and other crop plants for identifying heat-tolerant genotypes needed to feed the growing global population. Additionally, phenotyping these biomolecular patterns will significantly expand the repertoire of traits for selection, facilitate the identification of genes controlling heat-induced patterns, and enable reliable breeding for these traits (Fig. 6.2). Their applications extend beyond developmental stages such as seed fill and germination to include indicators of stress responses and nutritional quality. These traits capture the complex interactions between crop physiology and environmental factors, underscoring the need to better understand protein, lipid, and starch dynamics in these legumes.

Integrating functional genomics with legume-focused breeding programs can maximize the potential of protein, oil, and starch as biomarkers. Combining molecular breeding techniques, genomic selection, and phenomics will further enhance the use of these traits in crop improvement (Fig. 6.6).

Moreover, advancing our understanding of the complex regulatory networks governing protein, oil, and starch metabolism will improve breeding efficiency, ultimately strengthening crop resilience and nutritional quality in the face of climate change. CRISPR/Cas9 represents a powerful gene-editing tool for modifying protein-, oil-, and starch-related traits in legumes. In addition, model organisms such as *Arabidopsis* can facilitate functional validation of candidate genes, enabling cross-verification before findings are applied to target crops. The growing genomic resources for *Arachis* species and other legumes will be instrumental in elucidating the genetic bases of these traits, thereby accelerating breeding efforts to enhance productivity, sustainability, and global food security.

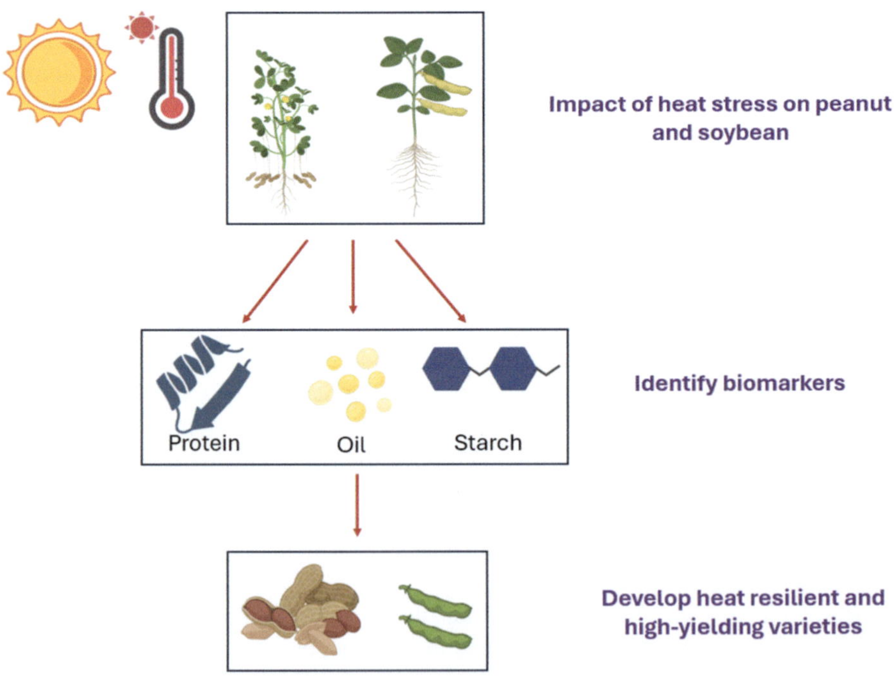

Fig. 6.6 Schematic representation of the strategy to enhance heat stress resilience in soybean and peanut

Acknowledgments This work was supported by the NIFA Hatch/Multi-state Grant (S009), National Peanut Board (Grant # 2012637 and 2014086), South Carolina Peanut Board (Grant # 2012613), the SC Soybean Board (2013753), and the USDA NIFA grant (2015061).

References

Abdellatif K, Zakaria A (2012) Drought stress tolerance of faba bean as studied by morphological traits and seed storage protein pattern. J Plant Stud 1(2):47–54. https://doi.org/10.5539/jps.v1n2p47

Abid G, Jebara M, Debode F, Vertommen D, Pyr dit Ruys S, Ghouili E, Jebara SH, Nefissi Ouertani R, El Ayed M, de Oliveira AC, Muhovski Y (2024) Comparative physiological, biochemical and proteomic analyses reveal key proteins and crucial regulatory pathways related to drought stress tolerance in faba bean (*Vicia faba* L.) leaves. Curr Plant Biol 37:100320. https://doi.org/10.1016/j.cpb.2024.100320

Abubakar MI (2022) Nutritional and health benefits of peanut (*Arachis hypogaea*)-based therapeutic foods as well as its effects on child health services delivery in North-West Nigeria. TIJPH 10(3):128–137. https://doi.org/10.21522/TIJPH.2013.10.03.Art012

Agüera E, De La Haba P (2021) Climate change impacts on sunflower (*Helianthus annus* L.) plants. Plants 10(12):2646. https://doi.org/10.3390/plants10122646

Akharume FU, Aluko RE, Adedeji AA (2021) Modification of plant proteins for improved functionality: a review. Compr Rev Food Sci Food Saf 20(1):198–224. https://doi.org/10.1111/1541-4337.12688

Akram NA, Shafiq F, Ashraf M (2018) Peanut (*Arachis hypogaea* L.): a prospective legume crop to offer multiple health benefits under changing climate. Compr Rev Food Sci Food Saf 17(5):1325–1338. https://doi.org/10.1111/1541-4337.12383

Aksouh-Harradj NM, Campbell LC, Mailer RJ (2006) Canola response to high and moderately high temperature stresses during seed maturation. Can J Plant Sci 86(4):967–980. https://doi.org/10.4141/P05-130

Aravind B, Shreeraksha RJ, Poornima R, Ravichandran D, Krishnaraj PU, Chimmad VP, Mirajkar KK, Bagewadi B, Janila P, Pandey MK, Varshney RK, Nayak SN (2024) Impact of heat stress on physiological characteristics and expression of heat shock proteins (HSPs) in groundnut (*Arachis hypogaea* L.). Physiol Mol Biol Plants 30(10):1691–1706. https://doi.org/10.1007/s12298-024-01520-y

Arya SS, Salve AR, Chauhan S (2016) Peanuts as functional food: a review. J Food Sci Technol 53(1):31–41. https://doi.org/10.1007/s13197-015-2007-9

Bagudam R, Mathew A, Kancherla E et al (2025) Transcriptome analysis of genes and carbon partitioning pathways involved in high temperature stress resilience in groundnut (*Arachis hypogaea* L.). Sci Rep 15:29939. https://doi.org/10.1038/s41598-025-15509-4

Bateman NR, Catchot AL, Gore J, Cook DR, Musser FR, Irby JT (2020) Effects of planting date for soybean growth, development, and yield in the southern USA. Agronomy 10(4):596. https://doi.org/10.3390/agronomy10040596

Bellaloui N, Smith JR, Ray JD, Gillen AM (2009) Effect of maturity on seed composition in the early soybean production system as measured on near-isogenic soybean lines. Crop Sci 49(2):608–620. https://doi.org/10.2135/cropsci2008.04.0192

Bellaloui N, Reddy KN, Mengistu A (2015) Drought and heat stress effects on soybean fatty acid composition and oil stability. In: Processing and impact on active components in food. Elsevier, pp 377–384

Bhattacharya A (2019) Effect of high-temperature stress on crop productivity. In: Effect of high temperature on crop productivity and metabolism of macro molecules. Elsevier, pp 1–114

Bonku R, Yu J (2020) Health aspects of peanuts as an outcome of its chemical composition. Food Sci Human Wellness 9(1):21–30. https://doi.org/10.1016/j.fshw.2019.12.005

Brentin RP (2014) Soy-based chemicals and materials: growing the value chain. In: Brentin RP (ed) ACS symposium series. American Chemical Society, Washington, DC, pp 1–23

Carrera C, Martínez MJ, Dardanelli J, Balzarini M (2009) Water deficit effect on the relationship between temperature during the seed fill period and soybean seed oil and protein concentrations. Crop Sci 49(3):990–998. https://doi.org/10.2135/cropsci2008.06.0361

Chakraborty K, Bishi SK, Singh AL, Zala PV, Mahatma MK, Kalariya KA, Jat RA (2018) Rapid induction of small heat shock proteins improves physiological adaptation to high temperature stress in peanut. J Agron Crop Sci 204(3):285–297. https://doi.org/10.1111/jac.12260

Chatterton NJ, Silvius JE (1979) Photosynthate partitioning into starch in soybean leaves: I. Effects of photoperiod *versus* photosynthetic period duration. Plant Physiol 64(5):749–753. https://doi.org/10.1104/pp.64.5.749

Chavarro C, Chu Y, Holbrook C, Isleib T, Bertioli D, Hovav R, Butts C, Lamb M, Sorensen R, A. Jackson S, Ozias-Akins P (2020) Pod and seed trait QTL identification to assist breeding for peanut market preferences. G3 Genes|Genomes|Genetics 10(7):2297–2315. https://doi.org/10.1534/g3.120.401147

Chen LM, Zhou XA, Li WB, Chang W, Zhou R, Wang C, Sha AH, Shan ZH, Zhang CJ, Qiu DZ, Yang ZL, Chen SL (2013) Genome-wide transcriptional analysis of two soybean genotypes under dehydration and rehydration conditions. BMC Genomics 14(1):687. https://doi.org/10.1186/1471-2164-14-687

Chen X, Li H, Pandey MK, Yang Q, Wang X, Garg V, Li H, Chi X, Doddamani D, Hong Y, Upadhyaya H, Guo H, Khan AW, Zhu F, Zhang X, Pan L, Pierce GJ, Zhou G, Krishnamohan KAVS, Chen M, Zhong N, Agarwal G, Li S, Chitikineni A, Zhang G, Sharma S, Chen N, Liu H, Janila P, Li S, Wang M, Wang T, Sun J, Li X, Li C, Wang M, Yu L, Wen S, Singh S, Yang Z, Zhao J, Zhang C, Yu Y, Bi J, Zhang X, Liu Z, Paterson AH, Wang S, Liang X, Varshney RK, Yu S (2016) Draft genome of the peanut A-genome progenitor (*Arachis duranensis*) provides insights into geocarpy, oil biosynthesis, and allergens. Proc Natl Acad Sci USA 113(24):6785–6790. https://doi.org/10.1073/pnas.1600899113

Chen S, Stefanova K, Siddique KHM, Cowling WA (2021) Transient daily heat stress during the early reproductive phase disrupts pod and seed development in *Brassica napus* L. Food Energy Secur 10(1):e262. https://doi.org/10.1002/fes3.262

Cheng K, Pan Y-F, Liu L-M, Zhang H-Q, Zhang Y-M (2021) Integrated transcriptomic and bioinformatics analyses reveal the molecular mechanisms for the differences in seed oil and starch content between Glycine max and Cicer arietinum. Front Plant Sci 12:743680. https://doi.org/10.3389/fpls.2021.743680

Chiou RY-Y (1990) Effects of heat treatments on peanut arachin and conarachin. J Food Biochem 14(3):219–232. https://doi.org/10.1111/j.1745-4514.1990.tb00835.x

Clemente TE, Cahoon EB (2009) Soybean oil: genetic approaches for modification of functionality and total content. Plant Physiol 151(3):1030–1040. https://doi.org/10.1104/pp.109.146282

Craufurd PQ, Prasad PVV, Kakani VG, Wheeler TR, Nigam SN (2003) Heat tolerance in groundnut. Field Crop Res 80:63–77

Coutinho ID, Facchinatto WM, Mertz-Henning LM, Viana AJC, Marin SRR, Santagneli SH, Nepomuceno AL, Colnago LA (2024) NMR fingerprinting of conventional and genetically modified soybean plants with AtAREB1 transcription factors. ACS Omega 9(30):32651–32661. https://doi.org/10.1021/acsomega.4c01796

Cox FR (1979) Effect of temperature treatment on peanut vegetative and fruit growth. Peanut Sci 6(1):14–17. https://doi.org/10.3146/i0095-3679-6-1-4

Cui S, McClements DJ, Xu X, Jiao B, Zhou L, Zhou H, Xiong L, Wang Q, Sun Q, Dai L (2023) Peanut proteins: extraction, modifications, and applications: a comprehensive review. Grain Oil Sci Technol 6(3):135–147. https://doi.org/10.1016/j.gaost.2023.07.001

Dias AS, Lidon FC (2009) Evaluation of grain filling rate and duration in bread and durum wheat, under heat stress after anthesis. J Agron Crop Sci 195(2):137–147. https://doi.org/10.1111/j.1439-037X.2008.00347.x

Djanaguiraman M, Prasad PVV, Boyle DL, Schapaugh WT (2011) High-temperature stress and Soybean leaves: leaf anatomy and photosynthesis. Crop Sci 51:2125–2131. https://doi.org/10.2135/cropsci2010.10.0571

Djanaguiraman M, Prasad PVV, Schapaugh WT (2013a) High day- or nighttime temperature alters leaf assimilation, reproductive success, and phosphatidic acid of pollen grain in Soybean [Glycine max (L.) Merr.]. Crop Sci 53:1594–1604. https://doi.org/10.2135/cropsci2012.07.0441

Djanaguiraman M, Prasad PVV, Boyle DL, Schapaugh WT (2013b) Soybean pollen anatomy, viability and pod set under high temperature stress. J Agro Crop Sci 199:171–177. https://doi.org/10.1111/jac.12005

Dong W, Xie Q, Li J, Liu Z, Wang Z, Wang C, Gao C (2025) BpMAPK3-mediated BpWRKY53 phosphorylation enhances Betula platyphylla drought stress tolerance by increasing flavonoid content. Plant J 121:e70089. https://doi.org/10.1111/tpj.70089

Dornbos DL, Mullen RE (1992) Soybean seed protein and oil contents and fatty acid composition adjustments by drought and temperature. J Am Oil Chem Soc 69(3):228–231. https://doi.org/10.1007/BF02635891

Dwivedi SL, Nigam SN, Jambunathan R, Sahrawat KL, Nagabhushanam GVS, Raghunath K (1993) Effect of genotypes and environments on oil content and oil quality parameters and their correlation in peanut (*Arachis hypogaea* L.). peanut. Science 20(2):84–89. https://doi.org/10.3146/i0095-3679-20-2-5

Egli DB, TeKrony DM, Heitholt JJ, Rupe J (2005) Air temperature during seed filling and soybean seed germination and vigor. Crop Sci 45(4):1329–1335. https://doi.org/10.2135/cropsci2004.0029

Eh T-J, Jiang Y, Jiang M, Li J, Lei P, Ji X, Kim H-I, Zhao X, Meng F (2024) The role of trehalose metabolism in plant stress tolerance. J Adv Res 76:57–72. https://doi.org/10.1016/j.jare.2024.07.013

El Sabagh A, Hossain A, Islam MS, Barutcular C, Ratnasekera D, Gormus O, Amanet K, Mubeen M, Nasim W, Fahad S, Tariq M, Llanes A, Meena RS, Ueda A, Saneoka H, Erman M, Hasanuzzaman M (2020) Drought and heat stress in cotton (*Gossypium hirsutum* L.): consequences and their possible mitigation strategies. In: Hasanuzzaman M (ed) Agronomic crops. Springer Singapore, Singapore, pp 613–634

Erdman JW (2000) Soy protein and cardiovascular disease: a statement for healthcare professionals from the nutrition committee of the AHA. Circulation 102(20):2555–2559. https://doi.org/10.1161/01.CIR.102.20.2555

Fan Y, Wang M, Li Z, Jiang H, Shi J, Shi X, Liu S, Zhao J, Kong L, Zhang W, Ma L (2022) Intake of soy, soy isoflavones and soy protein and risk of cancer incidence and mortality. Front Nutr 9:847421. https://doi.org/10.3389/fnut.2022.847421

Fehr WR, Welke GA, Hammond EG, Duvick DN, Cianzio SR (1992) Inheritance of reduced linolenic acid content in soybean genotypes A16 and A17. Crop Sci 32(4):903–906. https://doi.org/10.2135/cropsci1992.0011183X003200040013x

Ference BA, Ginsberg HN, Graham I, Ray KK, Packard CJ, Bruckert E, Hegele RA, Krauss RM, Raal FJ, Schunkert H, Watts GF, Borén J, Fazio S, Horton JD, Masana L, Nicholls SJ, Nordestgaard BG, Van De Sluis B, Taskinen M-R, Tokgözoğlu L, Landmesser U, Laufs U, Wiklund O, Stock JK, Chapman MJ, Catapano AL (2017) Low-density lipoproteins cause atherosclerotic cardiovascular disease. 1. Evidence from genetic, epidemiologic, and clinical studies. A consensus statement from the European Atherosclerosis Society Consensus Panel. Eur Heart J 38(32):2459–2472. https://doi.org/10.1093/eurheartj/ehx144

Galão OF, Carrão-Panizzi MC, Gontijo Mandarino JM, Santos Júnior OO, Maruyama SA, Figueiredo LC, Bonafe EG, Visentainer JV (2014) Differences of fatty acid composition in Brazilian genetic and conventional soybeans (*Glycine max* (L.) Merrill) grown in different regions. Food Res Int 62:589–594. https://doi.org/10.1016/j.foodres.2014.04.025

Ghoorchi T, Lund P, Larsen M, Hvelplund T, Hansen-Møller J, Weisbjerg MR (2013) Assessment of the mobile bag method for estimation of in vivo starch digestibility. Animal 7(2):265–271. https://doi.org/10.1017/S1751731112001504

Ghotbi-Ravandi AA, Sedighi M, Aghaei K, Mohtadi A (2021) Differential changes in D1 protein content and quantum yield of wild and cultivated barley genotypes caused by moderate and severe drought stress in relation to oxidative stress. Plant Mol Biol Report 39(3):501–507. https://doi.org/10.1007/s11105-020-01264-5

Gong A, Shi A, Liu H, Yu H, Liu L, Lin W, Wang Q (2018) Relationship of chemical properties of different peanut varieties to peanut butter storage stability. J Integr Agric 17(5):1003–1010. https://doi.org/10.1016/S2095-3119(18)61919-7

Goren A, Ashlock D, Tetlow IJ (2018) Starch formation inside plastids of higher plants. Protoplasma 255(6):1855–1876. https://doi.org/10.1007/s00709-018-1259-4

Govindasamy V, George P, Aher L, Ramesh SV, Thangasamy A, Anandan S, Raina SK, Kumar M, Rane J, Annapurna K, Minhas PS (2017) Comparative conventional and phenomics approaches to assess symbiotic effectiveness of Bradyrhizobia strains in soybean (*Glycine max* L. Merrill) to drought. Sci Rep 7(1):6958. https://doi.org/10.1038/s41598-017-06441-3

Hagely K, Konda AR, Kim J-H, Cahoon EB, Bilyeu K (2021) Molecular-assisted breeding for soybean with high oleic/low linolenic acid and elevated vitamin E in the seed oil. Mol Breed 41(1):3. https://doi.org/10.1007/s11032-020-01184-y

Hamidou F, Halilou O, Vadez V (2013) Assessment of groundnut under combined heat and drought stress. J Agron Crop Science 199(1):1–11. https://doi.org/10.1111/j.1439-037X.2012.00518.x

Hasler CM (2002) Functional foods: benefits, concerns and challenges—a position paper from the American council on science and health. J Nutr 132(12):3772–3781. https://doi.org/10.1093/jn/132.12.3772

Hildebrand DF, Hymowitz T (1981) Role of β-amylase in starch metabolism during soybean seed development and germination. Physiol Plant 53(4):429–434. https://doi.org/10.1111/j.1399-3054.1981.tb02726.x

Hou A, Chen P, Alloatti J, Li D, Mozzoni L, Zhang B, Shi A (2009) Genetic variability of seed sugar content in worldwide soybean germplasm collections. Crop Sci 49(3):903–912. https://doi.org/10.2135/cropsci2008.05.0256

Hsiang S, Kopp R, Jina A, Rising J, Delgado M, Mohan S, Rasmussen DJ, Muir-Wood R, Wilson P, Oppenheimer M, Larsen K, Houser T (2017) Estimating economic damage from climate change in the United States. Science 356(6345):1362–1369. https://doi.org/10.1126/science.aal4369

Huang B, Qi F, Sun Z, Miao L, Zhang Z, Liu H, Fang Y, Dong W, Tang F, Zheng Z, Zhang X (2019) Marker-assisted backcrossing to improve seed oleic acid content in four elite and popular peanut (*Arachis hypogaea* L.) cultivars with high oil content. Breed Sci 69(2):234–243. https://doi.org/10.1270/jsbbs.18107

Hughes GJ, Ryan DJ, Mukherjea R, Schasteen CS (2011) Protein Digestibility-Corrected Amino Acid Scores (PDCAAS) for soy protein isolates and concentrate: criteria for evaluation. J Agric Food Chem 59(23):12707–12712. https://doi.org/10.1021/jf203220v

Huth PJ, Fulgoni VL, Larson BT (2015) A systematic review of high-oleic vegetable oil substitutions for other fats and oils on cardiovascular disease risk factors: implications for novel high-oleic soybean oils. Adv Nutr 6(6):674–693. https://doi.org/10.3945/an.115.008979

Ingole HP, Saripalli G, Jones ZT, Parthiban M, Toyinbo J, Narayanan S, Mir RR, Rustgi S (2025) Genome-wide association analysis for pollen viability under heat stress in peanut. Plant Stress 15:100760. https://doi.org/10.1016/j.stress.2025.100760

Jagadish SVK, Way DA, Sharkey TD (2021) Plant heat stress: concepts directing future research. Plant Cell Environ 44(7):1992–2005. https://doi.org/10.1111/pce.14050

Jeammuangpuk P, Promchote P, Duangpatra J, Chaisan T, Onwimol D, Kvien CK (2020) Enhancement of Tainan 9 peanut seed storability and germination under low temperature. Int J Agron 2020:1–8. https://doi.org/10.1155/2020/8813285

Jeong W-H, Harada K, Yamada T, Abe J, Kitamura K (2010) Establishment of new method for analysis of starch contents and varietal differences in soybean seeds. Breed Sci 60(2):160–163. https://doi.org/10.1270/jsbbs.60.160

Jiang R (2002) Nut and peanut butter consumption and risk of type 2 diabetes in women. JAMA 288(20):2554. https://doi.org/10.1001/jama.288.20.2554

Jones JB, Provost M, Keaver L, Breen C, Ludy M-J, Mattes RD (2014) A randomized trial on the effects of flavorings on the health benefits of daily peanut consumption. Am J Clin Nutr 99(3):490–496. https://doi.org/10.3945/ajcn.113.069401

Joseph LM, Hymowitz T, Schmidt MA, Herman EM (2006) Evaluation of *glycine* germplasm for nulls of the immunodominant allergen P34/Gly m Bd 30k. Crop Sci 46(4):1755–1763. https://doi.org/10.2135/cropsci2005.12-0500

Juhász A, Belova T, Florides CG, Maulis C, Fischer I, Gell G, Birinyi Z, Ong J, Keeble-Gagnère G, Maharajan A, Ma W, Gibson P, Jia J, Lang D, Mayer KFX, Spannagl M, International Wheat Genome Sequencing Consortium, Tye-Din JA, Appels R, Olsen OA (2018) Genome mapping of seed-borne allergens and immunoresponsive proteins in wheat. Sci Adv 4(8):eaar8602. https://doi.org/10.1126/sciadv.aar8602

Jumrani K, Bhatia VS (2018) Impact of combined stress of high temperature and water deficit on growth and seed yield of soybean. Physiol Mol Biol Plants 24(1):37–50. https://doi.org/10.1007/s12298-017-0480-5

Kakani VG, Prasad PVV, Craufurd PQ, Wheeler TR (2002) Response of in vitro pollen germination and pollen tube growth of groundnut (*Arachis hypogaea* L.) genotypes to temperature. Plant Cell Environ 25:1651–1661. https://doi.org/10.1046/j.1365-3040.2002.00943.x

Kakati JP, Fallen B, Bridges W, Narayanan S (2022a) Characterization of a soybean (*Glycine max* L. Merr.) population for germination and seedling root traits under water stress. Agronomy 12(8):1944. https://doi.org/10.3390/agronomy12081944

Kakati JP, Lwe ZZ, Narayanan S (2022b) Heat stress during the early flowering stage did not affect seed fatty acid contents in conventional oleic peanut varieties. Peanut Sci 49(1):63–69. https://doi.org/10.3146/0095-3679-491-PS21-12

Kakati JP, Fallen B, Armstrong P, Yan S, Bridges W, Narayanan S (2024) High-protein soybean lines with stable seed protein content under heat and drought stresses. J Agric Food Res 18:101469. https://doi.org/10.1016/j.jafr.2024.101469

Katam R, Shokri S, Murthy N, Singh SK, Suravajhala P, Khan MN, Bahmani M, Sakata K, Reddy KR (2020) Proteomics, physiological, and biochemical analysis of cross tolerance mechanisms in response to heat and water stresses in soybean. PLoS One 15(6):e0233905. https://doi.org/10.1371/journal.pone.0233905

Korde LA, Wu AH, Fears T et al (2009) Childhood soy intake and breast cancer risk in Asian American women. Cancer Epidemiol Biomarkers Prev 18:1050–1059

Krishnan P, Ramakrishnan B, Reddy KR, Reddy VR (2011) High-temperature effects on rice growth, yield, and grain quality. In: Advances in agronomy. Elsevier, pp 87–206

Koti S, Reddy KR, Reddy VR, Kakani VG, Zhao D (2005) Interactive effects of carbon dioxide, temperature, and ultraviolet-B radiation on soybean (Glycine max L.) flower and pollen morphology, pollen production, germination, and tube lengths. J Exp Bot 56:725–736

Krishnan HB, Kim W-S, Oehrle NW, Smith JR, Gillman JD (2020) Effect of heat stress on seed protein composition and ultrastructure of protein storage vacuoles in the cotyledonary parenchyma cells of soybean genotypes that are either tolerant or sensitive to elevated temperatures. IJMS 21(13):4775. https://doi.org/10.3390/ijms21134775

Kumar S, Bhushan B, Wakchaure GC, Dutta R, Jat BS, Meena KK, Rakshit S, Pathak H (2023) Unveiling the impact of heat stress on seed biochemical composition of major cereal crops: implications for crop resilience and nutritional value. Plant Stress 9:100183. https://doi.org/10.1016/j.stress.2023.100183

Kunta S, Agmon S, Chedvat I, Levy Y, Chu Y, Ozias-Akins P, Hovav R (2021) Identification of consistent QTL for time to maturation in Virginia-type peanut (Arachis hypogaea L.). BMC Plant Biol 21(1):186. https://doi.org/10.1186/s12870-021-02951-5

Kunta S, Parimi P, Levy Y, Kottakota C, Chedvat I, Chu Y, Ozias-Akins P, Hovav R (2022) A first insight into the genetics of maturity trait in runner × Virginia types peanut background. Sci Rep 12(1):15267. https://doi.org/10.1038/s41598-022-19653-z

Lai H, Li X, Chen Y, Liu Z (2024) Mitigating heat-induced yield loss in peanut: insights into 24-epibrassinolide-mediated improvement in antioxidant capacity, photosynthesis, and kernel weight. Field Crop Res 316:109521. https://doi.org/10.1016/j.fcr.2024.109521

Lamaoui M, Jemo M, Datla R, Bekkaoui F (2018) Heat and drought stresses in crops and approaches for their mitigation. Front Chem 6:26. https://doi.org/10.3389/fchem.2018.00026

Li P, Zhang Y, Wu X, Liu Y (2018) Drought stress impact on leaf proteome variations of faba bean (Vicia faba L.) in the Qinghai-Tibet Plateau of China. 3 Biotech 8(2):110

Lobell DB, Asner GP (2003) Climate and management contributions to recent trends in U.S. agricultural yields. Science 299(5609):1032–1032. https://doi.org/10.1126/science.1078475

Lipton D, Rubenstein MA, Weiskopf SR, Carter SL, Peterson J, Crozier L, Fogarty M, Gaichas S, Hyde K, Morelli TL, Morisette J, Moustahfid H, Munoz R, Poudel R, Staudinger M, Stock C, Thompson L, Waples RS, Weltzin J (2018) Ecosystems, ecosystem services, and biodiversity. U.S. Global Change Research Program 268:321. https://doi.org/10.7930/NCA4.2018.CH7

Logan J, Mueller MA, Graves CR (1998) A comparison of early and recommended soybean production systems in Tennessee. J Prod Agric 11(3):319–325. https://doi.org/10.2134/jpa1998.0319

Machado J, Nunes da Silva M, Vasconcelos MW, Santos CS (2025) The impact of climate change-induced abiotic stresses on the nutritional quality of legume seeds. J Exp Bot 76(12):3288–3310. https://doi.org/10.1093/jxb/eraf085

Mathur S, Seo B, Jajoo A, Reddy KR, Reddy VR (2023) Chlorophyll fluorescence is a potential indicator to measure photochemical efficiency in early to late soybean maturity groups under changing day lengths and temperatures. Front Plant Sci 14:1228464. https://doi.org/10.3389/fpls.2023.1228464

Miransari M (2016) Soybean, protein, and oil production under stress. In: Environmental stresses in soybean production. Elsevier, pp 157–176

Morgillo S, Hill AM, Coates AM (2019) The effects of nut consumption on vascular function. Nutrients 11(1):116. https://doi.org/10.3390/nu11010116

Munck L (1992) The case of high-lysine barley breeding. In: Shewry PR (ed) Barley: genetics, biochemistry, molecular biology and biotechnology. CAB International, Wallingford, pp 573–601

Nakagawa ACS, Ario N, Tomita Y, Tanaka S, Murayama N, Mizuta C, Iwaya-Inoue M, Ishibashi Y (2020) High temperature during soybean seed development differentially alters lipid and protein metabolism. Plant Prod Sci 23(4):504–512. https://doi.org/10.1080/1343943X.2020.1742581

Narayanan S, Zoong-Lwe ZS, Gandhi N, Welti R, Fallen B, Smith JR, Rustgi S (2020) Comparative lipidomic analysis reveals heat stress responses of two soybean genotypes differing in temperature sensitivity. Plants 9(4):457. https://doi.org/10.3390/plants9040457

Nayak SN, Hebbal V, Bharati P, Nadaf HL, Naidu GK, Bhat RS (2020) Profiling of nutraceuticals and proximates in peanut genotypes differing for seed coat color and seed size. Front Nutr 7:45. https://doi.org/10.3389/fnut.2020.00045

NOAA (2025) Climate change: global temperature | NOAA. Climate.gov. https://www.climate.gov/news-features/understanding-climate/climate-change-global-temperature. Accessed 6 Sept 2025

Onemli F (2012) Impact of climate change on oil fatty acid composition of peanut (Arachis hypogaea L.) in three market classes. Chilean J Agric Res 72(4):483–488. https://doi.org/10.4067/S0718-58392012000400004

Patel JD, Wang ML, Dang P, Butts C, Lamb M, Chen CY (2022) Insights into the genomic architecture of seed and pod quality traits in the U.S. peanut mini-core diversity panel. Plants 11(7):837. https://doi.org/10.3390/plants11070837

Perveen A, Sheheryar S, Ahmad F, Mustafa G, Moura AA, Campos FAP, Domont GB, Nishan U, Ullah R, Ibrahim MA, Nogueira FCS, Shah M (2025) Integrative physiological, biochemical, and proteomic analysis of the leaves of two cotton genotypes under heat stress. PLoS One 20(1):e0316630. https://doi.org/10.1371/journal.pone.0316630

Prasad PVV, Craufurd PQ, Summerfield RJ (1999a) Sensitivity of peanut to timing of heat stress during reproductive development. Crop Sci 39(5):1352–1357. https://doi.org/10.2135/cropsci1999.3951352x

Prasad PVV, Craufurd PQ, Summerfield RJ (1999b) Fruit number in relation to pollen production and viability in groundnut exposed to short episodes of heat stress. Ann Bot 84(3):381–386. https://doi.org/10.1006/anbo.1999.0926

Prasad PVV, Craufurd PQ, Summerfield RJ (1999) Sensitivity of peanut to timing of heat stress during reproductive development. Crop Sci 39:1352–1357. https://doi.org/10.2135/cropsci1999.3951352x

Prasad PVV, Craufurd PQ, Summerfield RJ, Wheeler TR (2000) Effects of short episodes of heat stress on flower production and fruit-set of groundnut (Arachis hypogaea L.). J Exp Bot 51:777–784

Puteh A, Thuzar M, Mondal MMA, Abdullah NPB, Halim MA (2013) Soybean [Glycine max (L.) Merrill] seed yield response to high temperature stress during reproductive growth stages. Aust J Crop Sci 7:1472–1479

Prasad PVV, Boote KJ, Hartwell Allen L, Thomas JMG (2003) Super-optimal temperatures are detrimental to peanut (*Arachis hypogaea* L.) reproductive processes and yield at both ambient and elevated carbon dioxide. Glob Chang Biol 9(12):1775–1787. https://doi.org/10.1046/j.1365-2486.2003.00708.x

Qin P, Wang T, Luo Y (2022) A review on plant-based proteins from soybean: health benefits and soy product development. J Agric Food Res 7:100265. https://doi.org/10.1016/j.jafr.2021.100265

Rabara RC, Tripathi P, Rushton PJ (2017) Comparative metabolome profile between tobacco and soybean grown under water-stressed conditions. Biomed Res Int 2017:1–12. https://doi.org/10.1155/2017/3065251

Ray JD, Sinclair TR (1998) The effect of pot size on growth and transpiration of maize and soybean during water deficit stress. J Exp Bot 49(325):1381–1386. https://doi.org/10.1093/jxb/49.325.1381

Raza A, Bhardwaj S, Rahman MA, García-Caparrós P, Habib M, Saeed F, Charagh S, Foyer CH, Siddique KHM, Varshney RK (2024) Trehalose: a sugar molecule involved in temperature stress management in plants. Crop J 12(1):1–16. https://doi.org/10.1016/j.cj.2023.09.010

Ren Z, He S, Zhao N, Zhai H, Liu Q (2019) A sucrose non-fermenting-1-related protein kinase-1 gene, *IbSnRK1*, improves starch content, composition, granule size, degree of crystallinity and gelatinization in transgenic sweet potato. Plant Biotechnol J 17(1):21–32. https://doi.org/10.1111/pbi.12944

Rizzo G, Baroni L (2018) Soy, soy foods and their role in vegetarian diets. Nutrients 10(1):43. https://doi.org/10.3390/nu10010043

Ros E (2010) Health benefits of nut consumption. Nutrients 2(7):652–682. https://doi.org/10.3390/nu2070652

Rustgi S, Kakati JP, Jones ZT, Zoong Lwe ZS, Narayanan S (2021a) Heat tolerance as a function of membrane lipid remodeling in the major US oilseed crops (soybean and peanut). J Plant Biochem Biotechnol 30(4):652–667. https://doi.org/10.1007/s13562-021-00729-2

Rustgi S, Kashyap S, Deleu LJ, Delcour J (2021b) Reduced-immunogenicity wheat now coming to age. In: Wani SH, Mohan A, Singh GP (eds) Physiological, molecular, and genetic perspectives of wheat improvement. Springer Nature, Cham, pp 279–300

Salem MA, Kakani VG, Koti S, Reddy KR (2007) Pollen-based screening of soybean genotypes for high temperatures. Crop Sci 47:219–231. https://doi.org/10.2135/cropsci2006.07.0443

Schauberger B, Archontoulis S, Arneth A, Balkovic J, Ciais P, Deryng D, Elliott J, Folberth C, Khabarov N, Müller C, Pugh TAM, Rolinski S, Schaphoff S, Schmid E, Wang X, Schlenker W, Frieler K (2017) Consistent negative response of US crops to high temperatures in observations and crop models. Nat Commun 8(1):13931. https://doi.org/10.1038/ncomms13931

Sehgal A, Sita K, Siddique KHM, Kumar R, Bhogireddy S, Varshney RK, HanumanthaRao B, Nair RM, Prasad PVV, Nayyar H (2018) Drought or/and heat-stress effects on seed filling in food crops: impacts on functional biochemistry, seed yields, and nutritional quality. Front Plant Sci 9:1705. https://doi.org/10.3389/fpls.2018.01705

Selvaraj MG, Burow G, Burke JJ, Belamkar V, Puppala N, Burow MD (2011) Heat stress screening of peanut (*Arachis hypogaea* L.) seedlings for acquired thermotolerance. Plant Growth Regul 65(1):83–91. https://doi.org/10.1007/s10725-011-9577-y

Sharma V, Gangurde SS, Nayak SN, Gowda AS, Sukanth BS, Mahadevaiah SS, Manohar SS, Choudhary RS, Anitha T, Malavalli SS, Srikanth SN, Bajaj P, Sharma S, Varshney RK, Latha P, Janila P, Bhat RS, Pandey MK (2023) Genetic mapping identified three hotspot genomic

regions and candidate genes controlling heat tolerance-related traits in groundnut. Front Plant Sci 14:1182867. https://doi.org/10.3389/fpls.2023.1182867
Shea Z, Singer WM, Zhang B (2020) Soybean production, versatility, and improvement. In: Legume crops [Working Title]. IntechOpen
Shu XO, Jin F, Dai Q et al (2001) Soyfood intake during adolescence and subsequent risk of breast cancer among Chinese women. Cancer Epidemiol Biomarkers Prev 10:483–488
Smith JR, Mengistu A, Nelson RL, Paris RL (2009) Identification of soybean accessions with high germinability in high-temperature environments. Crop Sci 48(6):2279–2288. https://doi.org/10.2135/cropsci2008.01.0026
SoyStats®: a reference guide to important soybean facts & figures. 2025 ed., American Soybean Association, 2025, SoyStats®. https://soystats.com/wp-content/uploads/2025Soystats1.pdf
Spivey WW, Rustgi S, Welti R, Roth MR, Burow MD, Bridges WC, Narayanan S (2023) Lipid modulation contributes to heat stress adaptation in peanut. Front Plant Sci 14:1299371. https://doi.org/10.3389/fpls.2023.1299371
Stark MT (ed) (2006) Archaeology of Asia, 1st edn. Wiley
Sun Z, Qi F, Liu H, Qin L, Xu J, Shi L, Zhang Z, Miao L, Huang B, Dong W, Wang X, Tian M, Feng J, Zhao R, Zheng Z, Zhang X (2022) QTL mapping of quality traits in peanut using whole-genome resequencing. Crop J 10(1):177–184. https://doi.org/10.1016/j.cj.2021.04.008
Tahir ISA et al (2005) Remobilization of nitrogen and carbohydrate from stems of bread wheat in response to heat stress during grain filling. J Agric Sci 191:106–116
Ullah A, Nadeem F, Nawaz A, KHM S, Farooq M (2022) Heat stress effects on the reproductive physiology and yield of wheat. J Agron Crop Sci 208(1):1–17. https://doi.org/10.1111/jac.12572
United Nations (2012) Feeding the world sustainably. In: United Nations. https://www.un.org/en/chronicle/article/feeding-world-sustainably. Accessed 6 Sept 2025
United Nations (2022) World population to reach 8 billion on 15 November 2022. In: United Nations. https://www.un.org/en/desa/world-population-reach-8-billion-15-november-2022. Accessed 6 Sept 2025
Vaidya AS et al (2019) Dynamic control of plant water use using designed ABA receptor agonists. Science 366:eaaw8848. https://doi.org/10.1126/science.aaw8848
Van der Laan L, Elango D, Ferela A, O'Rourke JA, Singh AK (2025) Transcriptional insights into soybean genotypes under prolonged heat stress: Identification of key genes and soil influences for enhanced tolerance. Plant Stress 2025:101038. https://doi.org/10.1016/j.stress.2025.101038
Veas REA, Ergo VV, Vega CRC, Lascano RH, Rondanini DP, Carrera CS (2022) Soybean seed growth dynamics exposed to heat and water stress during the filling period under field conditions. J Agron Crop Sci 208(4):472–485. https://doi.org/10.1111/jac.12523
Vennam RR, Beard KM, Haak DC, Balota M (2025) Evaluation of peanut physiological responses to heat and drought stress across growth chamber and field environments. Plants 14(17):2687. https://doi.org/10.3390/plants14172687
Vidal RO, Nascimento LCD, Maurício Costa Mondego J, Amarante Guimarães Pereira G, Falsarella Carazzolle M (2012) Identification of SNPs in RNA-seq data of two cultivars of *Glycine max* (soybean) differing in drought resistance. Genet Mol Biol 35(1 suppl 1):331–334. https://doi.org/10.1590/S1415-47572012000200014
Wang F, Ye Y, Niu Y, Wan F, Qi B, Chen X, Zhou Q, Chen B (2016) A tomato plastidic ATP/ADP transporter gene SlAATP increases starch content in transgenic Arabidopsis. Physiol Mol Biol Plants 22(4):497–506. https://doi.org/10.1007/s12298-016-0389-4
Wang Y, Lyu B, Fu H, Li J, Ji L, Gong H, Zhang R, Liu J, Yu H (2023) The development process of plant-based meat alternatives: raw material formulations and processing strategies. Food Res Int 167:112689. https://doi.org/10.1016/j.foodres.2023.112689
Wang F, Miao H, Zhang S, Hu X, Li C, Yang W, Chen J (2025) Identification of a new major oil content QTL overlapped with FAD2B in cultivated peanut (*Arachis hypogaea* L.). Plants 14(4):615. https://doi.org/10.3390/plants14040615
Watanabe D, Adányi N, Takács K, Maczó A, Nagy A, Gelencsér É, Pachner M, Lauter K, Baumgartner S, Vollmann J (2017) Development of soybeans with low P34 allergen protein

concentration for reduced allergenicity of soy foods. J Sci Food Agric 97(3):1010–1017. https://doi.org/10.1002/jsfa.7827

Weaver CC, Monfort WS, Grey TL, Pilon C, Tubbs RS (2021) Effect of post shelling storage environments on seed germination and vigor of peanut. Agron J 113(5):4116–4131. https://doi.org/10.1002/agj2.20795

Wei J, Liu X, Li L, Zhao H, Liu S, Yu X, Shen Y, Zhou Y, Zhu Y, Shu Y, Ma H (2020) Quantitative proteomic, physiological and biochemical analysis of cotyledon, embryo, leaf and pod reveals the effects of high temperature and humidity stress on seed vigor formation in soybean. BMC Plant Biol 20(1):127. https://doi.org/10.1186/s12870-020-02335-1

Wijewardana C, Reddy KR, Bellaloui N (2019) Soybean seed physiology, quality, and chemical composition under soil moisture stress. Food Chem 278:92–100. https://doi.org/10.1016/j.foodchem.2018.11.035

Wood I (1968) The effect of temperature at early flowering on the growth and development of peanuts (*Arachis hypogaea*). Aust J Agric Res 19(2):241. https://doi.org/10.1071/AR9680241

Wu AH, Yu MC, Tseng CC, Pike MC (2008) Epidemiology of soy exposures and breast cancer risk. Br J Cancer 98:9–14

Wu P, Liu A, Zhang Y, Feng K, Zhao S, Li L (2021) NnABI4-mediated ABA regulation of starch biosynthesis in lotus (Nelumbo nucifera Gaertn). IJMS 22(24):13506. https://doi.org/10.3390/ijms222413506

Yu R, Huppertz T, Vasiljevic T (2025) Effect of shear and pH on heat-induced changes in faba bean proteins. Foods 14(16):2906. https://doi.org/10.3390/foods14162906

Zhang FF, Haslam DE, Terry MB et al (2017a) Dietary isoflavone intake and all-cause mortality in breast cancer survivors: the breast cancer family registry. Cancer 123:2070–2079

Zhang X, Zhang J, He X, Wang Y, Ma X, Yin D (2017b) Genome-wide association study of major agronomic traits related to domestication in peanut. Front Plant Sci 8:1611. https://doi.org/10.3389/fpls.2017.01611

Zhang S, Hu X, Miao H, Chu Y, Cui F, Yang W, Wang C, Shen Y, Xu T, Zhao L, Zhang J, Chen J (2019) QTL identification for seed weight and size based on a high-density SLAF-seq genetic map in peanut (*Arachis hypogaea* L.). BMC Plant Biol 19(1):537. https://doi.org/10.1186/s12870-019-2164-5

Zhang X, Zhu L, Ren M, Xiang C, Tang X, Xia Y, Song D, Li F (2023) Genome-wide association studies revealed the genetic loci and candidate genes of pod-related traits in peanut (*Arachis hypogaea* L.). Agronomy 13(7):1863. https://doi.org/10.3390/agronomy13071863

Zoong Lwe ZS, Welti R, Anco D, Naveed S, Rustgi S, Narayanan S (2020) Heat stress elicits remodeling in the anther lipidome of peanut. Sci Rep 10(1):22163. https://doi.org/10.1038/s41598-020-78695-3

Epigenetics in Crop Abiotic Stress Adaptation and Memory

7

Praveen Kumar and Vijay Gahlaut

Abstract

Crop productivity is negatively affected by various environmental stresses such as cold, heat, drought, salinity and heavy metal stress. Plants, due to their sessile nature, have thus developed various mechanisms to adapt themselves under abiotic stress conditions. Epigenetic mechanisms play a pivotal role in plant stress adaptations by inducing alterations in gene expression without modifying the DNA sequence. Epigenetic modifications are reversible and confer a dynamic plasticity to the plant's molecular mechanisms, thereby enhancing their growth, survival and adaptability via modulating gene expression in response to abiotic stressors. Epigenetic mechanisms include DNA methylation, histone modifications and non-coding RNA-mediated regulations. In this book chapter, we are elaborating these fundamental but essential epigenetic mechanisms in crop plants under optimal as well as common abiotic stress conditions, including heat, drought, salt and cold. We are also investigating some cereal crop studies to highlight epigenetic markers associated with agriculturally important traits. These markers can further help the researchers develop abiotic stress-resilient crop varieties. Finally, we are highlighting the concept of stress memory and its agricultural implications, which can help researchers develop abiotic stress-resilient crop varieties.

P. Kumar
Department of Biotechnology, University Institute of Biotechnology, Chandigarh University, Mohali, Punjab, India

V. Gahlaut (✉)
Department of Biotechnology, University Institute of Biotechnology, Chandigarh University, Mohali, Punjab, India

University Centre for Research and Development, Chandigarh University, Mohali, Punjab, India

V. Gahlaut et al. (eds.), *Smart Crop Development*,
https://doi.org/10.1007/978-981-95-7680-7_7

Keywords
DNA methylation · Heat stress · Transposable elements · Epigenetic memory · Stress-resilient crops

7.1 Introduction

Rapid climatic shifts increase the frequency and severity of abiotic stresses and negatively affect crop growth and yield. Epigenetic regulations allow the plant to modulate gene expression in response to abiotic stressors without changing its DNA sequence. These mechanisms modulate chromatin accessibility by modifying chromatin structure, thereby altering the transcriptional activity in response to environmental stimuli (Kaur et al. 2024; Liu et al. 2024a; Wang et al. 2023a; Li et al. 2025). These epigenetic mechanisms regulate almost all the fundamental genetic functions such as replication, transcription, translation, DNA repair and post-translational modifications under optimal as well as stress conditions. DNA methylation, histone modifications and RNA-dependent DNA methylations are the three major fundamental epigenetic mechanisms are involved in plant adaptations. DNA methylation has mostly been found to be involved in the gene repression and silencing of transposable elements (TEs) in the chromosomal heterochromatic regions, thus stabilising the genome integrity of plants under abiotic stress conditions (Zhang et al. 2015; Tran et al. 2005; Panda et al. 2016). The covalent insertion of a methyl group at the 5′ position of cytosine in symmetric (CG, CHG) and asymmetric (CHH) contexts function as an epigenetic marker in the genome, commonly associated with gene regulation under both optimal and adverse conditions (Rehman et al. 2021; Liu et al. 2024b; Wang and Wang 2022). These epigenetic marks are reversible and inheritable in nature, thus making a promising target for a sustainable crop improvement (Li et al. 2008). The regulation of DNA methylation is dependent on DNA methyltransferases (DNMTs) and demethylase enzymes (Zhong et al. 2014; Garg et al. 2014; Gouil and Baulcombe 2016). Unlike DNA methylation, histone modifications such as histone methylations and acetylation are majorly involved in chromatin remodelling by acting downstream to the DNA methylation and reinforce heterochromatin establishment under stressed conditions (Faivre et al. 2024; Tittel-Elmer et al. 2010; Yang et al. 2023). Histone modifications act as molecular signals that either relaxes or condenses the chromatin structure, thereby regulating gene expression via controlling the accessibility of transcription factors to the promoter regions present in the genome sites (Diallo et al. 2012; Zhou and Hu 2010; Zhou et al. 2021).

On the other hand, RNA-mediated DNA methylation mechanisms involve non-coding RNAs such as microRNAs (miRNAs), small interfering RNAs (siRNAs) and long non-coding RNAs (lncRNAs)–guided methylation, which facilitate the de novo methylation mechanisms and promote plants' acclimatisation under abiotic stress conditions (Panda et al. 2020; Chellappan et al. 2010; Bond and Baulcombe 2015). RNA-mediated methylation mechanisms play a pivotal role in stress adaptation and environmental receptiveness, rather than in the stable transmission of heritable traits. This chapter highlights the fundamental epigenetic mechanisms in plants under optimal conditions and their alterations in response to abiotic stressors. Extensively studied examples of crop plants, such as rice, wheat and maize, are used in this chapter for a deeper understanding of epigenetic mechanisms. Moreover, these stress-specific epigenomic alterations are not just discrete events but are intricately linked to "plant stress memory" or "epigenetic memory", which enables crops to escalate stronger responses upon re-exposure. Eventually, this chapter also emphasises the potential of epigenetic memory and priming in breeding strategies for climate-resilient crops. Finally, this chapter bridges the fundamental mechanisms at the molecular level by integrating molecular insights, thus providing a vision for sustainable crop improvement through epigenomic interventions.

7.2 Epigenetic Mechanisms in Plants: Gene Expression Regulation

Epigenetic mechanisms regulate the expression of genes involved in basic biological processes, such as DNA replication, transcription, translation and cell division, under optimal conditions without altering the DNA sequence. These mechanisms regulate biological processes by modifying genetic function through DNA methylation, histone modifications and RNA-directed DNA methylation pathways. In this section, we elaborate on these pathways and the key regulatory molecules involved under optimal and stressed conditions. DNA methylation is a fundamental but vital epigenetic mechanism in plants, enabling dynamic yet heritable regulation of gene expression in response to environmental stimuli. The addition of a methyl group to cytosine bases particularly occurs predominantly in CpG islands (sequences containing CG, CHG and CHH, where H = A, T or C) and influences gene expression (Guo et al. 2019; Sapna et al. 2020). DNA methylation at CpG islands is not a static mark but a dynamic and reversible regulatory layer. Cytosine methylation in plants usually occurs in symmetrical (CG, CHG) and asymmetrical (CHH) sequences, each guided by specialised methyltransferases (Fang et al. 2022; Gu et al. 2024). The percentage of cytosine methylation in CG context generally found higher compared to the non-CG context (CHG and CHH) in crops like rice (CG = 54%, CHG = 37.3% and CHH = 12%), wheat (CG = 53%, CHG = 3.48% and CHH = 1.41%) and *Arabidopsis*, further indicating the abundance and the significance of CG methylation in plant growth and development under optimal conditions (Gardiner et al. 2015; Li et al. 2012). In nuclear genome, gene expression and suppression usually depend on location and the amount of methylation. A higher cytosine methylation in the promoter region is usually linked with gene suppression, whereas hypermethylation in the gene body (introns, exons) is usually associated with gene overexpression (Gardiner et al. 2015). Distinct classes of DNA methyltransferases establish and maintain these epigenetic marks (or stress memory), dynamically responding to abiotic challenges like drought, salinity, heat and nutrient deficiency (Dai et al. 2005; Gahlaut et al. 2020). Three types of methyltransferases are generally involved in DNA methylation mechanism,

METHYLTRANSFERASE 1 (MET1), CHROMOMETHYLASE 3 (CMT3) and DOMAINS REARRANGED METHYLTRANSFERASE (DRM1/2), where MET1 and CMT3 are often involved in the inheritance of GC and CHG methylation contexts by simply replicating methylation patterns; conversely, DRM is mostly associated with CHH context and de novo methylation, which is essential for plant adaptability under stress conditions (Xun et al. 2025; Read et al. 2022; Chen et al. 2023). In the CG context, methylation is primarily maintained by METHYLTRANSFERASE 1 (MET1), a homolog of mammalian DNMT1 (DNA (cytosine-5)-methyltransferase 1), which restores methylation at hemi-methylated sites post-replication to preserve transcriptional patterns (Saze et al. 2003). In plants, de novo methylation is established by DRM1/2 via the RNA-directed DNA methylation (RdDM) pathway, where siRNAs guide DRM enzymes to specific genomic loci, particularly TEs and repetitive sequences (Gallego-Bartolomé et al. 2019; Palomar et al. 2020).

The ncRNA-mediated epigenetic regulation includes siRNAs, miRNAs and lncRNAs, which act as scaffolds, orchestrating chromatin modification via the RdDM pathway, thus promoting de novo methylation of TEs and stress-responsive genes (Xun et al. 2025; Huang et al. 2016). In the beginning of RdDM pathway, the heterochromatic region is detected by SHH1, which further recruits the RNA polymerase IV (POL IV) to transcribe the non-coding RNAs (Law et al. 2011; Law et al. 2013; Ye et al. 2016; Daxinger et al. 2009). These single-stranded non-coding RNAs are then converted to double-stranded RNAs (dsRNAs) by RNA-dependent RNA polymerase 2 (RDR2), which subsequently cuts into 24-nt siRNAs by dicer-like protein (DCL3). These 24-nt siRNAs are loaded onto the Argonaute proteins (AGO4/AGO6), forming a complex that interacts with the scaffold RNAs generated by RNA POL V at the target loci (Wang and Axtell 2016; Sigman et al. 2021; Palomar et al. 2020). This whole interaction facilitates the recruitment of DRM2, which finally catalyses the DNA methylation in CG, CHG and CHH contexts. RdDM pathway increases the genome integrity by transposon silencing and contributes to stress response and influences flowering, seed development and reproduction in crop plants (Wang et al. 2022a; Li et al. 2015) (Fig. 7.1).

Histone modifications are covalent post-translational changes to histone proteins, which include acetylation, methylation, phosphorylation and ubiquitination, and are catalysed by histone acetyltransferases (HATs), histone deacetylases (HDACs), histone methyltransferases (HMTs) and histone demethylases (HDMs) (Hou et al. 2021; Chen et al. 2024). Generally, histone modifications act as epigenetic signals which encourage or repress gene transcription, altering chromatin accessibility or recruiting chromatin remodelers in plants under abiotic stress conditions (Lv et al. 2021; Zheng et al. 2021; He et al. 2024). Histone acetylation is involved in loosening the chromatin architecture by neutralising the positive charges, which promotes transcriptional activation of stress-responsive genes under stressed conditions (Lin et al. 2022; Zheng et al. 2021a). Under drought conditions, histone methylations at H3K9ac and H3K4me3 sites have been found to increase at promoters of *OsDERB1* and *OsLEA* genes, which further accelerate the drought stress-responsive genes transcription in rice (Li et al. 2023; Wang et al. 2023b; Du

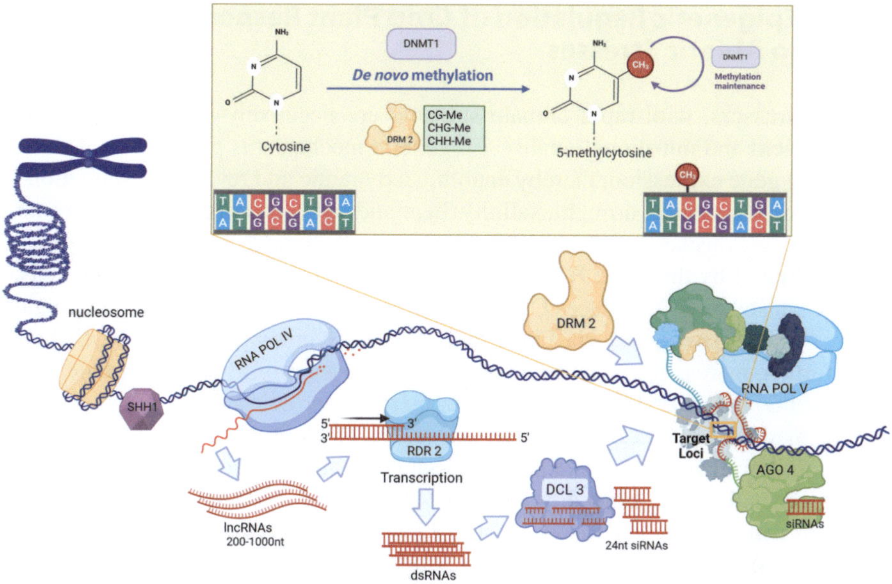

Fig. 7.1 The model illustrates the canonical RNA-directed DNA methylation (RdDM) pathway in plants. The RNA polymerase IV (Pol IV) generates precursor transcripts that are converted into double-stranded RNAs (dsRNAs) by RNA-DEPENDENT RNA POLYMERASE 2 (RDR2). These dsRNAs are processed by DICER-LIKE 3 (DCL3) into 24-nucleotide short interfering RNAs (siRNAs), which are loaded onto ARGONAUTE 4 (AGO4). The AGO–siRNA complexes associate with RNA polymerase V (Pol V) transcripts, facilitating the recruitment of the DNA methyltransferase DOMAINS REARRANGED METHYLTRANSFERASE 2 (DRM2)

et al. 2013). Similarly, salt stress has been found to induce H3K27me3 demethylation at ion transporter genes, facilitating their activation in *Arabidopsis* (Yamaguchi and Ito 2021). Further, an increased H3K4me2 and H4K5ac at heat shock protein genes has been found to increase thermotolerance in *G. hirsutum* under heat-stressed conditions (He et al. 2022). However, drought-tolerant wheat cultivars showed a reduced H3K27me3 at ABA-responsive genes, correlating with enhanced expression and stress adaptation under drought conditions (Naderi et al. 2024; Wang et al. 2022b; Temel et al. 2017). Additionally, cold stress triggers H3K9ac enrichment at genes involved in membrane stabilisation and ROS detoxification, which shows variation in histone methylation patterns among tolerant and sensitive genotypes, suggesting epigenetic markers for breeding (Pradhan et al. 2019). Cumulatively, histone modifications are reversible and circumstantial, allowing rapid adaptation of crop plants via activating (H3K4me3, H3K9ac) or repressing (H3K27me3, H3K9me2) stress-responsive genes, thus contributing to epigenetic priming of crop plants. Additionally, DNA methylation, along with ncRNAs, forms a multi-layered epigenetic network that increases the stress tolerance in crop plants.

7.3 Epigenetic Regulation of Crop Plant Responses to Abiotic Stresses

Abiotic stressors, with rapid climate shifting, are negatively affecting the crops' growth, yield and nutritional quality. Epigenetic mechanisms play a crucial role in regulating gene expression, thereby enabling a dynamic and reversible adaptation to abiotic stresses such as drought, salinity, heat and cold. The overexpression of abiotic stress-responsive genes and the induction of DNA repair pathway enzymes are also facilitated by the same RdDM pathway under abiotic stress conditions. DNA damage caused by ROS induces DNA demethylation at TEs and the promoter region of stress-responsive genes via DNA repair pathway enzymes (DME, ROS, DML 2/3), further increasing genome instability and susceptibility at specific genomic loci. Further, SHH1 and CLSY1-4, chromatin reader proteins, detect and bind to the histone marks (H3K9me2/H3K4me3) and recruit RNA POL IV/V and other TFs in the chromatin region to initiate siRNA biogenesis after transcription (Huang et al. 2016; Law et al. 2011). Long non-coding double-stranded RNAs (lnc-ssRNAs) generated by single-stranded RNAs via RDR2 were used as precursor molecules to produce small 24nt siRNAs via DCL3 enzyme. Finally, the siRNAs-guided DRM1/2 induce methylation and further establish a self-sustaining de novo methylation loop to silence the stress-responsive genes and TEs at specific genomic loci, thus increasing abiotic tolerance and genome stability (Wei et al. 2014; Le et al. 2014) (Fig. 7.2).

7.3.1 Epigenetic Regulation in the Drought Stress Response

Drought is among the most detrimental environmental factors, severely limiting crop growth and yield. Many studies have revealed drought-induced DNA methylation dynamics in various crop species, including wheat, maize, rice and barley (Chwialkowska et al. 2016; Naderi et al. 2024; Wang et al. 2021). For instance, in maize, exposure to drought stress leads to an overall increase in DNA methylation levels, with drought-tolerant inbred lines exhibiting a more stable methylome compared to drought-sensitive ones (Wang et al. 2021). Similarly, in barley, drought induces a comparable number of methylation and demethylation events in leaves, but a higher frequency of methylation events in roots, indicating that organ-specific methylome modulation may play a crucial role in conferring drought stress tolerance (Chwialkowska et al. 2016).

Several epigenomic studies has demonstrated the prominent regulatory role of DNA and histone methylation pattern in rice (*Oryza sativa* L.) genome under drought conditions (Shaik and Ramakrishna 2012; Ou et al. 2009; Zheng et al. 2013a). Epimutations, alteration in methylation pattern, triggered by the short-term drought stress, are transgenerational and play a critical role in drought tolerance (Zheng et al. 2017). Studies have found epimutations mostly in promoter regions, intergenic regions and TEs in the plants under drought conditions, where drought triggers the hypomethylation in CG and CHG context in intergenic regions while hypermethylation in TEs (Ou et al. 2009). This might be due to the increased

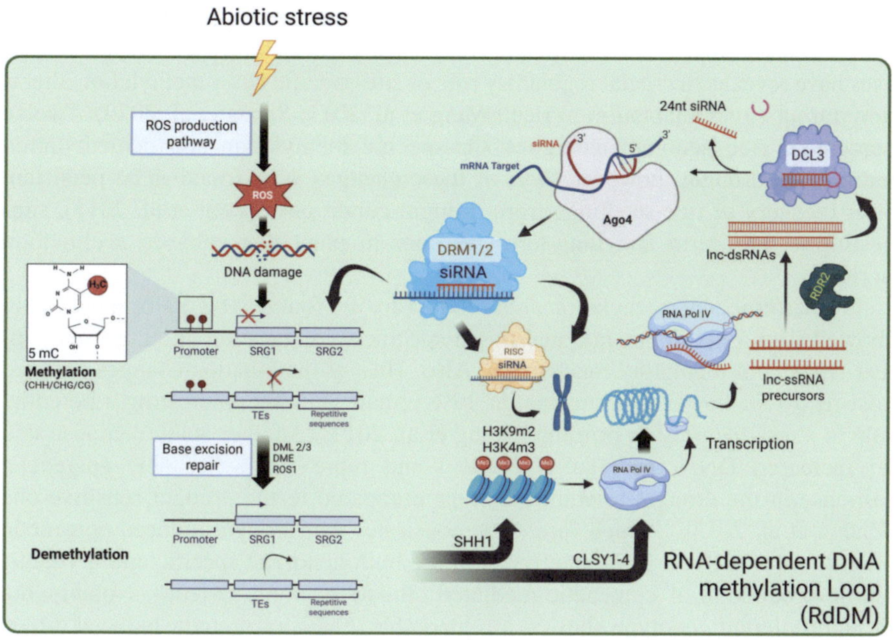

Fig. 7.2 The schematic representation illustrates a stress-responsive RNA-directed DNA methylation (RdDM) loop in plants under abiotic stress conditions. Plants initiate an epigenetic defence against abiotic stress through a tightly coordinated interplay of DNA methylation, demethylation and the RdDM pathway. ROS-induced DNA damage triggers demethylation via repair enzymes, while chromatin readers (SHH, CLSY1-4) recruit RNA POL IV/V to initiate siRNA production. These siRNAs guide DRM1/2 to re-establish methylation, silencing stress-responsive genes and TEs, thereby enhancing genome stability and abiotic stress tolerance in crop plants

sensitivity of genes associated with DNA methyltransferases and the proteins related to siRNA, which further serve as transgenerational epigenetic markers involving in drought adaptations (Ou et al. 2009; Kumar et al. 2023). Bioinformatics approach has shown the downregulation of drought responsive genes involve in chromatin remodelling in rice plants under drought which further implies the suppression of chromatin plasticity and TEs to maintain genome integrity (Shaik and Ramakrishna 2012). Further studies have revealed the role of CG island in the promoter region of drought-responsive genes. Certain genes such as HSP70 and RH25 lack the CpG islands in their promoter regions and are found to be upregulated, which further suggests the acclimation response of rice in drought conditions (Sapna et al. 2020). In contrast, histone methylation (such as H3K4me3) has found to be inversely involved in gene expression and affects a small proportion of drought-responsive genes (Zong et al. 2013).

These altered methylation patterns at 5mC sites accumulate directionally in the intergenic regions and TEs sites across generations and provide a heritable long-term adaptation and evolution under changing environmental conditions (Zheng et al. 2017). Contrasting rice cultivar studies highlight the pivotal role of genotype-specific DNA methylation modulations in rice drought tolerance (Wang et al. 2011; Karan et al. 2012). Methylation-sensitive amplified polymorphism analyses have revealed a crucial regulatory role of site-specific DNA methylation alteration during stress adaptation in rice (Wang et al. 2011; Sapna et al. 2020). Twelve percent of rice genome undergoes site-specific methylation and demethylation induced by drought; however, 29% of these changes were found to be persistent after recovery of rice seedlings from drought conditions (Wang et al. 2011), suggesting a short-term and long-term memory in plant that enhance acclimation in rice.

Furthermore, the methylation alterations were not found to be uniform but rather showed strong developmental and tissue specificity, suggesting an optimised context-dependent regulatory mechanism. Also, 70% of methylation changes reverted post-recovery; however, a substantial 29% remained stable, indicating a potential role in stress memory or priming (Wang et al. 2011). Another study demonstrated an increased DNA methylation (~39%) and more stable heritable epigenetic response in the drought-tolerant genotype compared to the drought-sensitive one (Zheng et al. 2013b), which further demonstrates that drought-induced epigenetic alterations, mainly in DNA methylation, are both genotype specific and heritable, aiding production of epigenetic-mediated climate-resilient varieties. Notably, the drought-tolerant genotype showed a remarkable 29.8% of drought-induced differentially methylated loci (DML) transmitted across generations compared to the sensitive genotype (3.2%), indicating more robust entrainment capacity in the drought-tolerant genotype (Zheng et al. 2013b; Wang et al. 2016).

These studies emphasise DNA methylation as a predominant and multifaceted epigenetic mechanisms in rice under drought stress responses. Notably, cytosine methylation patterns serve as reliable molecular imprints that discriminate between drought-tolerant and drought-sensitive genotypes, with hypomethylation consistently associated with enhanced tolerance. Moreover, these studies also fortify the concept of epigenetic memory; rather than transient methylation changes, they serve as a reliable molecular marker in selecting or engineering drought-tolerant phenotypes. Moreover, these findings support the concept that environmentally responsive epigenetic modifications, especially those that persist beyond stress exposure, are integral to adaptive responses and may serve as promising targets for breeding drought-resilient rice varieties.

7.3.2 Epigenetic Regulation in the Salt Stress Response

Soil salinity is a major abiotic stressor severely impacting global agricultural productivity, with over 20% of irrigated land affected worldwide. Epigenetic regulation has emerged as a central mechanism in crop adaptation to salt stress, with recent studies converging on its role in transcriptional reprogramming, stress memory and phenotypic plasticity. Studies have demonstrated significant alterations in DNA methylation patterns and histone marks, which cumulatively regulate gene expression in wheat and rice under salinity stress conditions (Hoseini et al. 2024; Wang et al. 2014; Kumar et al. 2017). Studies have also demonstrated CHH/CHG methylation in TEs upstream of the *OsHKT1;5* gene in rice which enhances the ion homeostasis under salt stress (Wang et al. 2020) and further identified a methylation reader chaperone TF complex which activates this key transporter (Wang et al. 2020), suggesting a predominant role of DNA methylation-mediated epigenetic regulation mechanism among the other epigenetic mechanisms. Similarly, soybean roots exhibit methylation dynamics under salinity, with stress-induced demethylation linked to transcriptional activation. However, studies have also demonstrated non-CG methylation patterns in barley and suggesting tissue-level epigenetic plasticity (Konate et al. 2018). The study also highlighted the critical role of cytosine methylation in modulating the expression of WRKY TFs (WRKY50 and WRKY72) in rice (*Oryza sativa* L.) seedlings under salt stress conditions. Statistical analysis of DNA methylation pattern revealed that methylation levels of these genes exhibited a negative correlation with gene expression at higher salt concentrations (90 mM and 150 mM), indicating a possible epigenetic repression mechanism (Çelik et al. 2019). However, the methylation-pattern correlation at 210 mM NaCl appeared more complex, but overall findings suggest that differential methylation of WRKY genes may optimise plant responses to salt stress (Çelik et al. 2019). However, studies also emphasise that many epigenetic alterations (cytosine methylation) are not directed and may arise as non-targeted responses to environmental stimuli. Furthermore, the inherent genetic variations for salt tolerance in rice germplasm appear to function independently of DNA methylation patterns, implying that epigenetic marks induced by abiotic stress are subsequently fixed through natural selection, rather than completely on the basis of stress adaptation (Karan et al. 2012). However, studies have demonstrated that methylome flexibility is genotype and tissue specific and is a key determinant of salt tolerance in crop plants. By virtue of ELISA-based quantification of 5mC and in situ imaging, studies have demonstrated that salt stress elicited tissue-specific methylation changes exhibiting comparatively rapid demethylation in the tolerant variety in rice, apparently facilitated by upregulation of DNA demethylases (Ferreira et al. 2015). However, sensitive varieties have shown lower methylation plasticity, probably due to synchronised induction of both methyltransferases and demethylases, which further leads to the attenuation of epigenomic alterations.

Additionally, studies revealed that sugar beet cultivars show genotype-specific miRNA responses to salt stress, implicating sRNAs in cultivar-dependent salt tolerance. Therefore, similar to drought stress, the epigenetic mechanism of DNA methylation pattern variation, along with the engagement of ncRNA molecules, is predominant as well as is tissue and genotype specific. Further, DGE analysis has shown that cytochrome P450 genes in cotton are epigenetically regulated under salinity, with methylation and ncRNA interactions modulating stress responses (Sun et al. 2021). Additionally, lncRNAs regulate salt tolerance by recruiting HDACs, which have been found to repress the transcription of *INDETERMINATE SPIKELET1* gene in (Cheng et al. 2018). Besides, studies have also emphasised the role of small RNAs, particularly miRNAs and siRNAs, in transcriptional- and post-transcriptional-level regulations of salt-responsive genes across multiple crop species (Kumar et al. 2017).

In contrast to DNA methylation, histone modifications have found to play a significant role in crops under salt stress conditions. The overexpression of the H3K4 demethylase JMJ15 has been found to enhance salt tolerance in *Arabidopsis* by repressing stress-inducible genes marked by H3K4me3 (Shen et al. 2014). Studies have demonstrated that genes involved in cellulose biosynthesis and cell wall integrity in maize were found to be regulated by histone acetyltransferase GCN5 (Hu et al. 2022; Zheng et al. 2019). In *Arabidopsis*, the light-dependent transcriptional memory is primarily maintained by a consistent histone modification, further establishing a link between environmental factors and epigenetic memory. In *Arabidopsis*, the light-dependent transcriptional memory is primarily maintained by a consistent histone modification, further establishing a link between environmental factors and epigenetic memory (Feng et al. 2016). Epigenetic stress memory in the form of histone modifications, DNA methylations and small RNAs facilitates a long-term and a transgenerational acclimation to the crop plants.

These epigenomic studies further elucidate the significant role of epigenetic mechanisms, DNA methylations and ncRNA regulation in plant adaptation to climatic stressors. Moreover, studies involved in "priming" in response to abiotic stressors showing transgenerational resilience thus offer a prospective framework for epigenome-assisted breeding of salt-tolerant crops.

7.3.3 Epigenetic Regulation in Heat Stress Response

Heat stress negatively impacts the reproduction and causes sterility during the anthesis stage in crop plants, which decreases the plant's survival and, most importantly, yield. Epigenetic regulations help to maintain homeostasis and ensure reproductive success in crop plants in warming conditions. Epigenetic mechanisms involved in heat stress response include histone modifications, DNA methylation and small non-coding RNAs, which interactively alter the gene expression in crops under heat stress conditions. Studies have found alterations in DNA methylation patterns associated to the genes involved in heat tolerance and adaptive response. For instance, a temporarily decreased methylation level associated with increased gene activation has identified in the genes involved with heat response in *Arabidopsis* under heat stress (Korotko et al. 2021; Naydenov et al. 2015). These modifications help plants acclimate and further transfer the methylation pattern as a memory to the next generation, giving offsprings a higher heat tolerance.

Histone modifications lead to gene activation or suppression via acetylation or demethylation of the chromatin regions under environmental stimuli. Studies have found the overexpression of acetyltransferases and demethylases genes in plants under heat stress conditions, which further causes the overexpression of heat-responsive genes and thus leads to plant survival (Chen et al. 2024; Yang et al. 2023). Studies have identified the overexpression of heat shock factor, i.e. HSF2A, after the removal of the methyl group from the histone mark, H3K27me3, which is known as the key regulator of stress memory under heat stress conditions (Liu et al. 2019; He et al. 2022).

Non-coding RNAs such as small RNAs, miRNAs and siRNAs are also involved in maintaining homeostasis via silencing or degrading the mRNAs, thus interrupting the production of certain heat-sensitive proteins under heat stress conditions and enhancing thermotolerance (He et al. 2019; Liu et al. 2017). For instance, studies have identified the over-accumulation of certain miRNAs, which help regulate gene activity under higher temperature conditions (Li et al. 2021). Studies using the model crop species like *Arabidopsis* help researchers deeply understand the molecular mechanisms of epigenetic regulations in crop plants under heat stress conditions, thus enabling climate-smart breeding strategies in food crops. Overall, epigenetic regulations support the long-term survival of crop plants under integrative abiotic stressors (heat, cold, salinity and drought) with rapidly changing climatic conditions, thus ensuring food security in the face of climate change.

Conclusively, histone modifications, DNA methylation and small RNA molecules interactively and solely enhance thermotolerance and generate stress memory, which is a resilient trait passed to the next generation. Deciphering these mechanisms might help develop robust crop varieties in the face of rising global warming climatic conditions.

7.3.4 Epigenetic Regulation in the Cold Stress Response

Epigenetic mechanisms, DNA methylation, histone modifications and non-coding RNAs also play a pivotal role in cold stress responses by regulating gene expression and also contribute to generating stress memory in plants under cold stress. Plants' exposure to non-lethal low temperatures triggers physiological as well as molecular alterations which affect membrane fluidity, gene expression patterns and the production of cold-responsive antifreeze-proteins. Chromatin remodelling and histone modifications are key components of this stress memory, playing roles in both heat and cold stress responses (Park et al. 2018). A deeper understanding of how these epigenetic processes function under cold conditions is crucial for breeding or engineering crops with enhanced cold tolerance, an urgent need in the context of climate change and global food security. Targeted genetic strategies that modify epigenetic regulators or enhance the expression of protective proteins like antifreeze proteins show great promise in improving plant resilience. Recent review studies, such as by Abdulraheem et al. (2024), have comprehensively discussed the role of TFs, secondary messengers like calcium ions and ROS and protective proteins, including antifreeze proteins and dehydrins. They also highlight the significance of physiological responses such as changes in membrane fluidity, accumulation of osmoprotectants and hormone signalling in building cold tolerance.

DNA methylation pattern alterations are specific to genotype, tissue and developmental stage of the plant under cold stress conditions. Moreover, studies have demonstrated the transgenerational inheritance of these altered patterns in rice and suggested that pattern alterations might be inherited through mitosis (Pan et al.

2011). Another study demonstrated a higher methylation pattern alteration in petioles than in the leaf blades in cassava, which further showed the tissue specificity of methylation pattern alterations under cold stress conditions (Yu et al. 2024). These cold-induced differentially methylated regions (DMRs) are associated with gene expression alteration and activation of transcription factors. Studies have shown a complex regulatory network of DNA methylation-mediated epigenetic regulations in plants under multifactor abiotic stressors (Latowski et al. 2011). These alterations are heritable and provide a long-term adaptation to the crop plant and therefore are critical for improving tolerance and crop productivity.

Histone modifications significantly affect the plant's response to cold stress by enhancing chromatin accessibility at target loci and inducing bivalent histone modifications, particularly H3K4me3 and H3K27me3, in actively transcribed genic regions in crop plants under cold stress (Zheng et al. 2019). Key histone-modifying enzymes that regulate stress-responsive genes have been identified (Lim et al. 2020), and further studies demonstrate that histone variants and post-translational modifications can alter chromatin conformation, thereby impacting transcription under stress conditions.

Non-coding RNAs (ncRNAs) also contribute to cold stress regulation via epigenetic mechanisms. In *Arabidopsis*, miRNAs such as miR397, miR408 and miR402, and in rice, Osa-miR156, Osa-miR319 and Osa-miR528, have been identified as key regulators during cold stress responses (Maryan et al. 2023; Kumar et al. 2024). Long non-coding RNAs (lncRNAs) like SVALKA have also been implicated in cold-responsive signalling pathways. These ncRNAs influence gene expression at multiple levels, viz. transcriptional, post-transcriptional and translational, and can also modify histones and chromatin organisation (Di et al. 2014). As a result, they represent promising targets for developing climate-resilient crops through molecular breeding or biotechnological approaches.

7.4 Epiregulation of Abiotic Stress Memory and Acclimation

Epigenetic alterations are stored in the form of stress memory, a process also called "Priming", which enables the plants to acclimatise to pre-exposed stressed conditions (Sun et al. 2021). Priming enables the plants to respond faster or stronger during subsequent stress exposures. These alterations, DNA methylation, histone modifications and gene expressions in the form of memory, are short term and permanent, depending on the stress exposure and duration (Forestan et al. 2019). Therefore, researchers have categorised stress memory into three groups: (1) transcriptional memory, which maintains altered gene expression levels during the plant recovery phase; (2) epigenetic memory, which is regulated by epigenetic or methylation marks; and (3) delayed memory, which shows alterations only upon similar stress re-exposure (Forestan et al. 2019). All these priming strategies trigger biochemical, physiological and molecular and epigenetic alterations for short term or might be persistent in nature. Molecular mechanisms such as the RNA-directed DNA methylation pathway and alternative splicing have been found to play a key role in the drought memory response in plants. For instance, flowering time in maize was affected by drought stress, particularly involving a combination of epigenetic, transcriptional and delayed memory mechanisms (Forestan et al. 2019).

Recent studies on maize have shown the epigenetic basis of drought stress memory in crop plants. For instance, histone modifications (H3K4me3) led to the enrichment of heat responsive genes, *ZmHSFA2* and *ZmDREB2A*, which are associated with enhanced thermotolerance in maize under heat stress (Ruan et al. 2024). Similarly, priming to the cold stress led to the sustained expression of *ZmcBF* gene after histone acetylation. The phytohormone signalling pathway is also regulated via stress priming. Stress-induced epimutations in the regulatory regions of *ZmPP2C* and *ZmSnRK2* have also been found to play a key role in ABA signalling pathways. Similarly, drought priming induced in sunflower (*Helianthus annuus*) was linked to the DNA methylation alterations in the genes such as *HaDREB2* and *HaNCED*, which further enhanced the ABA-mediated response. Furthermore, the repressive histone mark H3K27me3 induced after priming has been found to contribute to long-term stress memory and maintain drought-responsive gene expression. Studies have also shown the genotype-specific methylation pattern in olive (*Olea europaea*) under salt stress, which has been found to alter the *OeNHX1* and *OeSOS1* gene expression. Finally, these epigenetic alterations align with physiological traits such as stomatal conductance and facilitate the integration of epigenomic and phenotypic data for breeding programs.

Studies have observed a persistent histone acetylation and DNA demethylation pattern in *Arabidopsis* at loci such as HSFA2 and RD29A, along with transgenerational memory effects across three generations under combined heat and drought stress conditions (Li et al. 2021). Similarly, a heritable demethylation in the *FWA* locus, which plays a role in regulating flowering time, was observed under stress conditions. This exemplifies how epigenetic modifications can influence developmental timing in response to environmental cues. ncRNAs also play a vital role in regulating plant stress responses through epigenetic pathways. For instance, miR398 is downregulated during oxidative stress, allowing increased expression of its targets CSD1 and CSD2, which are involved in detoxifying reactive oxygen species. During drought, miR169 regulates stress responses by modulating NF-YA transcription factors. Additionally, siRNAs guide DNA methylation via the RdDM pathway, as seen in rice, where siRNAs target *OsSRO1c*, a gene involved in reactive oxygen species signalling. lncRNAs like LDMAR in rice also influence important traits such as photoperiod-sensitive male sterility under heat stress, demonstrating their involvement in chromatin remodelling and transcriptional memory (Ding et al. 2012).

Despite the potential of epigenetic memory to improve crop stress resilience, several challenges must be addressed. These include the stability of epigenetic marks across different environments, the need for high-resolution mapping techniques like bisulfite sequencing and the importance of integrating epigenomic data with transcriptomic and metabolomic analyses. Overcoming these hurdles is essential to fully harness the potential of epigenetics in crop improvement. In conclusion, epigenetic stress memory offers a powerful avenue for enhancing plant adaptation to abiotic stresses.

7.5 Agricultural Implications of Epigenetic Memory in Crop Plants

Epigenetic stress memory in crop plants offers a transformative approach to improve crop tolerance under abiotic stress conditions. This concept helps in understanding the plant's survival and adaptation to future stress conditions. The concept of priming and transgenerational stress memory helps understand the epigenetic pattern alterations in the subsequent generations, hence can help target the molecular mechanisms for the production of abiotic stress-tolerant crop varieties. Through priming, we can enhance the physiological and transcriptional flexibility of future generations under abiotic stress conditions. The transgenerational memory presents a robust opportunity for producing climate-resilient crop varieties. These modifications persist across generations, contributing to improved water-use efficiency and ABA-regulated stress responses (Kumar et al. 2024). In maize (*Zea mays*), similar epigenetic responses have been observed. Under heat stress, enrichment of the H3K4me3 histone mark at genes like *ZmHSFA2* and *ZmDREB2A* enhances thermotolerance. Meanwhile, cold priming helps maintain the expression of *ZmCBF* genes through histone acetylation, preparing the plant for future cold events (Ruan et al. 2024). These epigenetic marks not only modulate gene expression but also influence key agronomic traits such as flowering time, root architecture and ion transport, as seen in olive under salt stress condition, where genotype-specific methylation of *OeNHX1* and *OeSOS1* correlates with stomatal conductance and osmotic balance. The integration of epigenomic data with physiological and transcriptomic profiles enables the identification of stress-resilient genotypes, accelerating marker-assisted selection and epigenome-assisted breeding. Moreover, non-coding RNAs such as *miR398* and *miR169* and lncRNAs like *LDMAR* play a pivotal role in optimisation of stress responses and establishing transcriptional memory, particularly under drought, oxidative and heat stress (Ruan et al. 2024; Ding et al. 2012). The advent of CRISPR/dCas9-based epigenome editing offers a targeted approach to modulate stress-responsive loci without altering the DNA sequence, exemplified by demethylation of *OsDREB1* to enhance cold tolerance. However, challenges remain in ensuring the stability and heritability of epigenetic marks across diverse environments and generations.

From an agronomic perspective, leveraging epigenetic memory can reduce reliance on chemical inputs, improve yield stability under fluctuating climates and support sustainable agriculture. It also opens roadways for developing crops with enhanced nutrient use efficiency, disease resistance and stress-adaptive phenotypes without transgenic interventions. Significantly, the temporal and stress-specific nature of epigenetic memory weathering through cis-priming or trans-priming offers flexibility in designing priming strategies tailored to regional agroecological conditions. While model plants like *Arabidopsis* have laid the groundwork for understanding epigenetic regulation, crop-specific studies are crucial for translating these insights into field applications. In conclusion, epigenetic stress memory represents a transformative tool in modern agriculture, bridging molecular biology with practical crop improvement. Its integration into the breeding pipeline, supported by robust molecular diagnostics and phenotypic screening holds the key to cultivating resilient crops capable of thriving in the face of climate variability and resource constraints.

7.6 Conclusions

Epigenetic mechanisms, specifically DNA methylation, histone modifications and non-coding RNAs, play a dynamic regulatory role that supports crop stress response. These processes also contribute to transgenerational stress memory and mediate rapid transcriptional reprogramming in response to biotic and abiotic stresses, providing a potential pathway for heritable resilience. Although there has been considerable progress, we still don't fully understand how specific epigenetic changes lead to particular traits, especially in open field conditions. To unravel the complex, context-dependent roles of epigenetic regulators, we need integrated multi-omics studies combined with detailed epigenome mapping and experimental validation. It is also crucial to explore how epigenetic flexibility interacts with the plant's genetic makeup to effectively use these mechanisms for crop improvement. Going forward, research should focus on developing reliable epigenetic markers and precise editing tools that can fine-tune stress-related genes without negatively affecting yield or quality. In summary, the epigenetic regulation represents a promising frontier in plant stress biology; its translation into agronomic innovation demands a deeper mechanistic understanding, robust field validation and ethical consideration of epigenome editing technologies.

Acknowledgements This work was supported by the Anusandhan National Research Foundation (ANRF) under the Startup Research Grant (SRG) with award number SRG/2023/000320.

References

Abdulraheem MI, Xiong Y, Moshood AY, Cadenas-Pliego G, Zhang H, Hu J (2024) Mechanisms of plant epigenetic regulation in response to plant stress: recent discoveries and implications. Plants 13(2):163

Bond DM, Baulcombe DC (2015) Epigenetic transitions leading to heritable, RNA-mediated de novo silencing in *Arabidopsis thaliana*. Proc Natl Acad Sci USA 112(3):917–922

Çelik Ö, Meriç S, Ayan A, Atak Ç (2019) Epigenetic analysis of WRKY transcription factor genes in salt stressed rice (*Oryza sativa* L.) plants. Environ Exp Bot 159:121–131

Chellappan P, Xia J, Zhou X, Gao S, Zhang X, Coutino G, Vazquez F, Zhang W, Jin H (2010) siRNAs from miRNA sites mediate DNA methylation of target genes. Nucleic Acids Res 38(20):6883–6894

Chen J, Lu J, Liu J, Fang J, Zhong X, Song J (2023) DNA conformational dynamics in the context-dependent non-CG CHH methylation by plant methyltransferase DRM2. J Biol Chem 299(12):105433

Chen Z, Xu Q, Wang J, Zhao H, Yue Y, Liu B, Xiong L, Zhao Y, Zhou DX (2024) A histone deacetylase confers plant tolerance to heat stress by controlling protein lysine deacetylation and stress granule formation in rice. Cell Rep 43(9):114642

Cheng X, Zhang S, Tao W, Zhang X, Liu J, Sun J, Zhang H, Pu L, Huang R, Chen T (2018) INDETERMINATE SPIKELET1 recruits histone deacetylase and a transcriptional repression complex to regulate rice salt tolerance. Plant Physiol 178(2):824–837

Chwialkowska K, Nowakowska U, Mroziewicz A, Szarejko I, Kwasniewski M (2016) Water-deficiency conditions differently modulate the methylome of roots and leaves in barley (*Hordeum vulgare* L.). J Exp Bot 67(4):1109–1121

Dai Y, Ni Z, Dai J, Zhao T, Sun Q (2005) Isolation and expression analysis of genes encoding DNA methyltransferase in wheat (*Triticum aestivum* L.). Biochim Biophys Acta Gene Struct Expr 1729(2):118–125

Daxinger L, Kanno T, Bucher E, van der Winden J, Naumann U, Matzke AJM, Matzke M (2009) A stepwise pathway for biogenesis of 24-nt secondary siRNAs and spreading of DNA methylation. EMBO J 28(1):48–57

Di C, Yuan J, Wu Y, Li J, Lin H, Hu L, Zhang T, Qi Y, Gerstein MB, Guo Y, Lu ZJ (2014) Characterization of stress-responsive lncRNAs in *Arabidopsis thaliana* by integrating expression, epigenetic and structural features. Plant J 80(5):848–861

Diallo AO, Ali-Benali MA, Badawi M, Houde M, Sarhan F (2012) Expression of vernalization responsive genes in wheat is associated with histone H3 trimethylation. Mol Gen Genomics 287:575–590

Ding J, Lu Q, Ouyang Y, Zhang Q (2012) A long noncoding RNA regulates photoperiod-sensitive male sterility, an essential component of hybrid rice. Proc Natl Acad Sci USA 109(7):2654–2659

Du Z, Li H, Wei Q, Zhao X, Wang C, Zhu Q, Yi X, Xu W, Liu XS, Jin W, Su Z (2013) Genome-wide analysis of histone modifications: H3K4me2, H3K4me3, H3K9ac, and H3K27ac in *Oryza sativa* L. Japonica. Mol Plant 6(5):1463–1472

Faivre L, Kinscher NF, Kuhlmann AB, Xu X, Kaufmann K, Schubert D (2024) Cold stress induces rapid gene-specific changes in the levels of H3K4me3 and H3K27me3 in *Arabidopsis thaliana*. Front Plant Sci 15:1390144

Fang J, Jiang J, Leichter SM et al (2022) Mechanistic basis for maintenance of CHG DNA methylation in plants. Nat Commun 13:3877

Feng XJ, Li JR, Qi SL, Lin QF, Jin JB, Hua XJ (2016) Light affects salt stress-induced transcriptional memory of P5CS1 in *Arabidopsis*. Proc Natl Acad Sci USA 113(51):E8335–E8343

Ferreira LJ, Azevedo V, Maroco J, Oliveira MM, Santos AP (2015) Salt tolerant and sensitive rice varieties display differential methylome flexibility under salt stress. PLoS One 10(5):e0124060

Forestan C, Farinati S, Zambelli F, Pavesi G, Rossi V, Varotto S (2019) Epigenetic signatures of stress adaptation and flowering regulation in response to extended drought and recovery in *Zea mays*. Plant Cell Environ 43(1):55–75

Gahlaut V, Samtani H, Khurana P (2020) Genome-wide identification and expression profiling of cytosine-5 DNA methyltransferases during drought and heat stress in wheat (*Triticum aestivum*). Genomics 112(6):4796–4807

Gallego-Bartolomé J, Liu W, Kuo PH, Park SY, Chory J, Jacobsen SE (2019) Co-targeting RNA polymerases IV and V promotes efficient de novo DNA methylation in *Arabidopsis*. Cell 176(5):1068–1082.e19

Gardiner LJ, Quinton-Tulloch M, Olohan L, Price J, Hall N, Hall A (2015) A genome-wide survey of DNA methylation in hexaploid wheat. Genome Biol 16(1):273

Garg R, Kumari R, Tiwari S, Goyal S (2014) Genomic survey, gene expression analysis and structural modeling suggest diverse roles of DNA methyltransferases in legumes. PLoS One 9(2):e88947

Gouil Q, Baulcombe DC (2016) DNA methylation signatures of the plant chromomethyltransferases. PLoS Genet 12(12):e1006526

Gu C, Pei MS, Guo ZH et al (2024) Multi-omics provide insights into the regulation of DNA methylation in pear fruit metabolism. Genome Biol 25:70

Guo H, Wu T, Li S, He Q, Yang Z, Zhang W, Gan Y, Sun P, Xiang G, Zhang H, Deng H (2019) The methylation patterns and transcriptional responses to chilling stress at the seedling stage in rice. Int J Mol Sci 20(20):5089

He J, Jiang Z, Gao L, You C, Ma X, Wang X, Xu X, Mo B, Chen X, Liu L (2019) Genome-wide transcript and small RNA profiling reveals transcriptomic responses to heat stress. Plant Physiol 181(2):609–629

He S, Zhang Y, Wang J, Wang Y, Ji F, Sun L, Zhang G, Hao F (2022) H3K4me2, H4K5ac and DNA methylation function in short- and long-term heat stress responses through affecting the expression of the stress-related genes in *Gossypium hirsutum*. Environ Exp Bot 194:104699

He C, Bi S, Li Y, Song C, Zhang H, Xu X, Li Q, Saeed S, Chen W, Zhao C, Lan C, Su H, Mao H, Yan W (2024) Dynamic atlas of histone modifications and gene regulatory networks in endosperm of bread wheat. Nat Commun 15(1):9572

Hoseini M, Arzani A, Saeidi G, Araniti F (2024) Agro-physiological and DNA methylation responses to salinity stress in wheat (*Triticum aestivum* L.), *Aegilops cylindrica* host, and their introgressed lines. Plants 13(19):2673

Hou J, Ren R, Xiao H, Chen Z, Yu J, Zhang H, Shi Q, Hou H, He S, Li L (2021) Characteristic and evolution of HAT and HDAC genes in Gramineae genomes and their expression analysis under diverse stress in *Oryza sativa*. Planta 253(3):72

Hu D, Li R, Dong S, Zhang J, Zhao B, Ren B, Ren H, Yao H, Wang Z, Liu P (2022) Maize (*Zea mays* L.) responses to salt stress in terms of root anatomy, respiration and antioxidative enzyme activity. BMC Plant Biol 22(1):602

Huang W, Xian Z, Hu G, Li Z (2016) SlAGO4A, a core factor of RNA-directed DNA methylation (RdDM) pathway, plays an important role under salt and drought stress in tomato. Mol Breed 36:28

Karan R, DeLeon T, Biradar H, Subudhi PK (2012) Salt stress induced variation in DNA methylation pattern and its influence on gene expression in contrasting rice genotypes. PLoS One 7(6):e40203

Kaur R, Singh D, Singh PK, Kaur G, Shah K, Pandey H, Devi SG, Pandey AK, Rishi V (2024) Identifying the genetic and epigenetic basis for asymmetric bZIP expression in temperature-stressed bread wheat. Plant Growth Regul 103:581–595

Konate M, Wilkinson MJ, Mayne BT, Pederson SM, Scott ES, Berger B, Rodriguez Lopez CM (2018) Salt stress induces non-CG methylation in coding regions of barley seedlings (*Hordeum vulgare*). Epigenomes 2(2):12

Korotko U, Chwiałkowska K, Sańko-Sawczenko I, Kwasniewski M (2021) DNA demethylation in response to heat stress in Arabidopsis thaliana. Int J Mol Sci 22(4):1555

Kumar S, Beena AS, Awana M, Singh A (2017) Salt-induced tissue-specific cytosine methylation downregulates expression of HKT genes in contrasting wheat (*Triticum aestivum* L.) genotypes. DNA Cell Biol 36(4):283–294

Kumar S, Seem K, Mohapatra T (2023) Biochemical and epigenetic modulations under drought: remembering the stress tolerance mechanism in rice. Life 13(5):1156

Kumar S, Seem K, Kumar S, Singh A, Krishnan SG, Mohapatra T (2024) DNA methylome analysis provides insights into gene regulatory mechanism for better performance of rice under fluctuating environmental conditions: epigenomics of adaptive plasticity. Planta 259:4

Latowski D, Kuczyńska P, Strzałka K (2011) Xanthophyll cycle—a mechanism protecting plants against oxidative stress. Redox Rep 16(2):78–90

Law JA, Vashisht AA, Wohlschlegel JA, Jacobsen SE (2011) SHH1, a homeodomain protein required for DNA methylation, as well as RDR2, RDM4, and chromatin remodeling factors, associate with RNA polymerase IV. PLoS Genet 7(7):e1002195

Law JA, Du J, Hale CJ, Feng S, Krajewski K, Palanca AMS, Strahl BD, Patel DJ, Jacobsen SE (2013) Polymerase IV occupancy at RNA-directed DNA methylation sites requires SHH1. Nature 498:385–389

Le TN, Schumann U, Smith NA et al (2014) DNA demethylases target promoter transposable elements to positively regulate stress responsive genes in *Arabidopsis*. Genome Biol 15:458

Li X, Wang X, He K, Ma Y, Su N, He H, Stolc V, Tongprasit W, Jin W, Jiang J et al (2008) High-resolution mapping of epigenetic modifications of the rice genome uncovers interplay between DNA methylation, histone methylation, and gene expression. Plant Cell 20(2):259–276

Li X, Zhu J, Hu F, Ge S, Ye M, Xiang H, Zhang G, Zheng X, Zhang H, Zhang S, Li Q, Luo R, Yu C, Yu J, Sun J, Zou X, Cao X, Xie X, Wang J, Wang W (2012) Single-base resolution maps of cultivated and wild rice methylomes and regulatory roles of DNA methylation in plant gene expression. BMC Genomics 13(1):300

Li Q, Gent JI, Zynda G, Song J, Hirsch CN, Dawe RK, Madzima TF, McGinnis KM, Lisch D, Schmitz RJ, Vaughn MW, Springer NM (2015) RNA-directed DNA methylation enforces boundaries between heterochromatin and euchromatin in the maize genome. Proc Natl Acad Sci USA 112(47):14728–14733

Li N, Euring D, Cha JY, Lin Z, Lu M, Huang LJ, Kim WY (2021) Plant hormone-mediated regulation of heat tolerance in response to global climate change. Front Plant Sci 11:627969

Li G, Cao R, Ma L, Jiao G, Chen P, Dong N, Li X, Duan Y, Li X, Zhu M, Shao G, Sheng Z, Hu S, Tang S, Wei X, Yu Y, Hu P (2023) OsLEA1b modulates starch biosynthesis at high temperatures in rice. Plants 12(23):4070

Li J, Liu H, Qian H, Lu S, Wu Y, Hua J, Zou B (2025) Chromatin accessibility mediated by CHROMATIN REMODELING 11 promotes chilling tolerance in rice. Plant Physiol 197(1):kiaf018

Lim CJ, Park J, Shen M, Park HJ, Cheong MS, Park KS, Baek D, Bae MJ, Ali A, Jan M (2020) The histone-modifying complex PWR/HOS15/HD2C epigenetically regulates cold tolerance. Plant Physiol 184(2):1097–1111

Lin J, Song N, Liu D, Liu X, Chu W, Li J, Chang S, Liu Z, Chen Y, Yang Q, Liu X, Yao Y, Guo W, Xin M, Peng H, Ni Z, Sun Q, Hu Z (2022) Histone acetyltransferase TaHAG1 interacts with TaNACL to promote heat stress tolerance in wheat. Plant Biotechnol J 20(9):1645–1647

Liu Q, Yang T, Yu T, Zhang S, Mao X, Zhao J, Wang X, Dong J, Liu B (2017) Integrating small RNA sequencing with QTL mapping for identification of miRNAs and their target genes associated with heat tolerance at the flowering stage in rice. Front Plant Sci 8:43

Liu J, Feng L, Gu X, Deng X, Qiu Q, Li Q, Zhang Y, Wang M, Deng Y, Wang E, He Y, Bäurle I, Li J, Cao X, He Z (2019) An H3K27me3 demethylase-HSFA2 regulatory loop orchestrates transgenerational thermomemory in *Arabidopsis*. Cell Res 29(5):379–390

Liu B, Yang D, Wang D, Liang C, Wang J, Lisch D, Zhao M (2024a) Heritable changes of epialleles near genes in maize can be triggered in the absence of CHH methylation. Plant Physiol 194(4):2511–2532

Liu Y, Zhu T, Zhou X, Chen W, He C, Wang X, Chen C, Wei J, Lan C, Liu M, Su H, Li Q, Hu X, Bi S, Ouyang W, Li X, Mao H, Kishi M, Kaufmann K, Fernie AR, Chen D, Yan W (2024b) Integration of the DD-genome reshapes gene transcription, chromatin architecture and metabolome of allohexaploid wheat leading to enhanced adaptability. bioRxiv Preprint

Lv Z, Li Z, Wang M, Zhao F, Zhang W, Li C, Gong L, Zhang Y, Mason AS, Liu B (2021) Conservation and trans-regulation of histone modification in the A and B subgenomes of polyploid wheat during domestication and ploidy transition. BMC Biol 19:42

Maryan KE, Farrokhi N, Samizadeh Lahiji H (2023) Cold-responsive transcription factors in *Arabidopsis* and rice: a regulatory network analysis using array data and gene co-expression network. PLoS One 18(6):e0286324

Naderi S, Maali-Amiri R, Sadeghi L, Hamidi A (2024) Physio-biochemical and DNA methylation analysis of the defense response network of wheat to drought stress. Plant Physiol Biochem 209:108516

Naydenov M, Baev V, Apostolova E, Gospodinova N, Sablok G, Gozmanova M, Yahubyan G (2015) High-temperature effect on genes engaged in DNA methylation and affected by DNA methylation in *Arabidopsis*. Plant Physiol Biochem 87:102–108

Ou X, Long L, Zhang Y, Xue Y, Liu J, Lin X, Liu B (2009) Spaceflight induces both transient and heritable alterations in DNA methylation and gene expression in rice (*Oryza sativa* L.). Mutat Res Fundam Mol Mech Mutagen 662(1–2):44–53

Palomar VM, Garciarrubio A, Garay-Arroyo A, Martínez-Martínez C, Rosas-Bringas O, Reyes JL, Covarrubias AA (2020) The canonical RdDM pathway mediates the control of seed germination timing under salinity. Plant J 104(5):1253–1268

Pan Y, Wang W, Zhao X, Zhu L, Fu B, Li Z (2011) DNA methylation alterations of rice in response to cold stress. Plant Omics 4(7):364–369

Panda K, Ji L, Neumann DA, Daron J, Schmitz RJ, Slotkin RK (2016) Full-length autonomous transposable elements are preferentially targeted by expression-dependent forms of RNA-directed DNA methylation. Genome Biol 17:170

Panda K, McCue AD, Slotkin RK (2020) *Arabidopsis* RNA Polymerase IV generates 21–22 nucleotide small RNAs that can participate in RNA-directed DNA methylation and may regulate genes. Philos Trans R Soc Lond Ser B Biol Sci 375(1795):20190417

Park J, Lim CJ, Shen M, Yun DJ (2018) Epigenetic switch from repressive to permissive chromatin in response to cold stress. Proc Natl Acad Sci USA 115(23):E5400–E5409

Pradhan SK, Pandit E, Nayak DK, Behera L, Mohapatra T (2019) Genes, pathways and transcription factors involved in seedling stage chilling stress tolerance in indica rice through RNA-Seq analysis. BMC Plant Biol 19:352

Read A, Weiss T, Crisp PA, Liang Z, Noshay J, Menard CC, Wang C, Song M, Hirsch CN, Springer NM, Zhang F (2022) Genome-wide loss of CHH methylation with limited transcriptome changes in *Setaria viridis* DOMAINS REARRANGED METHYLTRANSFERASE (DRM) mutants. Plant J 110(2):403–419

Rehman Z, Baig A, Shabir G, Aslam K, Shahzad M, Rehman S, Shah M, Khan AR (2021) Drought stress induces differential DNA methylation shift at symmetric and asymmetric cytosine sites in the promoter region of ZmEXPB2 gene in maize. Int J Agric Biol 25(2):319–326

Ruan M, Zhao H, Wen Y, Chen H, He F, Hou X, Song X, Jiang H, Ruan YL, Wu L (2024) The complex transcriptional regulation of heat stress response in maize. Stress Biol 4:24

Sapna H, Ashwini N, Ramesh S, Nataraja KN (2020) Assessment of DNA methylation pattern under drought stress using methylation-sensitive randomly amplified polymorphism analysis in rice. J Agric Sci 158(8):609–617

Saze H, Kakutani T, Miura A, Choi Y, Jacobsen SE et al (2003) Maintenance of CpG methylation is essential for epigenetic inheritance during plant gametogenesis. Nat Genet 34(1):65–69

Shaik R, Ramakrishna W (2012) Bioinformatic analysis of epigenetic and microRNA mediated regulation of drought responsive genes in rice. PLoS One 7(11):e49331

Shen Y, Conde e Silva N, Audonnet L, Servet C, Wei W, Zhou DX (2014) Over-expression of histone H3K4 demethylase gene JMJ15 enhances salt tolerance in *Arabidopsis*. Front Plant Sci 5:290

Sigman MJ, Panda K, Kirchner R, McLain LL, Payne H, Peasari JR, Husbands AY, Slotkin RK, McCue AD (2021) An siRNA-guided ARGONAUTE protein directs RNA polymerase V to initiate DNA methylation. Nat Plants 7:1461–1474

Sun K, Fang H, Chen Y, Zhuang Z, Chen Q, Shan T, Khan MKR, Zhang J, Wang B (2021) Genome-wide analysis of the cytochrome P450 gene family involved in salt tolerance in *Gossypium hirsutum*. Front Plant Sci 12:685054

Temel A, Janack B, Humbeck K (2017) Drought stress-related physiological changes and histone modifications in barley primary leaves at HSP17 gene. Agronomy 7(2):43

Tittel-Elmer M, Bucher E, Broger L, Mathieu O, Paszkowski J, Vaillant I (2010) Stress-induced activation of heterochromatic transcription. PLoS Genet 6(10):e1001175

Tran RK, Zilberman D, de Bustos C, Ditt RF, Henikoff JG, Lindroth AM, Delrow J, Boyle T, Kwong S, Bryson TD, Jacobsen SE, Henikoff S (2005) Chromatin and siRNA pathways cooperate to maintain DNA methylation of small transposable elements in *Arabidopsis*. Genome Biol 6:R90

Wang F, Axtell MJ (2016) AGO4 is specifically required for heterochromatic siRNA accumulation at Pol V-dependent loci in *Arabidopsis thaliana*. Plant J 89(4):771–781

Wang M, Wang J (2022) Transcriptome and DNA methylome analyses provide insight into the heterosis in flag leaf of inter-subspecific hybrid rice. Plant Mol Biol 108:105–125

Wang WS, Pan YJ, Zhao XQ, Dwivedi D, Zhu LH, Ali J, Fu BY, Li ZK (2011) Drought-induced site-specific DNA methylation and its association with drought tolerance in rice (*Oryza sativa* L.). J Exp Bot 62(6):1951–1960

Wang M, Qin L, Xie C, Li W, Yuan J, Kong L, Yu W, Xia G, Liu S (2014) Induced and constitutive DNA methylation in a salinity-tolerant wheat introgression line. Plant Cell Physiol 55(7):1354–1365

Wang W, Qin Q, Sun F, Wang Y, Xu D, Li Z, Fu B (2016) Genome-wide differences in DNA methylation changes in two contrasting rice genotypes in response to drought conditions. Front Plant Sci 7:1675

Wang J, Nan N, Li N, Liu Y, Wang TJ, Hwang I, Liu B, Xu ZY (2020) A DNA methylation reader chaperone regulator transcription factor complex activates OsHKT1;5 expression during salinity stress. Plant Cell 32(11):3535–3558

Wang Q, Xu J, Pu X, Lv H, Liu Y, Ma H et al (2021) Maize DNA methylation in response to drought stress is involved in target gene expression and alternative splicing. Int J Mol Sci 22(15):8285

Wang L, Zheng K, Zeng L, Xu D, Zhu T, Yin Y, Zhan H, Wu Y, Yang DL (2022a) Reinforcement of CHH methylation through RNA-directed DNA methylation ensures sexual reproduction in rice. Plant Physiol 188(2):1189–1209

Wang X, Pan C, Long J, Bai S, Yao M, Chen J, Sun G, Fan Y, Wang Z, Liu F, Liu C, Li Q (2022b) Genome-wide identification of the jumonji C domain-containing histone demethylase gene family in wheat and their expression analysis under drought stress. Front Plant Sci 13:987257

Wang Y, Wang F, Lu H, Lin R, Liu J, Liu Y, Xu J, Wu Y, Wang Z, Zhou M, Mo X, Wu Z, Shou H, Zheng S, Mao C (2023a) Rice chromatin protein OsHMGB1 is involved in phosphate homeostasis and plant growth by affecting chromatin accessibility. New Phytol 240(2):460–475

Wang H, Fan Z, Shliaha PV, Miele M, Hendrickson RC, Jiang X, Helin K (2023b) H3K4me3 regulates RNA polymerase II promoter-proximal pause-release. Nature 615(7951):339–348

Wei L, Gu L, Song X, Cui X, Lu Z, Zhou M, Wang L, Hu F, Zhai J, Meyers BC, Cao X (2014) Dicer-like 3 produces transposable element-associated 24-nt siRNAs that control agricultural traits in rice. Proc Natl Acad Sci USA 111(10):3877–3882

Xun H, Lian L, Yuan J, Hong J, Hao S, Zhao H, Liu S, Feng W, Yin H, Liu B, Wang X (2025) Domains rearranged methyltransferases (DRMs)-mediated DNA methylation plays key roles in modulating gene expression and maintaining transposable element silencing in soybean. J Integr Plant Biol 67:1501

Yamaguchi N, Ito T (2021) Expression profiling of H3K27me3 demethylase genes during plant development and in response to environmental stress in *Arabidopsis*. Plant Signal Behav 16(11):1950445

Yang F, Sun Y, Du X, Chu Z, Zhong X, Chen X (2023) Plant-specific histone deacetylases associate with ARGONAUTE4 to promote heterochromatin stabilization and plant heat tolerance. New Phytol 238(2):802–817

Ye R, Chen Z, Lian B, Rowley MJ, Xia N, Chai J, Li Y, He XJ, Wierzbicki AT, Qi Y (2016) A Dicer-independent route for biogenesis of siRNAs that direct DNA methylation in *Arabidopsis*. Mol Cell 61(2):222–235

Yu G, Zhang B, Chen Q, Huang Z, Zhang B, Wang K, Han J (2024) Dynamic DNA methylation modifications in the cold stress response of cassava. Genomics 116(4):110871

Zhang J, Liu Y, Xia EH, Gao LZ (2015) Autotetraploid rice methylome analysis reveals methylation variation of transposable elements and their effects on gene expression. Proc Natl Acad Sci USA 112(50):E7022–E7029

Zheng X, Chen L, Li M, Lou Q, Xia H, Wang P, Li T, Liu H, Luo L (2013a) Transgenerational variations in DNA methylation induced by drought stress in two rice varieties with distinguished difference to drought resistance. PLoS One 8(11):e80253

Zheng X, Chen L, Lou Q, Xia H, Li M, Luo L (2013b) Changes in DNA methylation pattern at two seedling stages in water-saving and drought-resistant rice variety after drought stress domestication. J Genet Genomics 40(11):601–610

Zheng X, Chen L, Xia H, Wei H, Lou Q, Li M, Li T, Luo L (2017) Transgenerational epimutations induced by multi-generation drought imposition mediate rice plant's adaptation to drought condition. Sci Rep 7:39843

Zheng M, Liu X, Lin J, Liu X, Wang Z, Xin M, Yao Y, Peng H, Zhou D, Ni Z, Sun Q, Hu Z (2019) Histone acetyltransferase GCN5 contributes to cell wall integrity and salt stress tolerance by altering the expression of cellulose synthesis genes. Plant J 97(3):587–602

Zheng L, Ma S, Shen D, Fu H, Wang Y, Liu Y, Shah K, Yue C, Huang J (2021) Genome-wide identification of Gramineae histone modification genes and their potential roles in regulating wheat and maize growth and stress responses. BMC Plant Biol 21:543

Zheng M, Lin J, Liu X, Chu W, Li J, Gao Y, An K, Song W, Xin M, Yao Y (2021a) Histone acetyltransferase TaHAG1 acts as a crucial regulator to strengthen salt tolerance of hexaploid wheat. Plant Physiol 186(4):1951–1969

Zhong X, Du J, Hale CJ, Gallego-Bartolome J, Feng S, Vashisht AA, Chory J, Wohlschlegel JA, Patel DJ, Jacobsen SE (2014) Molecular mechanism of action of plant DRM de novo DNA methyltransferases. Cell 157(5):1050–1060

Zhou DX, Hu Y (2010) Regulatory function of histone modifications in controlling rice gene expression and plant growth. Rice 3:103–111

Zhou C, Yuan Z, Ma X, Zheng L, Zhang Y, Liu X (2021) Accessible chromatin regions and their functional interrelations with gene transcription and epigenetic modifications in sorghum genome. Plant Commun 2(1):100140

Zong W, Zhong X, You J, Xiong L (2013) Genome-wide profiling of histone H3K4-tri-methylation and gene expression in rice under drought stress. Plant Mol Biol 81:175–188

Harnessing the Potential of Transcription Factors in Developing Drought-Resilient Rice

8

Chetana Hasnu and Pratikshya Borah

Abstract

Climatic variability and unpredictable rainfall patterns have led to the frequency of drought worldwide. Rice is the primary cereal for much of the global population, yet dehydration stress remains a critical factor limiting its productivity and quality. In response to drought stress, rice plants undergo several adaptive modifications, including stomatal closure, cellular adaptation, and alterations in root development, induced by stress-responsive genes. These genes are controlled by different transcription factors like AP2/ERF, DREB, bZIP, MYB, WRKY, WOX, etc., mediating the upregulation or downregulation of their respective genes, enabling the plants to withstand drought and highlighting them as promising genomic candidates for crop improvement programs. These transcription factors (TFs) act as pivotal molecular switches orchestrating plant developmental processes under various stress environments. This book chapter seeks to elucidate the role of TFs in regulating rice responses to drought stress and to explore potential strategies for enhancing drought tolerance in rice cultivation.

Keywords

Drought stress · Transcription factors · Genes · Gene expression · Stress signalling

C. Hasnu
Department of Botany, The Assam Royal Global University, Guwahati, Assam, India

Department of Botany, Pandit Deendayal Upadhyaya Adarsha Mahavidyalaya, Tulungia, Bongaigaon, Assam, India

P. Borah (✉)
Department of Botany, The Assam Royal Global University, Guwahati, Assam, India
e-mail: pborah3@rgu.ac

V. Gahlaut et al. (eds.), *Smart Crop Development*,
https://doi.org/10.1007/978-981-95-7680-7_8

8.1 Introduction

Rice (*Oryza sativa* L.) is a monocotyledonous plant of the *Poaceae* family, which is consumed by approximately half the global population (Mohapatra et al. 2022). It is a semi-aquatic plant, which requires large amounts of water in its life cycle (Pandey and Shukla 2015). Rice is grown in environmental conditions exposed to various forms of biotic and abiotic stresses. Of these stresses, drought or dehydration is the most calamitous, resulting in an overall reduction in rice production (Nawaz et al. 2023). Owing to the population dynamics, increasing demand for rice production and depleting water resources due to climatic variability, the generation of drought-resistant rice varieties is the need of the hour (Pandey and Shukla 2015; Kim et al. 2020).

Drought refers to the depletion of soil moisture, which is necessary for plant growth, development, and completion of its life cycle (Sinton et al. 2023). As described by Ghosh et al. (2014), drought is a condition that arises when a region experiences prolonged deficiency in water availability—whether surface or underground—over months or years, typically due to persistently low precipitation. In plants, drought stress is assessed through morphological, physiological, and biochemical indicators, including reduced tissue water content, inhibition of cell elongation and growth, stomatal closure, diminished gas exchange, and impairment of enzyme-mediated reactions (Oladosu et al. 2019) (Fig. 8.1).

Flow chart illustrating the effects of drought stress on plants. Exposure to drought during any stage of the rice plant's growth affects it at morpho-physiological and

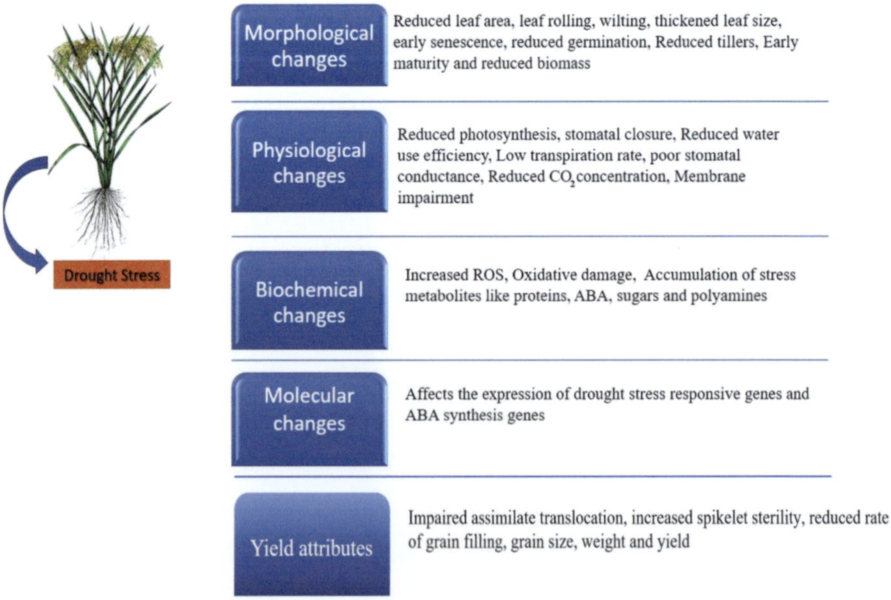

Fig. 8.1 Impact of drought in rice plants. (Modified from Kumar et al. 2023)

biochemical levels, generating complex traits for drought tolerance in rice (Hassan et al. 2023). These traits are governed by the activation of specific genes that encode transcription factors (TFs), playing a central role in regulating gene expression at the transcriptional level (Gao et al. 2006; Khong et al. 2008). Even subtle changes in TF activity, their sequences, or the DNA regions they target can markedly reshape gene regulatory networks, thereby influencing plant morphology or physiology (Khong et al. 2008). Acting as early stress-responsive regulators, TFs modulate the expression of numerous downstream target genes. Harnessing and manipulating TF-encoding genes has emerged as a widely adopted strategy to enhance plant resilience under stress. This approach increases the potential for developing new cultivars with improved tolerance to both individual and multiple stress conditions (Lan Thi Hoang et al. 2017).

8.2 Transcription Factors in Rice (*Oryza sativa* L.) Involved in Drought Stress

Transcription factors (TFs) are regulatory proteins that recognize specific DNA sequences and either promote or inhibit the transcription of numerous target genes (Wu et al. 2005). These proteins interact with cis-acting elements in the promoter region of eukaryotic genes, functioning as monomers or homodimers, to ensure precise spatial, temporal, and intensity-specific gene expression, thereby playing a crucial role in mediating plant responses to fluctuations (Katiyar et al. 2012; Mathew et al. 2016). Approximately 60 kinds of plant-specific TFs have been identified to date, with several—such as AP2/ERF, NAC, bZIP, MYB, and WRKY—being central to rice drought responses (Joshi et al. 2016; Xie et al. 2022). TFs are typically classified based on their DNA-binding domains. For example, the WRKY family derives its name from a highly conserved 60–amino acid WRKY domain (Agarwal et al. 2011). Although each TF family comprises numerous genes, only a subset functionally contributes to abiotic stress responses, orchestrating complex transcriptional networks that regulate large numbers of downstream genes and pathways (Joshi et al. 2016). Enhancement of the abiotic stress response in rice may be significantly achieved by either overexpression or knockout of specific transcription gene; for example, overexpression of *OsNAC17* showed drought tolerance by upregulating lignin biosynthetic genes and promoting lignin deposition in leaves and roots (Jung et al. 2022), whereas knockout of *OsNAC092* improves dehydration tolerance by modulating processes related to ROS response, photosynthesis, light signalling, and hormone- and MAPK-mediated pathways (Wang et al. 2022).

The AP2/ERF (APETELA2/ethylene response factor) family represents one of the largest groups of plant transcription factors, characterized by highly conserved AP2/ERF domains consisting of approximately 60–70 amino acid residues (Okamuro et al. 1997; Riechmann and Meyerowitz 1998). Structurally, this domain forms a typical three-dimensional configuration comprising of three β-folds and one α-helix, with YRG and RAYD elements in the AP2 domain, that are critical for DNA-binding activity (Sun et al. 2022; Ma et al. 2024). Functional studies have highlighted the role of specific members in rice drought tolerance. For instance, *OsERF48* regulates *OsCML16*, a calmodulin-like protein gene that promotes root growth and enhances drought resilience. Overexpression of *OsERF48* resulted in longer, denser roots, and under in vitro drought conditions (40% polyethylene glycol medium), and plants exhibited more vigorous root development and improved grain yield compared to controls (Jung et al. 2022). Similarly, the overexpression of *OsERF71* directly targets the promoter of *OsCINNAMOYL-COENZYME A REDUCTASE1*, a key lignin biosynthetic gene, conferring drought resistance during vegetative growth. Root-specific upregulation at the reproductive stage proved more effective in eliciting drought resistance, increasing grain yield by 23% to 42%. Enhanced root *OsERF71* activity facilitates cell wall modifications, including loosening and lignin biosynthesis, enlarged aerenchyma formation, and elevated lignification, thereby supporting adaptive root morphology under drought stress (Lee et al. 2016).

DREB (Dehydration-Responsive Element Binding) transcription factors belong to the ERF family and are classified into two major subclasses: DREB1/CBF, which are activated by low temperature, and DREB2, which respond to dehydration (Agarwal et al. 2006). In a study by Dubouzet et al. (2003), five rice DREB homolog cDNAs—*OsDREB1A*, *OsDREB1B*, *OsDREB1C*, *OsDREB1D*, and *OsDREB2A*—were isolated, with *OsDREB1A* expression induced by cold and *OsDREB2A* by dehydration, respectively. In rice protoplasts, both *OsDREB1A* and *OsDREB2A* were shown to activate DRE-driven GUS reporter gene transcription. In transgenic *Arabidopsis*, overexpression of *OsDREB2B* enhanced the expression of *DREB2A* target genes, identifying *OsDREB2B* as a key stress-inducible DREB2-type transcription factor involved in drought-responsive gene expression in rice (Matsukura et al. 2010). Furthermore, simultaneous overexpression of DREB and PIF transcription factors improved drought tolerance and promoted cell elongation in transgenic plants (Kudo et al. 2017). Overexpression of *OsDRAP1*, a DREB2-like gene, also conferred drought tolerance in the transgenic rice, positively influencing water balance, redox homeostasis, and vascular development under drought stress (Huang et al. 2018).

The NAC (NAM, ATAF1/2 and CUC2) family represents one of the largest groups of plant transcription factors, playing a pivotal role in transcriptional reprogramming associated with diverse abiotic stress responses (Jung et al. 2022). Genome-wide analysis has identified 151 NACs in rice, several of which are directly involved in regulating drought stress response (Shim et al. 2018). Overexpression of NAC genes such as *OsNAC14*, *OsNAC15*, and *OsNAC17* has shown to enhance drought tolerance in rice through multiple mechanisms, including increased panicle number, including increased panicle number, elevated accumulation of hydrogen peroxide (H_2O_2) and malondialdehyde (MDA), modulation of reactive oxygen species (ROS), and enhanced lignin deposition in leaves and root (Shim et al. 2018; Jung et al. 2022; Ao et al. 2024). Additionally, *OsNAC52* activates the promoters of key abscisic acid (ABA) biosynthesis genes (*OsNCED1*, *OsNCED2*, and *OsNCED5*), further contributing to drought resilience.

Rice contains 89 bZIP (Basic Leucine Zipper Motif) transcription factors, among which members of the third subfamily—such as *OsbZIP23*, *OsbZIP46*, and *OsbZIP72*—play key roles in stress responses by regulating abscisic acid signalling (Yang et al. 2019). *OsbZIP46* has been identified as an abiotic stress-responsive gene within this group (Tang et al. 2016). Chang et al. (2017) in their study found that the co-overexpression of two genes, encoding a constitutively active form of a bZIP transcription factor (*OsbZIP46CA1*) and a protein kinase (*SAPK6*) involved in the abscisic acid signalling pathway, showed significantly enhanced drought resistance compared with the single-gene transgenic lines.

The MYB (Myeloblastosis) transcription factor family is defined by the presence of a highly conserved MYB domain, which is ubiquitous across eukaryotes. This domain is located at the N-terminus, while the variable C-terminus functions as a transactivation domain, regulating diverse biological processes within MYB proteins (Tang et al. 2019). MYB proteins typically consist of 52 amino acid residues and contain 1, 2, or 3 imperfect repeats. The MYB TF family members are classified into four groups based on the repeats in their MYB domain, viz., R2R3-MYB, MYB1-R, 4R-MYB, and R1R2R3-MYB (MYB3R) (Butt et al. 2017; Tiwari et al. 2020). Functional studies in rice highlight the role of MYB transcription factors in drought tolerance. Overexpression of *OsMYB48–1* reduced water loss, lowered malondialdehyde (MDA) levels, and increased proline accumulation under stress conditions. At both germination and post-germination stages, transgenic plants demonstrated elevated endogenous ABA levels and regulated the expression of some ABA biosynthesis genes (*OsNCED4*, *OsNCED5*), early signalling genes (*OsPP2C68*, *OSRK1*), and late responsive genes (*RAB21*, *OsLEA3*, *RAB16C*, and *RAB16D*) under drought conditions (Xiong et al. 2014). Similarly, the R2R3-Type MYB transcription factor OsFLP (FLP- FOUR LIPS) positively regulates drought responses by directly binding to the promoters of *OsNAC1* and *OsNAC6*, thereby mediating stomatal closure and activating stress-responsive genes (Qu et al. 2022). Overexpression of *OsMYB6* has also been shown to enhance tolerance to both drought and salt stress, as evidenced by higher proline content, increased catalase (CAT) and superoxide dismutase (SOD) activities, and reduced relative electrolyte leakage (REL) and MDA levels in transgenic plants (Tang et al. 2019). In contrast, OsMYB1R1 acts as a negative regulator of drought resistance; its downregulation via RNA interference (RNAi) resulted in lower REC and MDA levels and higher proline accumulation, thereby improving drought tolerance in rice (Peng et al. 2023).

Recognizing and analyzing the expression of transcription factor genes is crucial as these are master regulators governing the activity of numerous downstream genes. Exploring their molecular mechanisms provide deeper insights into the stress signalling network, thereby advancing strategies to enhance rice resilience against drought. The table below outlines the roles of different transcription factors in regulating plant development under drought stress (Table 8.1).

Table 8.1 The transcription factors family expressed in response to drought stress in rice

Transcription factor family	Domain	Gene name	Expression	References
AP2/ERF (APETALA2/ Ethylene Responsive Factor)	AP2/ERF domain contains 60–70 amino acid residues, which include the YRG and RAYD structural domains	*OsERF48*, *OsERF71*, *OsERF922*, *OsERF83*	Enhances root growth in of seedling, modification of cell wall and root structure in root meristem, ABA level modulation in shoot, higher photochemical efficiency	Licausi et al. (2013), Lee et al. (2016), Jung et al. (2017), Ma et al. (2024) and Jung et al. (2021)
DREB (Dehydration-responsive element binding)	AP2 domain and can bind to the cis-acting elements DRE/ CRT	*OsDREB2A*	Protect plant cell improving survival rates	Cui et al. (2011), Zhang and Xia (2023)
NAC (NAM, ATAF1/2 and CUC2)	NAC Domain is situated in N-terminus about 150 amino acid residues in length and consists of five (A–E) subdomains	*ONAC095*, *OsNAC52*, *ONAC066*, *SNAC3*, *OsNAC17*, *OsNAC14*	Decreased water loss rate, increased proline and soluble sugar contents; activates the promoters of key ABA biosynthesis genes; modulation of ROS; modulating lignin accumulation in leaves and root; higher number of panicle and rice filling rate	Gao et al. (2010), Nakashima et al. (2012), Fang et al. (2015), Nuruzzaman et al. (2015), Huang et al. (2016), Shim et al. (2018), Jung et al. (2022), and Ao et al. (2024)
MYB (Myeloblastosis)	MYB domain is highly conserved across all eukaryotes and is located at the N terminus, whereas the C terminus is variable and acts as a transacting domain	*OsMYB2*, *OsMYB6*, *OsMYB48-1*, *OsMYBR1*, *OsMYB511*	Mediated by regulating the cell cycle at seedling stage; accumulate greater amount of proline and soluble sugars; regulate the expression of some ABA biosynthesis genes, early signalling genes and late responsive genes; early panicle development	Yang et al. (2012), Xiong et al. (2014), Huang et al. (2015), Yin et al. (2017), Tang et al. (2019), and Biswas et al. (2023)

(continued)

Table 8.1 (continued)

Transcription factor family	Domain	Gene name	Expression	References
WRKY	WRKY protein domain—60 amino acid DNA-binding domain (DBD) characterized by a highly conserved N-terminal core, WRKYGQK motif	*OsWRKY47*, *OsWRKY55*, *ZmWRKY58*, *OsWRKY114*	Photosynthetic reduction and high yields; accumulation of ROS; positive regulation of drought stress tolerance; negative regulation of drought tolerance by restriction of stomatal closure in rice	Cai et al. (2014), Raineri et al. (2015), Jiang et al. (2017), Huang et al. (2021), and Song et al. (2022)
WOX (WUSCHEL homeobox)	WOX homeodomain consists of two extended loops between helices 1 and 2, in between a highly conserved structure	*OsWOX13*, *OsWOX4*	High yields and reduction in photosynthesis; triggers early flowering; primary root elongation	Minh-Thu et al. (2018), Chen et al. (2020)
bZIP (basic Leucine Zipper Motif)	bZIP domain, a leucine zipper domain, and a basic DBD	*OsbZIP72*, *OsbZIP62*, *OsABF2*, *OsbZIP46*, *OsbZIP46CA1*	Regulator of ABA response	Lu et al. (2009), Hossain et al. (2010), Das et al. (2019), Yang et al. (2019), Tang et al. (2016), and Chang et al. (2017)
bHLH (basic Helix-Loop-Helix)	~60 amino acids long comprising of a DNA-binding basic region (b) at N-terminus followed by two α-helices separated by a variable loop region (HLH) at C-terminus	*OsbHLH148*	Increased drought tolerance by involving in the jasmonate (JA) pathway	Seo et al. (2011), Thoben and Pucker (2023)

(continued)

Table 8.1 (continued)

Transcription factor family	Domain	Gene name	Expression	References
GRAS {Gibberellin-acid insensitive (GAI), Repressor of GA1 (RGA) and Scarecrow (SCR)}	Contains 400–770 amino acid residues with sequence homology in C-terminal domains highly divergent N-terminal amino acid sequences	*OsGRAS23*	Drought resistance by regulating expression of several stress-responsive genes	Xu et al. (2015)

8.3 Stress Perception and Downstream Events

Plants modulate the expression pattern of TFs and other genes in response to abiotic stress, thereby reshaping signalling pathways and regulatory networks to enhance stress tolerance (Ramadoss et al. 2020). Stress perception occurs through membrane-localized receptors, and adaptation is mediated via signal transduction pathways that regulate the stress-responsive gene expression. These pathways function either through the accumulation or release of Ca^{2+} ions or through the de novo synthesis of secondary messengers such as reactive oxygen species (ROS), inositol triphosphates, hexaphosphates, and diacylglycerols or a combination of both (Hrmova and Lopato 2013). Elevated cytosolic Ca^{2+} levels are detected by calcium sensors or Ca^{2+}-binding proteins, which in turn activate protein kinases. This leads to the phosphorylation of multiple regulatory proteins, including transcription factors, ultimately triggering the expression of stress-responsive genes. Such gene activation supports stress tolerance and resistance while restoring cellular and tissue homeostasis (Tuteja and Mahajan 2007).

Several TFs genes have been reported to function under drought stress via both abscisic acid (ABA)–dependent and independent pathways (Joshi et al. 2016). ABA is a key plant hormone that regulates diverse developmental processes, including embryo maturation, seed germination and dormancy, floral initiation, and root growth. Beyond its developmental roles, ABA plays a critical part in mitigating drought damage by activating protective responses (Hussain et al. 2021). Under drought conditions, ABA levels rise sharply in vegetative tissues, leading to stomatal closure, accumulation of stress-related metabolites and proteins, and increased H_2O_2 productions in guard cells, which collectively signal and limit water loss from the plant (Dash et al. 2018).

A wide range of TFs regulate gene expression in response to drought stress. Among them, bZIP, AP2/EREBP, NAC, WORKY, HD-ZIP, and MYB families are the most prominent contributors to drought response. These TFs are activated through either ABA-dependent (*OsNAC2*, *OsWORKY50*, *OsbZIP72*, etc.) or ABA-independent (*SNAC1*, *OsWRKY5*, *OsWRKY114*, etc.) signalling pathways during stress conditions. Transgenic rice plants overexpressing these TFs consistently exhibit enhanced drought-tolerant phenotypes (Hrmova and Lopato 2013; Park and Jeong 2023). Within the NAC TF family, *OsNAC2* directly binds to the promoters of *LATE EMBRYOGENESIS ABUNDANT 3* (*OsLEA3*) (a marker of abiotic stress) and *Stress-Activated Protein Kinases 1* (*OsSAPK1*) (a marker gene of ABA signalling), indicating its dual role in coordinating both abiotic stress responses and ABA-mediated pathways, thereby functioning at the intersection of these networks (Shen et al. 2017). Similarly, activation of *OsbZIP42* acts as a positive regulator of ABA signalling and drought tolerance—its activity dependent on phosphorylation by stress-/ABA-activated protein kinase 4 (SAPK4) and an additional ABA-dependent modification of (Joo et al. 2019). Collectively, these interactions represent key molecular mechanisms underlying drought stress tolerance in rice (Fig. 8.2).

8.4 Conclusion and Future Potential

Rice quality and yield are highly dependent on their stress tolerance strategies adapted overtime due to continuous environmental variability. Rice production and productivity are severely constrained when plants fail to adapt to adverse environmental conditions, particularly drought stress, leading to significant yield losses. In recent years, substantial progress has been made in elucidating the molecular mechanisms underlying drought and salinity tolerance. Through genetic

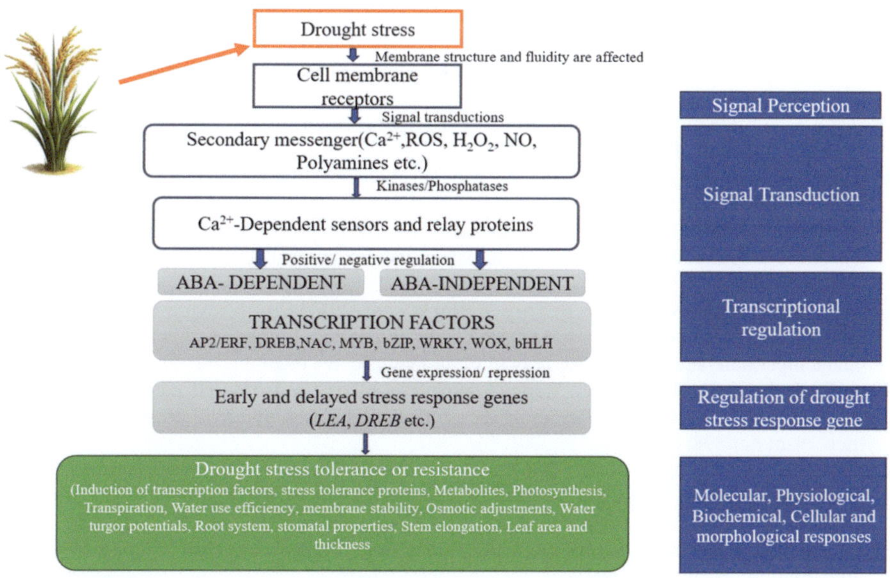

Fig. 8.2 Illustration of stress signal perception followed by downstream events—signal transduction and subsequent signalling cascades—that activate diverse transcription factors (TFs), leading to the expression of multiple stress-responsive genes and mediating drought stress responses in rice. (Adapted from Hrmova and Hussain 2021)

engineering and genome editing approaches, multiple regulatory pathways associated with drought stress in rice have been identified. Future research on transcription factors in rice drought stress can be attended toward more deep understanding of complex regulatory networks, identifying novel TFs with specific functions and their off-target effects, and studying the interplay between different TF families to optimize stress response mechanisms.

These insights pave the way for designing modified transcription factors (TFs) with enhanced DNA-binding efficiency. More importantly, this knowledge provides a foundation for applying TFs in crop bioengineering to improve drought tolerance. Accordingly, this chapter highlights current advances in understanding the roles of diverse TFs and genes in rice drought stress, emphasizing their connections to both ABA-dependent and ABA-independent pathways, and identifying potential transcription factors and regulatory networks that could be exploited to develop drought-resilient rice varieties.

Acknowledgments P. B. acknowledges The Assam Royal Global University for the financial support as a Seed Grant (RGU/Ch (Acad)/(07)/Botany (29)).

References

Agarwal PK, Agarwal P, Reddy MK, Sopory SK (2006) Role of DREB transcription factors in abiotic and biotic stress tolerance in plants. Plant Cell Rep 25:1263–1274

Agarwal P, Reddy MP, Chikara J (2011) WRKY: its structure, evolutionary relationship, DNA-binding selectivity, role in stress tolerance and development of plants. Mol Biol Rep 38:3883–3896

Ao CW, Xiang GJ, Wu YF, Wen Y, Zhu ZL, Sheng F, Du X (2024) OsNAC15 regulates drought and salt tolerance in rice. Physiol Mol Biol Plants 30:1–11

Biswas D, Gain H, Mandal A (2023) MYB transcription factor: a new weapon for biotic stress tolerance in plants. Plant Stress 10:100252

Butt HI, Yang Z, Gong Q, Chen E, Wang X, Zhao G et al (2017) GaMYB85, an R2R3 MYB gene, in transgenic Arabidopsis plays an important role in drought tolerance. BMC Plant Biol 17:1–17

Cai R, Zhao Y, Wang Y, Lin Y, Peng X, Li Q et al (2014) Overexpression of a maize WRKY58 gene enhances drought and salt tolerance in transgenic rice. Plant Cell Tissue Organ Cult (PCTOC) 119:565–577

Chang Y, Nguyen BH, Xie Y, Xiao B, Tang N, Zhu W et al (2017) Co-overexpression of the constitutively active form of OsbZIP46 and ABA-activated protein kinase SAPK6 improves drought and temperature stress resistance in rice. Front Plant Sci 8:1102

Chen R, Xu N, Yu B, Wu Q, Li X, Wang G, Huang J (2020) The WUSCHEL-related homeobox transcription factor OsWOX4 controls the primary root elongation by activating OsAUX1 in rice. Plant Sci 298:110575

Cui M, Zhang W, Zhang Q, Xu Z, Zhu Z, Duan F, Wu R (2011) Induced over-expression of the transcription factor OsDREB2A improves drought tolerance in rice. Plant Physiol Biochem 49(12):1384–1391

Das P, Lakra N, Nutan KK, Singla-Pareek SL, Pareek A (2019) A unique bZIP transcription factor imparting multiple stress tolerance in Rice. Rice 12:1–16

Dash PK, Rai R, Rai V, Pasupalak S (2018) Drought induced signaling in rice: delineating canonical and non-canonical pathways. Front Chem 6:264

Dubouzet JG, Sakuma Y, Ito Y, Kasuga M, Dubouzet EG, Miura S et al (2003) OsDREB genes in rice, Oryza sativa L., encode transcription activators that function in drought-, high-salt-and cold-responsive gene expression. Plant J 33(4):751–763

Fang Y, Liao K, Du H, Xu Y, Song H, Li X, Xiong L (2015) A stress-responsive NAC transcription factor SNAC3 confers heat and drought tolerance through modulation of reactive oxygen species in rice. J Exp Bot 66(21):6803–6817

Gao G, Zhong Y, Guo A, Zhu Q, Tang W, Zheng W et al (2006) DRTF: a database of rice transcription factors. Bioinformatics 22(10):1286–1287

Gao F, Xiong A, Peng R, Jin X, Xu J, Zhu B et al (2010) OsNAC52, a rice NAC transcription factor, potentially responds to ABA and confers drought tolerance in transgenic plants. Plant Cell Tissue Organ Cult (PCTOC) 100:255–262

Ghosh T, Mukhopadhyay A, Ghosh T, Mukhopadhyay A (2014) Drought Hazard in Bihar. In: Natural Hazard zonation of Bihar (India) using Geoinformatics: a schematic approach. Springer, pp 23–37

Hassan MA, Dahu N, Hongning T, Qian Z, Yueming Y, Yiru L, Shimei W (2023) Drought stress in rice: morpho-physiological and molecular responses and marker-assisted breeding. Front Plant Sci 14:1215371

Hossain MA, Cho JI, Han M, Ahn CH, Jeon JS, An G, Park PB (2010) The ABRE-binding bZIP transcription factor OsABF2 is a positive regulator of abiotic stress and ABA signaling in rice. J Plant Physiol 167(17):1512–1520

Hrmova M, Lopato S (2013) Enhancing abiotic stress tolerance in plants by modulating properties of stress responsive transcription factors. In: Genomics of plant genetic resources: volume 2. Crop productivity, food security and nutritional quality. Springer Netherlands, Dordrecht, pp 291–316

Hrmova M, Hussain SS (2021) Plant transcription factors involved in drought and associated stresses. Int J Mol Sci 22(11):5662

Huang P, Chen H, Mu R, Yuan X, Zhang HS, Huang J (2015) OsMYB511 encodes a MYB domain transcription activator early regulated by abiotic stress in rice. Genet Mol Res 14:9506–9517

Huang L, Hong Y, Zhang H, Li D, Song F (2016) Rice NAC transcription factor ONAC095 plays opposite roles in drought and cold stress tolerance. BMC Plant Biol 16:1–18

Huang L, Wang Y, Wang W, Zhao X, Qin Q, Sun F et al (2018) Characterization of transcription factor gene OsDRAP1 conferring drought tolerance in rice. Front Plant Sci 9:94

Huang K, Wu T, Ma Z, Li Z, Chen H, Zhang M et al (2021) Rice transcription factor OsWRKY55 is involved in the drought response and regulation of plant growth. Int J Mol Sci 22(9):4337

Hussain Q, Asim M, Zhang R, Khan R, Farooq S, Wu J (2021) Transcription factors interact with ABA through gene expression and signaling pathways to mitigate drought and salinity stress. Biomolecules 11(8):1159

Jiang J, Ma S, Ye N, Jiang M, Cao J, Zhang J (2017) WRKY transcription factors in plant responses to stresses. J Integr Plant Biol 59(2):86–101

Joo J, Lee YH, Song SI (2019) OsbZIP42 is a positive regulator of ABA signaling and confers drought tolerance to rice. Planta 249(5):1521–1533

Joshi R, Wani SH, Singh B, Bohra A, Dar ZA, Lone AA et al (2016) Transcription factors and plants response to drought stress: current understanding and future directions. Front Plant Sci 7:1029

Jung H, Chung PJ, Park SH, Redillas MCFR, Kim YS, Suh JW, Kim JK (2017) Overexpression of Os ERF 48 causes regulation of Os CML 16, a calmodulin-like protein gene that enhances root growth and drought tolerance. Plant Biotechnol J 15(10):1295–1308

Jung SE, Bang SW, Kim SH, Seo JS, Yoon HB, Kim YS, Kim JK (2021) Overexpression of OsERF83, a vascular tissue-specific transcription factor gene, confers drought tolerance in rice. Int J Mol Sci 22(14):7656

Jung SE, Kim TH, Shim JS, Bang SW, Yoon HB, Oh SH et al (2022) Rice NAC17 transcription factor enhances drought tolerance by modulating lignin accumulation. Plant Sci 323:111404

Katiyar A, Smita S, Lenka SK, Rajwanshi R, Chinnusamy V, Bansal KC (2012) Genome-wide classification and expression analysis of MYB transcription factor families in rice and Arabidopsis. BMC Genomics 13:1–19

Khong GN, Richaud F, Coudert Y, Pati PK, Santi C, Perin C et al (2008) Modulating rice stress tolerance by transcription factors. Biotechnol Genet Eng Rev 25(1):381–404

Kim Y, Chung YS, Lee E, Tripathi P, Heo S, Kim KH (2020) Root response to drought stress in rice (Oryza sativa L.). Int J Mol Sci 21(4):1513

Kudo M, Kidokoro S, Yoshida T, Mizoi J, Todaka D, Fernie AR et al (2017) Double overexpression of DREB and PIF transcription factors improves drought stress tolerance and cell elongation in transgenic plants. Plant Biotechnol J 15(4):458–471

Ku et al. (2015), corrected as Xu et al. (2015) the reference of which has already been added in the bibliography

Kumar A, Sengar RS, Pathak RK, Singh AK (2023) Integrated approaches to develop drought-tolerant rice: demand of era for global food security. J Plant Growth Regul 42(1):96–120

Lan Thi Hoang X, Ngoc Hai Nhi D, Binh Anh Thu N, Phuong Thao N, Phan Tran LS (2017) Transcription factors and their roles in signal transduction in plants under abiotic stresses. Curr Genomics 18(6):483–497

Lee DK, Jung H, Jang G, Jeong JS, Kim YS, Ha SH et al (2016) Overexpression of the OsERF71 transcription factor alters rice root structure and drought resistance. Plant Physiol 172(1):575–588

Licausi F, Ohme-Takagi M, Perata P (2013) APETALA 2/Ethylene Responsive Factor (AP 2/ERF) transcription factors: mediators of stress responses and developmental programs. New Phytol 199(3):639–649

Lu G, Gao C, Zheng X, Han B (2009) Identification of OsbZIP72 as a positive regulator of ABA response and drought tolerance in rice. Planta 229:605–615

Ma Z, Hu L, Jiang W (2024) Understanding AP2/ERF transcription factor responses and tolerance to various abiotic stresses in plants: a comprehensive review. Int J Mol Sci 25(2):893

Mathew IE, Das S, Mahto A, Agarwal P (2016) Three rice NAC transcription factors heteromerize and are associated with seed size. Front Plant Sci 7:1638

Matsukura S, Mizoi J, Yoshida T, Todaka D, Ito Y, Maruyama K et al (2010) Comprehensive analysis of rice DREB2-type genes that encode transcription factors involved in the expression of abiotic stress-responsive genes. Mol Gen Genomics 283:185–196

Minh-Thu PT, Kim JS, Chae S, Jun KM, Lee GS, Kim DE et al (2018) A WUSCHEL homeobox transcription factor, OsWOX13, enhances drought tolerance and triggers early flowering in rice. Mol Cells 41(8):781–798

Mohapatra PK, Sahu BB, Mohapatra PK, Sahu BB (2022) Botany of rice plant. In: Panicle architecture of rice and its relationship with grain filling, pp 27–48. https://doi.org/10.1007/978-3-030-67897-5

Nakashima K, Takasaki H, Mizoi J, Shinozaki K, Yamaguchi-Shinozaki K (2012) NAC transcription factors in plant abiotic stress responses. Biochim Biophys Acta (BBA)-Gene Regul Mech 1819(2):97–103

Nawaz M, Sun J, Shabbir S, Khattak WA, Ren G, Nie X et al (2023) A review of plants strategies to resist biotic and abiotic environmental stressors. Sci Total Environ 900:165832

Nuruzzaman M, Sharoni AM, Satoh K, Karim MR, Harikrishna JA, Shimizu T et al (2015) NAC transcription factor family genes are differentially expressed in rice during infections with Rice dwarf virus, Rice black-streaked dwarf virus, Rice grassy stunt virus, Rice ragged stunt virus, and Rice transitory yellowing virus. Front Plant Sci 6:676

Okamuro JK, Caster B, Villarroel R, Van Montagu M, Jofuku KD (1997) The AP2 domain of APETALA2 defines a large new family of DNA binding proteins in Arabidopsis. Proc Natl Acad Sci 94(13):7076–7081

Oladosu Y, Rafii MY, Samuel C, Fatai A, Magaji U, Kareem I et al (2019) Drought resistance in rice from conventional to molecular breeding: a review. Int J Mol Sci 20(14):3519

Pandey V, Shukla A (2015) Acclimation and tolerance strategies of rice under drought stress. Rice Sci 22(4):147–161

Park SY, Jeong DH (2023) Comprehensive analysis of rice seedling transcriptome during dehydration and rehydration. Int J Mol Sci 24(9):8439

Peng Y, Tang N, Zou J, Ran J, Chen X (2023) Rice MYB transcription factor OsMYB1R1 negatively regulates drought resistance. Plant Growth Regul 99(3):515–525

Qu X, Zou J, Wang J, Yang K, Wang X, Le J (2022) A rice R2R3-type MYB transcription factor OsFLP positively regulates drought stress response via OsNAC. Int J Mol Sci 23(11):5873

Raineri J, Wang S, Peleg Z, Blumwald E, Chan RL (2015) The rice transcription factor OsWRKY47 is a positive regulator of the response to water deficit stress. Plant Mol Biol 88:401–413

Ramadoss BR, Gangola MP, Gurunathan S (2020) NAC transcription factor family in rice: recent advancements in the development of stress-tolerant rice. In: Transcription factors for abiotic stress tolerance in plants. Academic, pp 47–61

Riechmann JL, Meyerowitz EM (1998) The AP2/EREBP family of plant transcription factors. Biol Chem 379:633–646

Seo JS, Joo J, Kim MJ, Kim YK, Nahm BH, Song SI et al (2011) OsbHLH148, a basic helix-loop-helix protein, interacts with OsJAZ proteins in a jasmonate signaling pathway leading to drought tolerance in rice. Plant J 65(6):907–921

Shen J, Lv B, Luo L, He J, Mao C, Xi D, Ming F (2017) The NAC-type transcription factor OsNAC2 regulates ABA-dependent genes and abiotic stress tolerance in rice. Sci Rep 7(1):40641

Shim JS, Oh N, Chung PJ, Kim YS, Choi YD, Kim JK (2018) Overexpression of OsNAC14 improves drought tolerance in rice. Front Plant Sci 9:310

Sinton FV, Lim RAH, Sam LM (2023) Evaluation of the growth and yield of upland rice varieties to drought stress. Asian J Plant Sci 22:465–472

Song G, Son S, Lee KS, Park YJ, Suh EJ, Lee SI, Park SR (2022) OsWRKY114 negatively regulates drought tolerance by restricting stomatal closure in rice. Plants 11(15):1938

Sun S, Liang X, Chen H, Hu L, Yang Z (2022) Identification of AP2/ERF transcription factor family genes and expression patterns in response to drought stress in Pinus massoniana. Forests 13(9):1430

Tang N, Ma S, Zong W, Yang N, Lv Y, Yan C et al (2016) MODD mediates deactivation and degradation of OsbZIP46 to negatively regulate ABA signaling and drought resistance in rice. Plant Cell 28(9):2161–2177

Tang Y, Bao X, Zhi Y, Wu Q, Guo Y, Yin X et al (2019) Overexpression of a MYB family gene, OsMYB6, increases drought and salinity stress tolerance in transgenic rice. Front Plant Sci 10:168

Thoben C, Pucker B (2023) Automatic annotation of the bHLH gene family in plants. BMC Genomics 24(1):780

Tiwari P, Indoliya Y, Chauhan AS, Pande V, Chakrabarty D (2020) Over-expression of rice R1-type MYB transcription factor confers different abiotic stress tolerance in transgenic Arabidopsis. Ecotoxicol Environ Saf 206:111361

Tuteja N, Mahajan S (2007) Calcium signaling network in plants: an overview. Plant Signal Behav 2(2):79–85

Wang B, Wang Y, Yu W, Wang L, Lan Q, Wang Y et al (2022) Knocking out the transcription factor OsNAC092 promoted rice drought tolerance. Biology 11(12):1830

Wu KL, Guo ZJ, Wang HH, Li J (2005) The WRKY family of transcription factors in rice and Arabidopsis and their origins. DNA Res 12(1):9–26

Xie W, Ding C, Hu H, Dong G, Zhang G, Qian Q, Ren D (2022) Molecular events of rice AP2/ERF transcription factors. Int J Mol Sci 23(19):12013

Xiong H, Li J, Liu P, Duan J, Zhao Y, Guo X et al (2014) Overexpression of OsMYB48-1, a novel MYB-related transcription factor, enhances drought and salinity tolerance in rice. PLoS One 9(3):e92913

Xu K, Chen S, Li T, Ma X, Liang X, Ding X et al (2015) OsGRAS23, a rice GRAS transcription factor gene, is involved in drought stress response through regulating expression of stress-responsive genes. BMC Plant Biol 15:1–13

Yang A, Dai X, Zhang WH (2012) A R2R3-type MYB gene, OsMYB2, is involved in salt, cold, and dehydration tolerance in rice. J Exp Bot 63(7):2541–2556

Yang S, Xu K, Chen S, Li T, Xia H, Chen L et al (2019) A stress-responsive bZIP transcription factor OsbZIP62 improves drought and oxidative tolerance in rice. BMC Plant Biol 19:1–15

Yin X, Cui Y, Wang M, Xia X (2017) Overexpression of a novel MYB-related transcription factor, OsMYBR1, confers improved drought tolerance and decreased ABA sensitivity in rice. Biochem Biophys Res Commun 490(4):1355–1361

Zhang Y, Xia P (2023) The DREB transcription factor, a biomacromolecule, responds to abiotic stress by regulating the expression of stress-related genes. Int J Biol Macromol 243:125231

Climate Resilient Agricultural Practices to Optimize the Therapeutic and Nutritional Potential of Microgreens for Sustainable Development of Urban Society

9

Varun Bhardwaj, Pragati Yadav, Kumari Prabha Singh, and Astha Gupta

Abstract

Microgreens are young seedlings of edible plants that have gained recognition due to their heightened nutrient content and bioactive compounds. Medicinal microgreens are rich in secondary metabolites such as flavonoids, alkaloids, phenolic compounds, and terpenoids, which contribute to their therapeutic properties. The chapter examines the cultivation methods, nutritional composition, and medicinal potential of medicinal microgreens. Optimal growing conditions, including soil and hydroponic and aeroponic systems, along with harvesting techniques are explored. In light of climate change, the chapter touches upon microgreen cultivation as part of climate-resilient agriculture. Despite several advantages, several challenges to microgreen cultivation remain, including sensitivity to environmental factors and the need for standardization of cultivation protocols. The chapter also discusses the therapeutic potential of medicinal microgreens, namely, as pharmaceuticals, in immune support, anti-aging and skin health, digestive health, and cognitive function. While their benefits are promising, further research is warranted to increase bioactive compounds and validate health claims through clinical trials. As consumer interest in health-

V. Bhardwaj · P. Yadav
Department of Botany, University of Delhi, New Delhi, India

K. P. Singh
College of Basic Sciences and Humanities, School of Business Studies, Punjab Agricultural University, Ludhiana, Punjab, India

A. Gupta (✉)
Department of Biotechnology, Sharda School of Biosciences and Technology, Sharda University, Greater Noida, Uttar Pradesh, India
e-mail: astha.gupta@sharda.ac.in

V. Gahlaut et al. (eds.), *Smart Crop Development*,
https://doi.org/10.1007/978-981-95-7680-7_9

promoting foods grows, medicinal plant microgreens offer a plant-based approach to nutrition and natural medicine. This chapter therefore provides a review of their potential, highlighting current research gaps and future directions for their broader application in health and wellness.

Keywords

Climate-resilient agriculture · Medicinal microgreens · Bioactive compounds · Microgreens cultivation · Health-promoting foods

9.1 Introduction

Ever since the introduction of "vegetable confetti" to high-end restaurants in California during the 1980s–1990s, microgreens have amassed immense fame in both culinary and nutritional spheres. Due to their rich aroma, taste, colour, and flavour, they are now commonly used as garnishing in soups, salads, and sandwiches. Studies in peer-reviewed journals citing their importance have grown exponentially, and their introduction to common people through dietary magazines and articles has helped microgreens become a household name, being dubbed as "superfoods" or "functional foods". Microgreens are developed from various food crops such as grains, vegetables, and herbs, which consist of young cotyledonary-leaved seedlings, harvested at the 7–21 days stage, sized 1–3 inches. Beyond vibrant colours, delicate textures, and intense flavours, microgreens harbour are a treasure trove of nutritional and therapeutic compounds (Patil et al. 2024). Upgradation of nutritional value in daily diet has always been a part of human evolution. The focus of our daily food has shifted from providing mere nutrition and prevention of deficiencies to the prevention of chronic diseases such as obesity. As a result, there is a growing emphasis on foods that are rich in antioxidants, anti-inflammatory compounds, and other bioactive substances that can promote health and prevent disease. Medicinal microgreens, with their concentrated nutrient profiles and potent phytochemicals, have emerged as a powerful ally in this new paradigm of nutrition (Sven Verlinden 2020; Choe et al. 2018; Zhang et al. 2021).

Since ancient times, medicinal plants have been used in the prevention and cure of diseases. About 80% of the modern world's medicines are derived from traditionally acclaimed medicinal plants (Vaou et al. 2021). Medicinal plants are abundant accumulators of secondary metabolites such as terpenoids, phenols, steroids, flavonoids, tannins, and aromatic compounds, which are widely used in the pharmaceutical, cosmetics, and food industry. Approximately 30,000 antimicrobial compounds have been isolated from over 1340 plants, so far (Pant et al. 2021; Vaou et al. 2021). Some notable mentions include caper (*Capparis decidua*), cranberry juice (*Vaccinium macrocarpon*), lemon balm (*Melissa officinalis*), garlic (*Allium sativum*), and tea tree (*Melaleuca alternifolia*). Indian Ayurveda mentions the use of plants such as Neem (*Azadirachta indica*), Indian snake root (*Rauwolfia serpentina*), Gotu Kola (*Centella asiatica*), Myrobolan (*Terminalia* species), Sandalwood (*Santalum album*), Camphor (*Cinnamomum camphora*), *ela* or cardamomum (*Elettaria cardamomum*), and Aswargandha (*Withania somnifera*) for their remarkable medicinal properties. These plants have been revered for centuries for their ability to treat a wide range of ailments and promote overall health and well-being (Vaou et al. 2021; Gurib-Fakim 2006).

The allure of medicinal microgreens lies in their simplicity and accessibility. Minimal space and resource requirements allow for greater availability to masses. Unlike their mature counterparts, microgreens are harvested at an early stage of development, when their nutritional content is at its peak. This concentrated nutrient profile makes them a potent source of pharmaco-active compounds, which are essential for promoting overall health as well as ameliorating ailments. Different plants show differences in the richness of bioactive compounds, therefore, a customized diet, according to the target disease, is preferable. By selecting specific microgreens tailored to target particular health conditions, individuals can maximize the therapeutic benefits and support their unique health needs effectively. This chapter provides an overview of the major class of medicinal microgreens under study across the world, further expanding on their nutritional and medicinal properties, cultivation practices and maintenance, and their applications in therapeutic and various health benefits.

9.2 Microgreens as Medicinal Plants: Medi-microgreens

Medicinal plants grown as microgreens refer to the tiny, immature form of the plants that are harvested when they have true leaves, usually within 1–3 weeks of germination. Microgreens are nutrient-dense greens rich in minerals, vitamins, and beneficial compounds. Microgreens are famous for their versatility in culinary applications, and their potential as medicinal plants has increased their attention. Many studies have found that microgreens have more nutritional value than mature plants (Giordano et al. 2022).

Due to their colourful appeal, crisp texture, and taste, juvenile, tender greens are used in salads, chutneys, and garnishes. These greens can also be used to make smoothies, cookies, powders, supplements, etc. (Maru 2024). Medicinal microgreens include species from varied families, such as herbs, vegetables, cereals, roots, and edible flowers. Table 9.1 describes the growth duration and taste of medicinal plants grown as microgreens.

9.3 Nutritional Properties of Medi-Microgreens

The increasing interest in health and wellness has led to an outpouring in demand for healthy and nutritional food options. Microgreens have become the most trending topic because of their high nutritional and medical properties among people

Table 9.1 Examples of microgreens, growth duration, and taste

S. no.	Microgreen plant	Family	Growth duration (days)	Taste	References
Vegetables					
1	*Amaranthus tricolor* (red amaranth)	Amaranthaceae	7–10	High intensity of grassy, with low astringency, and bitterness	Rozali and Rahim (2024)
2	*Eruca vesicaria* (Arugula)	Brassicaceae	20	Hot, bitter, astringent, and sour	Michell et al. (2020)
3	*Brassica juncea* (L.) *czern.* (Mustard)	Brassicaceae	14	Bitter, pungent, and sulphur like	Marchioni et al. (2021)
4	*Coriander sativum* (Cilantro)	Apiaceae	12–15	Fresh and citrusy	Shah and Rathod (2024)
Herbs					
1	*Foeniculum vulgare* Mill. (Fennel)	Apiaceae	10–14	Slightly sweet and anise like	Liu et al. (2024)
2	*Trigonella foenum-graecum* (Fenugreek)	Fabaceae	6–8	Strong bitter	Senevirathne et al. (2019)
3	*Perilla frutescens* var. *crispa* (Chinese red basil)	Lamiaceae	20	Citrusy, spicy, and woody	Dimita et al. (2022)
4	*Perilla frutescens* var. *frutescens* (Chinese green basil)	Lamiaceae	20	Fruity, sweet, spicy, and herbaceous	Dimita et al. (2022)
Roots					
1	*Beta vulgaris* L. (Red beet)	Amaranthaceae	14	Slightly earthy and sweet	Rizvi et al. (2024), Karuthedath (2024)
2	*Panax ginseng* (Korean ginseng)	Araliaceae	10–20	Mild bitter	Ebert (2022)
Cereals					
1	*Triticum aestivum* (Wheatgrass)	Poaceae	12–15	Acrid	Mishra et al. (2025)

who want healthy food without compromising their routine. Hence, it is easily cultivable for a very short period (7–21 days) (Xiao et al. 2016). Even some research has shown that it contains four to forty times more nutritional value counterpart fully grown vegetables (Weber 2017). Microgreens are rich sources of vitamins (A, C, and E), minerals, macronutrients (Ca, Mg, P, K, Na, etc.), and phenolic compounds such as flavonoids, carotenoids, and glucosinolates. These compounds possess antioxidant, anti-inflammatory, and anticancer properties, which can significantly aid in treating or preventing chronic diseases (Bhaswant et al. 2023). Microgreens from varied families show unique attributes that enhance their appeal and health benefits. For example, the Brassicaceae family, which includes choy sum and mustard greens, is well-known for its high bioactive chemical content and robust nutritional profile (Alloggia et al. 2023). These chemicals have a number of health benefits, including improved cardiovascular health and detoxification processes. Meanwhile, the Lamiaceae family, which includes herbs such as cilantro and basil, is known for its high concentration of phytonutrients and secondary metabolites. For decades, these natural substances have been recognized for their health benefits, such as their ability to relieve fever, alleviate body aches, and immune-boosting properties (Fayezizadeh et al. 2023). Likewise, every crop family differs in characteristics. Microgreens offer a variety of flavors, textures, and benefits, making them a healthy and tasty addition to any meal. With their myriads of health benefits and culinary possibilities, microgreens are poised to become a staple in the diets of individuals seeking to enhance their well-being through nutritious food choices (Karuthedath 2024). Table 9.2 comprises different microgreens of various plant families possessing vivid nutritional and medicinal properties.

9.4 Cultivation of Medi-microgreens

Two factors play most important role for medicinal microgreens to become superfood in future, the bundle of nutrition they provide and the ease of cultivating them. While the overall cultivation and handling of microgreens is much easier than growing full-fledged crops of the same species, there are proper techniques and procedures for making out the best germination rate and high yield. A wide range of parameters, such as the selection and pre-treatment of appropriate species seeds and growing mediums, soil and soil less, the indoor and outdoor growing conditions, illumination system, dosage of fertilizers, irrigation practices, harvesting techniques, and post-harvest storage and maintenance, need to be considered. The seeds of different species generally have different germination techniques to get the desired results. Other notable factors are the choice of chemicals to be used, such as the usage of fertilizers, or pesticides must be avoided if going for an organic microgreen product. Recently, the medium of cultivation in agriculture sector has seen a shift from soil-based cultivation to more advanced agricultural practices such as water-based aquaponics, hydroponics, and air-based aeroponics. The maintenance of a crop is as important as its germination. Hence, a closer eye is required for their growth and maintenance as the young leaves are very soft tissue and prone to microbial infections due to humidity and nutrients availability in the medium. The harvesting time of microgreens varies from 7 to 21 days; once harvested, their careful packaging and a suitable packaging material must be ensured. A shelf life maintenance-oriented packaging is preferred for maintaining their freshness and their transport. All these important factors contributing in successful cultivation of microgreens are summarized in Table 9.3 (Gunjal et al. 2024; Singh et al. 2024).

Table 9.2 Microgreens of various plants and their nutritional and medicinal properties

Microgreen	Family	Nutritional properties	Medicinal properties	References
Brassica rapa ssp. *chinensis* var. *parachinensis* (choy sum)	Brassicaceae	Essential amino acids: 15.8% Vitamin K1: 377 ± 29 μg/100 g FW Glucosinolates: Gluconapoleiferin 517 ± 125 μg/100 g FW) Potassium: 412 ± 47 mg/100 g FW Other macrominerals (magnesium, calcium, manganese, iron, nickel, copper, sodium, zinc) were in lower amounts than potassium	Effective in chronic diseases like cardiovascular, diabetes, nephropathy, and neuropathy It contains plenty of antioxidants and secondary metabolites	Zou et al. (2021)
Triticum aestivum L. (wheatgrass powder)	Poaceae	Protein: 25 g/100 g Chlorophyll: 513.50 mg/100 g β-Carotene: 182.46 μg/100 g Vitamin C: 14.4 mg/100 g Riboflavin: 2.620 μg/g Folic acid: 84 μg/100 g Total phenols: 17.95–38.49 mg GAE/g Total flavonoid: 10.56–30.64 mg QE/g DPPH: 62.12%–82.53%	Anti-inflammatory, anticancer, antidiabetic, immunomodulatory, cardiovascular, protective, and promotes good digestive health	Mishra et al. (2025)

Ocimum sp. (basil) Cultivar: Kapoor	Lamiaceae	Chl a: 0.93^{a} mg g – 1 FW Chl b: 0.24^{a} mg g – 1 FW Chl a + b: 1.18^{a} mg g – 1 FW Antioxidant capacity: 57.49^{b} Polyphenols: 1085.53^{b} mg GA 100 g – 1 FW Flavonoids: 1.58^{g} mg CAE g – 1 FW Anthocyanin: 12.83^{h} mg 100 g – 1 FW Nitrate: 721.22^{abc} mg kg – 1 FW Vitamin C: 2.40 ^{a} mg g – 1 FW	Anti-inflammatory, antioxidant activity, analgesic effect, antipyretic effect, stress-reducing effect, and immune-boosting properties	Fayezizadeh et al. (2023)
Linum usitatissimum (flaxseed)	Linaceae	Chlorophyll: 62.6% Carotenoid: 24.4% Phenol: 37.8% Flavonoids: 2.48 mg CAE g^{-1} FW Total phenols: 3.76 mg GAE g^{-1} FW Chlorogenic acid: 1.10 mg g^{-1} FW Antioxidant capacity: 8.06 μmol TEAC g^{-1} FW	Antioxidant, antidiabetic, and anticancer	Puccinelli et al. (2022)
Foeniculum vulgare Mill. (fennel)	Apiaceae	Protein: 4.4 g/100 g TDF and IDF: 4.28 and 3.98 g/100 g Alpha-tocopherol: 23.8 mg/100 g Beta-carotene: 9.1 mg/100 g Ascorbic acid: 52.5 ± 2.7b Oxalic acid: 68.2 mg/100 g FW Calcium: 94.7 ± 0.5f mg/ 100 g Potassium: 480.5 ± 6.0 h mg/ 100 g Sodium: 228.3 ± 0.6 mg/ 100 g	Weight loss, prevent and treat constipation, control of serum cholesterol level, lower the chance of diabetes mellitus and stimulate beneficial microorganisms	Ghoora et al. (2020)

(continued)

Table 9.2 (continued)

Microgreen	Family	Nutritional properties	Medicinal properties	References
Coriander sativum (coriander)	Apiaceae	Calcium: 10140.79 mg/kg Iron: 1503.58 mg/kg Potassium: 55530.98 mg/kg Phosphorus: 10504.07 mg/kg Sodium: 19004.15 mg/kg Zinc: 149.47 mg/kg Chlorophyll—13.36 mg/kg FW Lutein—98.6 mg/k DW Beta-carotene—325.1 mg/kg DW Ascorbic acid—121.40 mg/kg DW Polyphenols—15.25 mg/g DW	Antacids, anti-inflammatory or antifungal agent, relieve nausea, jaundice, diarrhoea, diabetes, abdominal pain, ulcers, eye infections, and headache	Shah and Rathod (2024), Bhaswant et al. (2023)
Moringa oleifera L. (moringa)	Moringaceae	Chlorophyll a: 1180.7 ± 329.0 μg g − 1 FW Chlorophyll b: 631.7 ± 35.1 μg g–1 FW Total chlorophyll: 1812.4 ± 340.4 μg g–1 FW Carotenoid: 458.3 ± 52.01 μg g–1 FW	Anticancer, treat cardiovascular and asthma-like diseases	Artadana and Pandji (2023)
Anethum graveolens (dill)	Apiaceae	Total carotenoids: 0.196 mg/g tw Total chlorophyll: 0.713 mg/g tw Chlorophyll a: 0.528 mg/g fw It also has a high amount of ascorbic acid, and flavonoids too.	Anti-inflammatory, antimicrobial, and relieve insomnia	Yordanova and Miroslavova (2024)
Eutrema japonicum (Wasabi)	Brassicaceae	Total glucosinolates content: 6.1 μmol/100 g FW TAA: 66.9 ± 3.5 mg/100 g FW Beta-carotene: 8.5 ± 0.5 mg/100 g FW Lutein/zeaxanthin: 4.1 ± 0.1 mg/100 g FW Volaxanthin: 1.8 ± 0.1 mg/100 g FW Total carotenoids: 14.4 ± 0.7 mg/100 g FW	Anticarcinogenic, antimutagenic, antidiabetic, and anticholinergic	Dereje et al. (2023)

Eruca sativa Mill. (arugula)	Brassicaceae	Total protein: 4.8 a% Total phenols: 3304.6 a mg kg – 1 of GAE Nitrates: 2442.2 a g L – 1 Beta-carotene: 12612 e mg equiv. 100 g – 1 Calcium: 380.7 bc mg kg^{-1} Iron:69.1 c mg kg^{-1} Magnesium: 202.6 bc mg kg^{-1} Sodium: 25.7 ab mg kg^{-1} Phosphorus: 0.8 a mg kg^{-1}	High amount of bioactive results in anti-inflammatory and antioxidant health benefits	Rebolledo et al. (2024)
Beta vulgaris L. (red beet)	Amaranthaceae	Phenolic content: 110.77 mg GA·100 g^{-1} FW Flavonoid content: 120.36 mg RU·100 g^{-1} FW Protein: 2.73 g 100 g^{-1} FW in agril Vitamin C: 18.35 mg 100 g^{-1} FW in vermiculite	Antioxidant and anti-inflammatory	Balik et al. (2024)
Echinacea purpurea L. (purple coneflower)	Asteraceae	Total chlorophyll: 23.22 μg g-1 Chlorophyll a: 15.13 μg g-1 TA Chlorophyll b: 6.75 μg g-1 TA Carotenoid: 3.31 μg g-1 TA Antioxidant capacity: 134.1 μmol TE g-1 Flavonoid: 23.83 mg QE 100 g-1 Phenolic content: 199.3 mg GAE g-1 Ascorbic acid: 66.11 mg LAA 100 g-1 Calcium: 16.07 g kg-1 Potassium: 46.12 g kg-1 Magnesium: 6.09 g kg-1 Sodium: 3.12 g kg-1	Reduces the risk of cancer, cardiovascular diseases, and degenerative diseases	Tunçtürk et al. (2024)

(continued)

Table 9.2 (continued)

Microgreen	Family	Nutritional properties	Medicinal properties	References
Amaranthus tricolor (red amaranth)	Amaranthaceae	Total phenolic content: 16.25 ± 0.41 mg GAE/g DW Total flavonoid content: 8.66 ± 0.22 mg Quer/g DW Percentage of DPPH inhibition: 49.14 ± 8.24%	Antioxidative, antimutagenic, anticancer, antifungal, anticarcinogenic	Rozali and Rahim (2024)
Brassica juncea (mustard)	Brassicaceae	Total phenolics: 49.3 mg GAE/100 g FW Ascorbic acid: 27.8-57 mg/100 g FW Alpha-tocopherol: 31.6 mg/100 g FW Beta-carotene: 7.4 mg/100 g FW Chlorophyll: 52.4–64.7 mg/100 g FW Calcium: 51 mg/100 g FW Magnesium: 35 mg/100 g FW Potassium: 322 mg/100 g FW Phosphorus: 24 mg/100 g FW Sodium: 74 mg/100 g FW Zinc: 1.21 mg/100 g FW Iron: 2.46 mg/100 g FW	Antioxidant, anticancer, antidiabetic, and anti-inflammatory	Dereje et al. (2023)

Table 9.3 Studies on microgreen cultivation practices for optimization of growth parameters

Cultivation step	Microgreen crop tested	Experimental design	Growing parameters	Findings	Reference
Pre-sowing seed treatment	Parsley (*Petroselinum hortense*)	Seeds were soaked in water at 10 °C for 12 hours	Growing medium: 2/3 peat mixed with 1/3 perlite; seeding density of 5 seeds per cm^2	Increased ascorbic acid content, germination rate and yield	Çakır et al. (2025)
	Chia (*Salvia hispanica* L.)	Seed soaking for 6 and 12 hrs, at room temperature and 70–80 °C	Coconut coir with sand and rice husk ash (1:1:1)	Higher percentage of germination, height, and fresh weight	Junpatiw and Sangpituk (2021)
	Table beet and chard (*Beta vulgaris* L.)	H_2O at 20 °C for 48 h; 0.3% H_2O_2 at 20 °C for 48 h; 0.3 M HCl at 20 °C for 2 h; or 4% NaOCl at 20 °C for 3 h	Matric priming (−1 MPa at 12 °C for 6 d in fine vermiculite)	Improvement in germination and growth	Lee et al. (2004)
Seed density	Chia (*Salvia hispanica* L.)	Seed sowing rate tested 56, 93, 130, 167, and 204 g/m^2	Coconut coir with sand and rice husk ash (1:1:1)	Highest fresh weight/sowing rate of 204 g/m2 and harvesting at 6–9 days	Junpatiw and Sangpituk (2021)

(continued)

Table 9.3 (continued)

Cultivation step	Microgreen crop tested	Experimental design	Growing parameters	Findings	Reference
	Rapini (*Brassica rapa*), kale (*B. oleracea*), and cress (*Lepidium sativum*)	3 to 5 seeds•cm^{-2} seed density and 11–14 days of harvesting	Temperature 20 °C (day/night), relative humidity (RH) ≥60%, photoperiod 18/6 (day/night), PPFD 30 μmol/m2/s	19% increase in yield at 5 seeds/cm2 density	Signore et al. (2024)
Type of growing media	Black radish (*Raphanus sativus*), broccoli (*Brassica oleracea*), and red beet (*Beta vulgaris* L.)	A peat-based substrate, perlite with an inorganic structure, vermiculite, a cotton-linen textile, agril, and filter paper	Climate-controlled growth chamber, at 20 °C and under 50% RH 16 h of light and 8 h of darkness	Vermiculite produced the overall highest yield	Balik et al. (2024)
	Chia (*Salvia hispanica* L.)	Five media: Coconut coir, rice husk ash, sand, coconut peat/rice husk ash (1:1), and coconut peat/sand/rice husk ash (1:1:1)	Seed soaking for 6 and 12 hrs, at room temperature and 70–80 °C	96–98% germination rate, improved height and fresh weight in medium coconut peat/sand/rice husk ash (1:1:1)	Junpatiw and Sangpituk (2021)
	Radish (*Raphanus sativus*)	75% white sphagnum peat +25% vermiculite and 75% coco coir dust +25% vermiculite	Grown in green house at 85% air humidity, germination at 20 °C, and maintenance below 34 °C during growth	Coco coirdust are low-cost and renewablealternative substrates	Thuong and Minh (2020)
Nutrient supplement	*B. oleracea*, *B. oleracea*, and *Brassica rapa* L.	NH_4/NO_3 molar ratios (5:95, 15:85, and 25:75)	Controlled hydroponic conditions	Increase in yield, bioactive compounds, and mineral concentrations	Palmitessa et al. (2020)

	Amaranth (*Amaranthus tricolor*), arugula (*Eruca sativa*), basil (*Ocimum basilicum*), broccoli (*Brassica oleracea*), radish (*Raphanus sativus*), kale (*Brassica napus*), mustard (*Brassica juncea*), pea (*Pisum sativum*)	Soaking in water at RT, fertilized with water-solublefertilizer 20 N-8.7P-16.6 K maintaining pH of 6.56 andelectrical conductivity of 0.41 $mS{\cdot}cm^{-1}$	Greenhouse cultivation at 25 °C with natural light	Increased accumulation of macro- and micronutrients and improved yield	Li et al. (2021)
Illumination system	Mustard (*Brassica juncea*) and kale (*Brassica napus*)	Blue (447 nm) and red (660 nm) light-emitting diodes (LEDs) with total photon flux density (TPFD) of 250 $\mu mol\ m^{-2}\ s^{-1}$	Nutrient-rich hydroponic solution, pH 5.5–6.5 and EC: 1.3–1.7 mS/cm	Blue light led to increase in elongation; B/R affect growth parameters and bioactive compounds	Brazaityte et al. (2021)

9.4.1 Pre-treatment of Seeds

The yield of a microgreen crop is heavily dependent on the quality of the seeds, which is determined majorly by three factors: seed germination (preferably >95%), seed density, and seed weight. Also, the seeds should be devoid of any disease-causing microorganisms on their surface. Proper instructions as directed by the US Food and Drug Administration (USFDA, 2022: https://www.fda.gov/media/127972/download?attachment) for carrying out sprout cultivation can be followed for the elimination of the pathogenic microorganisms. Thorough cleaning and washing of the seeds are mandatory post-chemical treatments (Galieni et al. 2020). In order to enhance germination and growth rate of medicinal microgreens, different soaking and priming methods have been studied by researchers. Matric priming applied to seeds of beet and chard microgreens resulted in 91% and 99% germination, respectively. The priming treatment included 6 days in vermiculite under −1 MPa and 12 °C temperature. Application of different soaking solutions, such water, H_2O_2, HCl, NaCl, and NaOCl in moderate amounts, were also tested in the same study, and it was observed that seeds soaked in 0.3% of H_2O_2 for 48 hours were the earliest to germinate (Lee et al. 2004). It was also observed that seed soaking in cold water is beneficial for large-sized seeds such as peas and radish. Seed size is also a determinant of the sowing range, it varies from 1 seed per cm^2 in larger seeds to 4 seeds per cm^2 in smaller seeds. A software called microgreen seed density calculation software has been developed for more precise sowing rates in a defined area (Microgreens seed density calculator 2020: https://extension.psu.edu/microgreens-seed-density-calculator). Application of such methods may significantly improve the germination rates of medicinal microgreens.

9.4.2 Growing Medium

Both soil-less and soil-based growing media are being used commercially for the production of microgreens, each with its own advantages and disadvantages. Certain parameters are required in the growing medium of microgreens for their proper germination and growth: pH ranging from 5.5 to 6.5, a low electrical conductivity, 55–70% (v/v) water holding capacity, and 20–30% (v/v) aeration. The classic soil-based medium is still popular among cultivars as it provides repertoire of micronutrients and retains moisture content as required. Substitutes of soil such as well-draining peat-moss mixes, coconut coir, synthetic or hemp seed germination mats, etc. are also gaining popularity. These soil-less substrates have enabled user-friendly kits where mats are available with pre-embedded seeds, thus making microgreens cultivation easier in households (Gunjal et al. 2024).

Other soil-less alternative growing mediums are taken up by commercial farmers for controlled and regulated environmental conditions. When a water-based nutrient-rich solution is used for the cultivation of crops, the technique is formally called hydroponics. There are two different types of hydroponic culture. First is the deep-water culture system. Here, roots develop in a deep water reservoir to gather

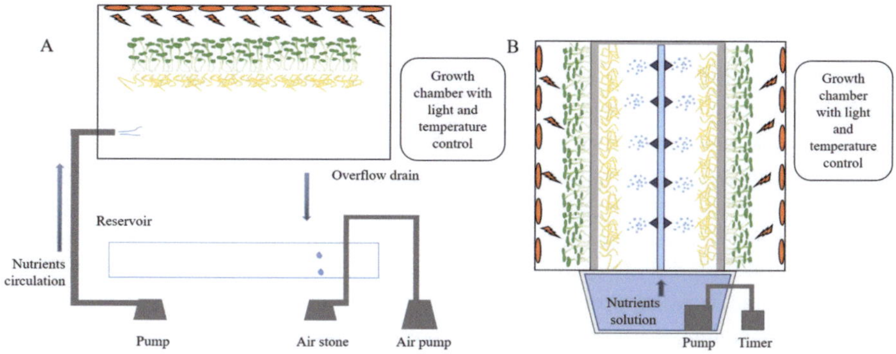

Fig. 9.1 Diagrammatic representation of (**a**) hydroponic/aquaponic culture, where roots are submerged in a circulating nutrients solution, and (**b**) aeroponic vertical farming set up for medicinal microgreen cultivation, where nutrient solution is taken up and sprayed in mist on roots from spray nozzles

nourishment, and reservoir is actively aerated to avoid root hypoxia. A special component called air stone or air pump is provided for oxygenation and aeration in the nutrient medium (Fig. 9.1). The second one is called the nutrient film technique. It involves continuously running water and provides a better aeration facility; however, as it relies on the continuous availability of water from a pump, its failure can cause rapid root damage due to dehydration. Another soil-less culture is called aeroponics, wherein consumption of water is reduced to the minimum. It is a process of growing plants by suspending their roots in the air, and nutrient-rich water solution is sprayed on the roots. Two major advantages are provided by this method, adequate oxygenation to the root zone and an efficient water usage system. Crops grown in aeroponic cultures are highly susceptible to unregulated nutrient supplies, disturbed nutrient recipes, power cuts, and other failures in instrumentation, making aeroponic a high-precision technique. Another alternate technique is aquaponics, where the culture media is water. It is generally combined with aquaculture (cultivation of fish) and depends on the breakdown of fish waste into micro- or macronutrients by the action of naturally occurring microbes for plant nourishment. As the natural supply of nutrients may fail to provide plant-specific requirements, a constant enrichment and adjustment of nutrient composition is required (Teng et al. 2023).

9.4.3 Key Factors in the Growing Medium

An additional nutrient supplement is not necessarily required by the medicinal microgreens for their survival; however, the application of fertilizers with different concentrations of macro- and micronutrients such as NH_4^+/NH_3^- ratio has shown to alter fresh weight and dry weight of broccoli and cauliflower microgreens. In addition, the levels of proteins, beta-carotene, and nitrates were also elevated (Palmitessa et al. 2020). Fertilizers such as calcium nitrate, in moderate concentration (150 mg/L), have been shown to increase the yield of microgreens (Gunjal et al. 2024). However, the usage of chemical fertilizers is generally discouraged as the medicinal microgreens are very fragile and susceptible to harsh environments. Organic bio-based fertilizers are preferred for sprinkling and spraying small amount at the juvenile stage, or a light fertilization may be achieved by mixing organic fertilizers in nutrient medium (Rizvi et al. 2023).

We also propose an idea here for the usage of protoplasts as a natural nutrient supplement in the growing medium. A mixture of micro- and macromolecules such as ions, polysaccharides, proteins, lipids, enzymes, etc. are packed in the protoplast, having a high potential for enhancing the growth of the microgreens if an optimized concentration is added to the culture. This aspect of research can be explored further by experimental designs on the effects of protoplast application on microgreen growth.

Other methods have also been employed by researchers to enhance the antioxidant levels in microgreen crops. One such study by Islam et al. (2019) exploited the natural mechanism of wheat microgreen to increase the content of flavonoids, beta carotenes, phenolic acid, anthocyanins, vitamin C, and other radical scavenging activities as a part of the response to salt stress in an optimized 12.5 mM to 25 mM NaCl concentration, where microgreen growth and yield were unaffected by salt stress. Similar studies performed in cabbage microgreens (Patras 2021) and *Brassica* microgreens (Maina et al. 2021) resulted in elevated antioxidant levels.

9.4.4 Usage of Advanced Technology

The high automation and high precision requirements of large-scale commercial microgreen cultivation necessitate the inclusion of the latest technological advancements of modern agriculture. The treatment combining the advanced oxidation process and micro-nano bubbles technique played a remarkable role in decontamination by showing anti-microbial activity along with the enhancement in phytonutrients in roselle medicinal microgreens (Phornvillay et al. 2022). Antimicrobial activity can also be treated by low-frequency ultrasound (20–40 kHz) and a $CaO/Ca(OCl)_2$ spray (Dong et al. 2022). Chemical treatments involve using chemicals as elicitors for triggering plant defence responses and using chemicals as preservatives for increasing post-harvest shelf life. Application of $CaCl_2$ in combination with UV-B in broccoli microgreens helped in the improvement of glucosinolate levels and shelf life (Lu et al. 2021). In a similar study by Gao et al. 2021, UV-A and Na_2SeO_3, in combination, enhanced antioxidant activity. This approach has also been applied during the storage of amaranth microgreens; improved anthocyanins, such as total betalain concentrations, were observed on immersion in 0.1 μM cyanocobalamin (vitamin B12) (Supapvanich et al. 2020). *IoT (Internet of things)* is another advanced technology being adopted in agricultural sector, while its application in microgreen cultivation is yet to expand. IoT provides basis for development and operation of real-time data and other environmental parameters that may be controlled remotely such as the nutrient solution, light intensity, humidifiers, etc. These IoT-based autonomous control systems have been adopted by some leading CEA (controlled environment agriculture)-based microgreen cultivation companies for commercial production (Teng et al. 2023).

9.4.5 Harvesting

The harvesting period of microgreens ranges from approximately 1 to 3 weeks, when the height of the plant reaches 5–10 cm depending on the growth of the species. The cut is made 1 cm above the medium; usage of blunt scissors or blades resulted in a reduction in storage life; hence, sharp blades should be preferred. The carefully harvested microgreens are washed and then stored in cold storage to repress the rate of senescence and spoilage. Removal of excess water and moisture from the washed microgreens is very important prior to packaging. Harvesting is also affected by the time of the day, especially in the spring-summer season. Harvesting of baby red chard and leaf roquette at the end of the day increased their shelf life by 2–6 days (Clarkson et al. 2005). The quality and storage life of baby spinach leaves were significantly improved on harvesting during early morning (Garrido et al. 2015).

The shelf life of microgreens ranges from 2 to 4 days at normal temperature, while it may extend to 10–15 days if stored at 5 °C; hence, the transportation of medicinal microgreens requires quick packaging and transport in a controlled and regulated environment. Other factors to be taken into consideration during packaging are relative humidity (90–95% optimum), pO_2/pCO_2 balance, and packaging material (Kyriacou et al. 2016). Moreover, in a study, a pre-harvest spray of calcium chloride improved the shelf life of broccoli microgreens along with the activation of ROS detoxification enzyme activities (Kou et al. 2015).

9.4.6 Challenges in Cultivation

The primary challenges encountered during cultivation of the medicinal microgreens are the factors affecting the growth and production, i.e. the pre-harvest parameters, post-harvest storage, and shelf life defining factors. Pre-sowing treatment of seeds generally involves the usage of chemicals for effective sterilization. Also, during the growth period of microgreen crops, fertilizers and pesticides are often required to be sprayed to increase production and avoid pests and microbial infections. Microgreens are sensitive to such harmful treatments; therefore, their application in an unregulated manner may result in problems associated with their production. Another challenge faced by microgreen producers in mass manufacturing is their shelf life and storage. The soft leaf tissue is very prone to damage during transport; therefore, their quality rapidly decreases during storage. The susceptibility of microgreens to pathogenic and microbial infections puts them at risk of harbouring food-borne diseases (Abaajeh et al. 2023; Lone et al. 2024). Therefore, the cultivation and harvesting of medicinal microgreens are a high-precision job.

9.4.7 Sustainability

The cultivation of microgreens in greenhouses has enabled their production and availability throughout the year without any hindrances from the changing weather and infertile soil conditions. Microgreen cultivation has emerged as a highly sustainable agricultural practice in the ever-increasing urban population. Some of the reasons are as follows: high on nutrition, compatible with both indoor and outdoor cultivation, short growth period enabling quick production, a minimum or no fertilization requirement, and their compatibility with vertical farming methods. A large land area and fertile soil are not a pre-requisite for growing microgreens; they can be easily grown in a well-maintained set-up both indoors and outdoors. Very little space, as compared to the full-grown crops, is required by these microgreens; moreover, space-efficient hydroponic, aquaponic, and aeroponic cultures are being adopted for vertical farming systems by many commercial start-ups and companies (Du et al. 2022).

9.5 Applications of Medi-microgreens in the Field of Nutrition and Medicine to Improve the Human Health

The young and tender microgreen seedlings are plausible quick and high-potential food especially for the urban population, due to their small size, a small area required for growth and maintenance, less resources required, short cultivation time, and easily amenable to kitchen gardening and indoor farming practices. So, they can be made available in every household. They have gained huge popularity among dieticians, nutritionists, other healthcare professionals, and researchers because of their richness in valuable physiologically active compounds. Medicinal microgreens are abundant in phytochemicals, bioactive compounds, carotenoids, anthocyanins, glucosinolates, tocopherols, antioxidants, vitamins, minerals, and other nutrients. It is already well-known that medicinal plants are a valuable source of active phytochemicals. They have been studied extensively for their usage in the treatment of various diseases. Medicinal microgreens are younger versions of their mature counterparts and contain higher concentrations of few active bio-compounds. Ease in growing, easy availability, and high nutrition—all these factors contribute in making medicinal microgreens a better herbal drug for the future. Some medicinal benefits of medicinal microgreens are listed in Fig. 9.2.

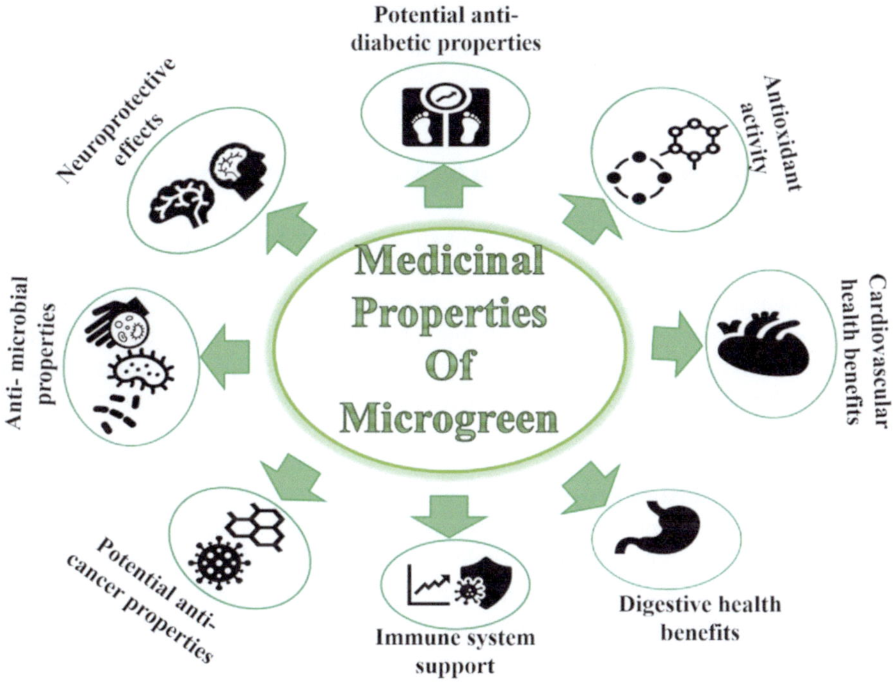

Fig. 9.2 Diagrammatic representation of the medicinal properties of microgreens

9.5.1 Chemical Composition and Nutritional Components of Medicinal Microgreens

Micro- and macro-nutrients, vitamins, and minerals are very important in a healthy diet. We obtain these essential elements from fruits, vegetables, and especially green leafy vegetables. It has now been well studied that microgreens provide a higher content of these essential elements than their mature counterparts. Vitamin C or ascorbic acid, an essential vitamin, required in daily human diet, also acts as an antioxidant, when tested in different stages of a plant. It was found to be highest in the young leafy microgreen stage when compared with sprout and mature plant stage (Di Bella et al. 2020). De la Fuente et al. (2019) also reported a high total ascorbic acid content in broccoli microgreen and a comparable range in kale, mustard, and radish microgreens. Phylloquinone or vitamin K was estimated in six microgreens, namely, Dijon mustard, opal basil, bull's blood beet, red amaranth, peppercress, and China rose radish, and phylloquinone content varied from 2.1 to 4 g/kg within the selected species (Xiao et al. 2015). Carotenoids, phenolic antioxidants, and vitamin A are important for eliminating free radicles and hence play a vital role in immune system development. The total content of tocopherol (vitamin A), carotenoids, and phenolic contents was also present in higher amount in different microgreens as tested by Michele Paradiso et al. (2018). Microgreens are also rich in nutrients such as Na, K, Cu, Mn, P, Ca, Mg, K, Zn, and Fe, which are involved in biosynthesis and production of antioxidant enzymes and hormones (Bhaswant et al. 2023; Rizvi et al. 2023).

9.5.2 Health Benefits Provided by Medicinal Microgreens

Potential health-promoting benefits of medicinal microgreens has been under the lens due to their richness in bioactive compounds. These bioactive compounds if taken in the regular diet have the potential to keep lifestyle and diet-related diseases at bay such as diabetes and other microbial and inflammatory diseases. As medicinal microgreens harbour a higher content of phytoactive compounds, minerals, and vitamins, they have the potential to showcase some disease-preventing properties that are discussed below.

9.5.2.1 Antioxidative Properties

In extreme circumstances of oxidative stress conditions, plants have evolved a mechanism to cope with the free radicles generated in cells. Antioxidant enzymes like ascorbate peroxidase, superoxide dismutase, catalase, etc. work as scavenging molecules to capture ROS (reactive oxygen species). Jiang et al. (2016) observed elevated activities of these enzymes in the red cabbage microgreens, which further helped in modulating levels of cholesterol and hypercholesterolemia. Methionine, L-asparagine, L-glutamine, and homocysteine are the amino acids that reduce oxidative stress and are found in high concentrations in radish, green bean, kale, and amaranth microgreens (Wojdyło et al. 2020). Recent report by de la Fuente et al. (2019) evaluated broccoli (*Brassica oleracea* L. var. *italica* Plenck), green curly kale (*Brassica oleracea* var. *sabellica* L.), red mustard (*Brassica juncea* (L.) Czern.), and radish (*Raphanus sativus* L.) hydroponic microgreens for ascorbic acid, total carotenoids, total soluble polyphenols, total isothiocyanates, total anthocyanins, and total antioxidant capacity and concluded that the four hydroponic Brassicaceae microgreens present a wide array of bioactive compounds with highest antioxidant activities. Fenugreek microgreens (FGM) were studied for their antioxidant and antimicrobial potential by Jayaraman and Ramasamy (2024). The study reported FGMs to be rich in flavonoids, tannins, and phenols. Methanolic extracts of FGMs were shown to scavenge ABTS, DPPH, and H_2O_2 radicals and exhibited antibacterial activity against *Pseudomonas aeruginosa*, *Aeromonas hydrophila*, and *Staphylococcus aureus*. Antioxidant compounds from microgreens also play a role in cardiovascular health and nervous system protection (Tallei et al. 2024).

9.5.2.2 Anticancerous Properties

There is an ever-increasing compendium of research targeted at microgreens harbouring compounds with putative anti-proliferative functions. Compounds such as ascorbic acid, tocopherol, glucosinolates, and polyphenols have been implicated in prevention and remediation of various types of cancers (Rizvi et al. 2023). An in vitro study reported anti-proliferative activity of microgreens against colon cancer cells. The microgreens taken for study were mustard, kale, radish, and broccoli. They were found to have accumulated total carotenoids and isothiocyanates at the microgreen stage, which probably helped in expressing this anti-proliferative property (Sehrish et al. 2023). The same set of microgreens grown hydroponically in an independent study proved that their effect on colon Caco-2 cells in vitro was comparable with that of 5-fluorouracil, a cancer medication, on exposure for up to 24 h (de la Fuente et al. 2019). HCT116 cell lines were subjected to extracellular vesicles (EVs) derived from *Raphanus sativus* L. var. *caudatus* Alef microgreens. EV treatment led to inhibition of cell proliferation in concentration- and time-dependent manner and consequently to apoptotic cell death. Bioactive macromolecules, particularly β-pleated sheet proteins were implicated in arresting the growth of HCT116 cells, thereby confirming the anticancerous activity of *R. sativus* microgreens (Kaimuangpak et al. 2022, 2024). Anti-proliferative/pro-oxidant efficacy of green pea, soybean, radish, Red Rambo radish, and rocket microgreens was evaluated by Truzzi et al. (2021), using Ewing sarcoma lines RD-ES and A673, respectively. The study showed that green pea microgreen extracts had anti-proliferative and pro-apoptotic effects in RD-ES and A673 spheroids due to high polyphenol faction of the extract.

9.5.2.3 Metabolic Disorders and Weight Management

A person's metabolism is highly dependent on his/her diet. A high-calorie and high-fat diet sometimes become the cause of metabolic disorders. When planning for a healthy diet, one must not leave medicinal microgreens behind for they have shown to be rich in bioactive compounds and also help in curing different inflammatory diseases. Hence, incorporating microgreens in the diet can help prevent diseases like obesity, diabetes, and CVD (cardiovascular diseases). Huang et al. (2016) reported diet supplemented with microgreen helped in reducing LDL levels, inflammatory cytokines in the liver, and hepatic cholesterol ester, thus modulating cholesterol metabolism and weight gain. Reduction in body weight by consumption of broccoli microgreens is also reported in another study by Li et al. (2021). They reported improvement in glucose tolerance and decreased insulin resistance.

9.5.2.4 Diabetes

Type 2 diabetes mellitus (T2DM) is a lifestyle-based, age-onset disease resulting from sedentary lifestyle and a diet rich in calories and fat and low in fruits and vegetables. Current treatment methods are either expensive or have uncontrollable side effects such as damage to the liver, hypoglycaemia, and symptoms of digestive tract and abnormal kidney function (Vestergaard 2021). As a consequence, microgreens are being sought as functional foods to relieve diabetes-related symptoms. A recent report on T2DM Wistar rats showed that orally administered *Brassica carinata* microgreen ethanolic extract (BMEE) lowered fasting blood glucose (FBG) levels, increased oral glucose tolerance, reduced insulin resistance, and increased insulin sensitization. Test subjects also showed elevated expression of lipid profile and mRNA expression levels of insulin receptor substrate (IRS-1) and glucose transporter 2 (GLUT-2) and reduced expression of nuclear factor kappa light-chain enhancer of activated B cells (NFKβ) in the liver. Overall, BMEE induced antidiabetic activity in experimental mice (Nakakaawa et al. 2025). Lyophilized broccoli (*Brassica oleracea* L.), in addition to improving gut microbiota, led to hypoglycaemic effects, along with improvement in blood lipid status, inflammatory factors level, and antioxidant indexes in mice (Ma et al. 2022). Persistent hyperglycaemic state in T2DM leads to production of advanced glycation end products, which bring about pathophysiology of diabetic complications. Microgreens of spinach (*Spinacia oleracea*) were shown to possess the highest antiglycaemic activity (Zhou et al. 2023).

9.5.2.5 Kidney Disorders

Kidney disorders come under the category of metabolic illness due to other related diseases such as diabetes, heart-related diseases, and hypertension, caused by sedentary lifestyle. Additionally, potassium-rich food is typically avoided by kidney and heart patients for maintaining blood pressure. Most of the vegetables are potassium rich, so nutritional research has focussed on cultivation of low potassium diet. Hydroponically grown lettuce and chicory microgreens with media lacking potassium resulted in reduction of total potassium content in the cultivated microgreens (Sehrish et al. 2023).

9.5.2.6 Cardiovascular Diseases

Incorporating fruits and vegetables into the diet of people suffering from cardiovascular diseases is suggested by doctors. An addition of medicinal microgreens is thus highly recommended to be taken in the daily diet as they are even richer in phytoactive compounds and phytochemicals. In a laboratory investigation by Huang et al. (2016), they used high-fat diet and microgreen-rich diet to check the accumulation of cholesterol ester, triacylglycerol, and LDL in mice. They observed that preventing hypercholesterolemia may help prevent CVD diseases.

9.5.3 Medicinal Microgreens in Gastronomy

A wide variety of microgreens are obtained from several botanical families such as Brassicaceae, Amaranthaceae, Asteraceae, Fabaceae, Cucurbitaceae, etc. which serve different purposes in adding creativity and flavours to dishes. Medicinal microgreens are even richer in phytoactive compounds, so they serve an array of strong tastes such as spicy (radish and arugula), sour (salicornia and beet), pungent (Brassicaceae), and bitter (Cucurbitaceae). Hence, their usage varies typically on the wide range of shape of leaves, colour, flavour, taste, etc. in the food items (Renna et al. 2017).

9.5.4 Usage of Protoplasts Isolated from Medicinal Microgreens

As medicinal microgreens are highly rich in nutritional compounds, we propose a novel exploitation by extraction of protoplasts from young leaves of microgreens. The protoplasts isolated from the rich phytoactive compound containing microgreens may serve as a package of multi-nutrients for the growth enhancement in plants growing in hydroponic culture, or they might prove subs for the water-based solutions used in aquaponic and aeroponic cultures.

9.6 Fitness, Well-Being Benefits, and Therapeutic Potentials of Medi-Microgreens

9.6.1 Immune Support

Microgreens are an emerging food source, associated with a large assemblage of micro- and macronutrients, such as Zn, Fe, Ca, K, P, N, S, Mn, Se, and Mo, among others. Armed with a stockpile of phytochemicals, microgreens have immense potential to improve human health and ameliorate diseases (Bhaswant et al. 2023). Sinusitis is an inflammation of the paranasal sinuses caused by bacterial, viral, and fungal pathogens. Antibiotics are the preferred treatment for bacterial sinusitis. However, due to increase in antimicrobial resistance, especially in immune-compromised patients, it is imperative to explore alternative therapies. *Echinacea* sp., a perennial herbaceous plant, is a well-known dietary supplement for common cold and other infectious diseases. Herbal extracts from aerial parts (such as flowers, stem, and leaves) and roots are known to stimulate the immune system against infections. *Echinacea purpurea* extract was shown to prevent respiratory tract infections (RTI) and secondary complications such as bronchitis, sinusitis, and pneumonia in children from 3 to 12 years old. The study indicated significant differences between control and test groups in the intensity of sinusitis in patients (Ogal et al. 2021; Poudineh et al. 2024; Vimalanathan et al. 2017). While studies on the effect of *Echinacea* extracts on RTIs are numerous, more research is warranted on *Echinacea* microgreens cultivation, their biochemical characterization, and their clinical efficacy. Tunçtürk et al. (2024) investigated changes in ascorbic acid, photosynthetic pigment, phenolic and flavonoid content, and antioxidant capacity of microgreens of *Echinacea purpurea*, *Calendula officinalis*, and *Silybum marianum*. The study concluded that *Echinacea purpurea* microgreens were the highest accumulators of photosynthetic pigment, total antioxidant substance, and flavonoids. The human gastrointestinal tract harbours a complex population of microorganisms such as bacteria, viruses, archaea, and fungi, which have a marked impact on general homeostasis and disease development (Clemente et al. 2012; Thursby and Juge 2017). Modulation of gut microbiome is an emerging preventive strategy to halt disease progression and promote overall well-being. Owing to bioactive compounds present in microgreens, the bacterial population in the gut can be regulated, for protection against a variety of chronic diseases. Using bioinformatic approaches, more than 400 phytocompounds which target gut microbiome were identified by Ni et al. (2015). The study also highlighted the role of flavonoids as a class of compounds that have a pronounced effect on gut microbiome with quercetin, catechin, and puerarin being key regulators. Cocoa-derived flavonoids were found to increase *bifidobacteria* and *lactobacilli* abundances and reduce *clostridia* counts paralleled by significant reduction in blood plasma triacylglycerols (Tzounis et al. 2011). Since microgreens are abundant in flavonoids, their consumption can help improve populations of healthy gut bacteria. Furthermore, improvement in gut microbiome enhances/alters bioavailability of essential nutrients. More research however is warranted to elucidate the interactions of microgreen compounds with the gut microbiome.

9.6.2 Anti-aging and Skin Health

Aging is a biological process marked by progressive loss of cellular and organ functions due to genetic and environmental factors, lifestyle, and nutrition. Oxidative stress, a principal mechanism of cellular aging, is caused by an imbalance between pro- and antioxidants and results in several age-associated, noncommunicable diseases (Martemucci et al. 2022).

Premature aging is closely linked to contemporary lifestyle patterns, such as more consumption of processed food, exposure to environmental pollutants, and chronic stress. Free radicals produced by environmental contaminants and UV radiation exposure contribute to oxidative stress and accelerate extrinsic skin aging. All of these elements work together to hasten the onset of age-related health problems (Bauer et al. 2009; Poljšak and Dahmane 2012). Antioxidants have the propensity to scavenge free radicals or interrupt their propagation, thereby lowering the level of oxidative stress and disrupt age-related dysregulated metabolic processes and deterioration in physiological functions (E Obrenovich et al. 2011). Microgreens contain a large amount of antioxidant compounds, such as carotenoids, phenolic antioxidants, anthocyanins, and glucosinolates (Bhaswant et al. 2023), suggesting their potential role in preventing oxidative stress and, subsequently, premature aging. Consumption of foods rich in carotenoids has unique benefits for general health, including reduction in wrinkles and skin aging (Darvin et al. 2011). Carotenoids promote the migration of nuclear factor erythroid 2-related factor 2 (Nrf2) to the nucleus, resulting in initiation of antioxidant enzyme cascades. In addition, carotenoids slow down the process of telomere shortening during DNA replication, counteracting the effect of oxidative stress (Babar et al. 2025; Kavalappa et al. 2019; Sheng et al. 2020; Varesi et al. 2022). Xanthophyllic carotenoids, namely, lutein and zeaxanthin, are used to treat skin conditions and administered topically or orally (Bin-Jumah et al. 2021). Lycopene scavenges free radicals in the skin more effectively than other carotenoids and is used to improve moisturization, texture, elasticity, and superficial structure (Caseiro et al. 2020). Dietary phenolic compounds are among the most easily accessible nutrients that exhibit antioxidant activity. Polyphenols, a subclass of phenols, are currently being highlighted to have anti-aging effects. Numerous studies have shown that polyphenols confer effects such as anti-inflammatory, promotion of cellular repair, and antioxidant activity. Polyphenols act as powerful scavengers for reactive oxygen species (ROS), reducing the risk of age-related diseases such as heart diseases, cancer, and neurodegenerative disorders (Liu et al. 2024). Phytochemicals, carotenoids, and phenolics are abundant in microgreens. Niroula et al. (2019) investigated the carotenoid profile of wheat and barley microgreens and found that the carotenoid content increased after germination and elongation. After 16 days of sowing, the carotenoid content in wheat rose significantly from 0.42 mg/100 g (dry weight) DW to 53.36 mg/100 g DW, while barley exhibited an increase from 0.78 mg/100 g DW to 56.08 mg/100 g DW. Sun et al. (2013) performed an extensive analysis of polyphenols in microgreens from five *Brassica* species. Their study identified 164 polyphenols, including 30 anthocyanins, 105 flavonol glycosides, and 29 derivatives of hydroxycinnamic and hydroxybenzoic acids.

9.6.3 Digestive Health

Basil (*Ocimum basilicum* L.) has many uses as a medicinal plant. Traditional and scientific literature is replete with accounts of basil for its health-promoting compounds and essential oils. Basil essential oils (eugenol) reduce inflammation and muscle spasms, have antibacterial and antifungal activity, and are useful for alleviating gastric discomfort. Owing to the presence of essential oils, antioxidants, and vitamins in basil microgreens, they can help combat oxidative stress, promote overall cellular health, and reduce inflammation, thereby supporting digestive function (Rusu et al. 2021). Mint (*Mentha* sp.) microgreens contain compounds such as menthol, menthone, and methyl esters, which possess muscle-relaxing and antispasmodic properties. These compounds can help relax the stomach and intestines, providing relief from bloating and stomach discomfort (Singh et al. 2024). Peppermint oil has been shown to lessen symptoms of irritable bowel syndrome (IBS) and other digestive conditions by relaxing the smooth muscles of the gastrointestinal tract, reducing inflammation, and decreasing pain sensation (Khanna et al. 2014). While the consumption of basil and mint microgreens appears to be beneficial for digestive health, further research will help fully understand their mechanisms and efficacy in various populations.

9.6.4 Mental Health

Medicinal microgreens influence mental health by virtue of their rich nutrient profile and the therapeutic aspects of their cultivation. Young tender greens are rich in vitamins, minerals, and antioxidants, which improve brain function and emotional well-being. Ginseng (*Panax ginseng*) is used to boost energy and increase mental and physical abilities. Principal component of ginseng are ginsenosides, which are steroidal glycosides or triterpene saponins. At least 28 ginsenosides have been reported from ginseng so far. Ginsenosides have been shown to act as anti-neuroinflammatory agents, combat oxidative stress, and modulate cholinergic signalling, thereby enhancing cognitive performance and reducing impairment. Regular consumption of ginseng has been shown to improve cognitive function in older adults (Coleman et al. 2003). Besides the nervous system, ginsenosides influence mental health through the gut-brain connection. Ginsenosides influence intestinal health by modifying gut microbiota and thereby indirectly promote mental well-being (Patel and Rauf 2017). While the consumption of ginseng microgreens appears promising for mental health support, further research is warranted to fully understand their mechanisms and efficacy across diverse populations.

9.7 Climate-Resilient Smart Agriculture and Medicinal Microgreens

The agriculture industry is focused on increasing the productivity and yield of nutritional crops to cope with food scarcity and meet the ever-increasing demand; however, the changing climate poses an unwavering challenge due to the limited land and other resources. The need for advanced climate-smart agricultural techniques over the age-old agricultural procedures has increased over the past few years. The climate-resilient measures must ensure the following points: reducing the emission of greenhouse gases, carbon sequestration, and resource management (Islam et al. 2022). According to a 2014 report of IPCC, it was estimated that the rising temperature will cause a reduction in yield by 25% by 2050 (IPCC, Intergovernmental Panel on Climate Change, 2014). Hence, smart agriculture aims to build climate-resilient farms that deal with impacts on soil, water, crop, human health, and GHG (greenhouse gas) emissions. Precision farming is an answer to such a question, which minimizes the usage of resources and is devoid of toxic chemicals, fertilizers, and pesticides. This has been successfully achieved in advanced farming techniques such as vertical farming, hydroponics, aquaponics, and aeroponics, which require limited land area, and the produce is available all year round, hence lowering the requirement of extensive storage facilities (Perera et al. 2025). Lately, the integration of advanced technologies such as IoT (Internet of things) and AI (artificial intelligence) has been recognized and is being adopted to carry out real-time monitoring of smaller and large-scale agricultural fields. These modern techniques show promising enhancement in productivity while also lowering the carbon footprint (Lakhiar et al. 2025).

Medicinal microgreens serve as a perfect nutrient solution for such low–area coverage smart agricultural techniques, offering a highly nutrient-rich composition. Their low land, water, and time requirement makes them highly suitable for integrating smart farming systems. Smart agriculture–led management of medicinal microgreens can be beneficial, especially to the urban population, due to guaranteed timely supply of nutrients, reduced availability of land area, and high consumer demand (Teng et al. 2023). Moreover, if the knowledge is wisely spread, the farmers and entrepreneurs in the rural area can also benefit from this low-investment,

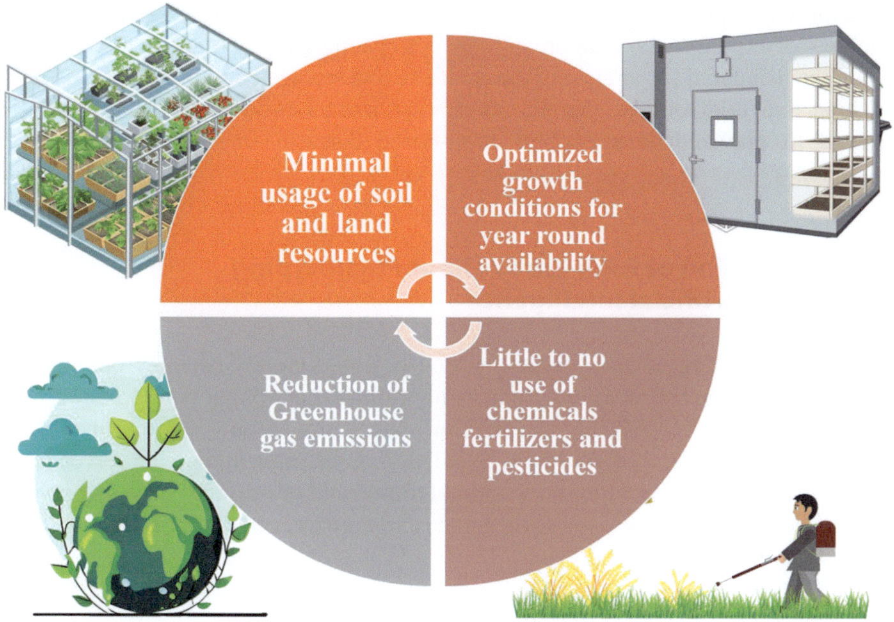

Fig. 9.3 Advantages of using climate-smart agriculture for medicinal microgreen cultivation

high-return idea of smart farming practices. The applications and benefits of smart agriculture in the cultivation of medicinal microgreens are summarized in Fig. 9.3.

9.8 Future Directions and Research Gaps

The rich nutritional profiles and possible health advantages of medicinal microgreens have attracted a considerable attention. They present promising options for sustainable agriculture and nutrition security in light of the world's population growth and urbanization. Several research gaps and future directions need to be addressed to harness their full potential.

9.8.1 Standardization of Cultivation Practices and Improvement of Shelf Life

An optimization of pre-harvest and post-harvest methods is crucial for maintaining the nutritional and medical properties while also improving their shelf life. More research is needed on sustainable, non-chemical methods for seed sterilization and antimicrobial treatment, as well as on pre-sowing and seed pre-germination techniques that can help standardize and accelerate the production cycle. Additionally, crop-specific standardization of seed sowing methods and the growing media will be helpful in improving cultivation practices (Kyriacou et al. 2016). Generation and analyses of experimental data on variable desiccating temperatures on microgreens would be beneficial for analysing the shelf life and storage conditions and loss/gain of bioactive compounds, thereby improving marketability. Further research is needed to prevent the spread of food-borne diseases in microgreens (Singh et al. 2024).

9.8.2 Inclusion of High-Throughput Phenotyping and Artificial Intelligence

Commercial scaling of microgreen production would greatly benefit from autonomous cutting-edge technologies, such as high-throughput phenotyping paired with artificial intelligence (AI), for instance, in image acquisition for determination of growth stage based on environment, calculation of leaf area and prediction of fresh weight at harvest, and prediction of optimal timing and position for product harvesting. In addition, AI can be used to detect the early emergence of human/plant pathogens (Teng et al. 2023).

9.8.3 Consumer Awareness

In recent years, microgreens have drawn a lot of attention due to their potential market as well as their nutritional advantages. However, customer demand and awareness play a major role in the microgreens industry's expansion. Factors such as taste, sugar content, texture, aroma, and visual appeal, all of which, play a crucial role in shaping consumer preferences where flavour and texture are key determinants of consumer acceptance. Studies have shown that microgreens with milder, less bitter, sour, or astringent flavours are more likely to be favoured by consumers. More research is needed to understand the phytochemical makeup and health advantages of microgreens in order to grow their market. Furthermore, it is essential to raise consumer education about their nutritional and therapeutic qualities, especially in developing countries like India, where knowledge is still lacking (Rizvi et al. 2023).

9.8.4 Need for Clinical Trials

Thorough clinical trials will be essential to understanding how microgreens influence human health and building a scientific database to study their potential benefits. In addition, research on medicinal microgreens' high nutritional profile will focus on developing preventive and therapeutic medicines. Further, more studies will assist their incorporation into evidence-based medical and dietary practices and validate their health-promoting qualities (Lone et al. 2024).

9.9 Conclusion

Medicinal microgreens are a "healthy" combination of nutrition and functional medicine. They have been shown to contain high concentrations of antioxidants, anti-inflammatory agents, and other phytochemicals, making them valuable for disease prevention and promotion of overall wellness. Cultivation of medicinal microgreens is simple and requires minimal space and resources, yet optimization of growth conditions remains a challenge. Limited consumer awareness also needs to be addressed. Further research is required on the phytochemical composition, bioactivity, and therapeutic applications of microgreens, achieved through clinical trials. As the demand for sustainable agriculture in light of global climate change comes to the forefront, microgreen cultivation has the potential to be a viable candidate for climate resilient agriculture. By bridging the gap between scientific research and consumer awareness, medicinal microgreens can emerge as a vital component of health and wellness strategies worldwide.

Declarations The authors declare that they have no conflict of interest.

References

Abaajeh AR, Kingston CE, Harty M (2023) Environmental factors influencing the growth and pathogenicity of microgreens bound for the market: a review. Renew Agric Food Syst 38:e12. https://doi.org/10.1017/S174217052300008X

Alloggia FP, Bafumo RF, Ramirez DA, Maza MA, Camargo AB (2023) Brassicaceae microgreens: A novel and promissory source of sustainable bioactive compounds. Curr Res Food Sci 6:100480

Artadana IBM, Pandji E (2023) Moringa oleifera L. microgreens and their antioxidant activity. In: E3S Web of Conferences, vol 374. EDP Sciences, p 00018

Babar M, Buzdar JA, Zaheer A, Nizam-ud-din N, Mustafa G, Khan BA, Hanif M, Asghar T, Qadeer A (2025) Carotenoids as a nutraceutical and health-promoting dietary supplement for human and animals: an updated review. Tradit Med Res 10(3):14. https://doi.org/10.53388/TMR20240831001

Balik S, Dasgan HY, Ikiz B, Gruda NS (2024) The performance of growing-media-shaped microgreens: the growth, yield, and nutrient profiles of broccoli, red beet, and black radish. Horticulturae 10(12):Article 12. https://doi.org/10.3390/horticulturae10121289

Bauer ME, Jeckel CMM, Luz C (2009) The role of stress factors during aging of the immune system. Ann N Y Acad Sci 1153:139–152. https://doi.org/10.1111/j.1749-6632.2008.03966.x

Bhaswant M, Shanmugam DK, Miyazawa T, Abe C, Miyazawa T (2023) Microgreens—a comprehensive review of bioactive molecules and health benefits. Molecules 28(2):867. https://doi.org/10.3390/molecules28020867

Bin-Jumah M, Alwakeel SS, Moga M, Buvnariu L, Bigiu N, Zia-Ul-Haq M (2021) Application of carotenoids in cosmetics. In: Zia-Ul-Haq M, Dewanjee S, Riaz M (eds) Carotenoids: structure and function in the human body. Springer International Publishing, pp 747–756. https://doi.org/10.1007/978-3-030-46459-2_24

Brazaitytė A, Miliauskienė J, Vaštakaitė-Kairienė V, Sutulienė R, Laužikė K, Duchovskis P, Małek S (2021) Effect of different ratios of blue and red led light on brassicaceae microgreens under a controlled environment. Plants 10(4):801

Çakır A, Sarıyer T, Özbay N (2025) Determination of quality characteristics in mature parsley (Petroselinum hortense) plants, parsley microgreens, and primed parsley microgreens. Black Sea J Agric 8(1):96–102. https://doi.org/10.47115/bsagriculture.1595301
Caseiro M, Ascenso A, Costa A, Creagh-Flynn J, Johnson M, Simões S (2020) Lycopene in human health. LWT 127:109323. https://doi.org/10.1016/j.lwt.2020.109323
Choe U, Yu LL, Wang TTY (2018) The science behind microgreens as an exciting new food for the 21st century. J Agric Food Chem. https://doi.org/10.1021/acs.jafc.8b03096
Clarkson G, Rothwell S, Taylor G (2005) End of day harvest extends shelf life. HortScience 40:1431–1435. https://doi.org/10.21273/HORTSCI.40.5.1431
Clemente JC, Ursell LK, Parfrey LW, Knight R (2012) The impact of the gut microbiota on human health: an integrative view. Cell 148(6):1258–1270. https://doi.org/10.1016/j.cell.2012.01.035
Coleman CI, Hebert JH, Reddy P (2003) The effects of Panax ginseng on quality of life. J Clin Pharm Ther 28(1):5–15. https://doi.org/10.1046/j.1365-2710.2003.00467.x
Darvin ME, Sterry W, Lademann J, Vergou T (2011) The role of carotenoids in human skin. Molecules 16(12):10491–10506. https://doi.org/10.3390/molecules161210491
de la Fuente B, López-García G, Máñez V, Alegría A, Barberá R, Cilla A (2019) Evaluation of the bioaccessibility of antioxidant bioactive compounds and minerals of four genotypes of Brassicaceae microgreens. Foods 8(7):7. https://doi.org/10.3390/foods8070250
Dereje B, Jacquier JC, Elliott-Kingston C, Harty M, Harbourne N (2023) Brassicaceae microgreens: phytochemical compositions, influences of growing practices, postharvest technology, health, and food applications. ACS Food Sci Technol 3(6):981–998
Di Bella MC, Niklas A, Toscano S, Picchi V, Romano D, Lo Scalzo R, Branca F (2020) Morphometric characteristics, polyphenols and ascorbic acid variation in Brassica oleracea L. novel foods: sprouts, microgreens and baby leaves. Agronomy 10(6):6. https://doi.org/10.3390/agronomy10060782
Dimita R, Min Allah S, Luvisi A, Greco D, De Bellis L, Accogli R, Mininni C, Negro C (2022) Volatile compounds and total phenolic content of Perilla frutescens at microgreens and mature stages. Horticulturae 8(1):71
Dong M, Park HK, Wang Y, Feng H (2022) Control Escherichia coli O157:H7 growth on sprouting brassicacae seeds with high acoustic power density (APD) ultrasound plus mild heat and calcium-oxide antimicrobial spray. Food Control 132:108482. https://doi.org/10.1016/j.foodcont.2021.108482
Du M, Xiao Z, Luo Y (2022) Advances and emerging trends in cultivation substrates for growing sprouts and microgreens toward safe and sustainable agriculture. Curr Opin Food Sci 46:100863. https://doi.org/10.1016/j.cofs.2022.100863
E Obrenovich M, Li Y, Parvathaneni K, B Yendluri B, H Palacios H, Leszek J, Aliev G (2011) Antioxidants in health, disease and aging. CNS Neurol Disord Drug Targets 10(2):192–207. https://doi.org/10.2174/187152711794480375
Ebert AW (2022) Sprouts and microgreens—novel food sources for healthy diets. Plants 11(4):571
Fayezizadeh MR, Ansari NA, Sourestani MM, Hasanuzzaman M (2023) Biochemical compounds, antioxidant capacity, leaf color profile and yield of basil (Ocimum sp.) Microgreens in Floating System. Plants 12(14):2652
Galieni A, Falcinelli B, Stagnari F, Datti A, Benincasa P (2020) Sprouts and microgreens: trends, opportunities, and horizons for novel research. Agronomy 10(9):9. https://doi.org/10.3390/agronomy10091424
Gao M, He R, Shi R, Li Y, Song S, Zhang Y et al (2021) Combination of selenium and UVA radiation affects growth and phytochemicals of broccoli microgreens. Molecules 26(15):4646
Garrido Y, Tudela JA, Gil MI (2015) Time of day for harvest and delay before processing affect the quality of minimally processed baby spinach. Postharvest Biol Technol 110:9–17. https://doi.org/10.1016/j.postharvbio.2015.07.002
Ghoora MD, Babu DR, Srividya N (2020) Nutrient composition, oxalate content and nutritional ranking of ten culinary microgreens. J Food Compos Anal 91:103495

Giordano M, Petropoulos SA, Kyriacou MC, Graziani G, Zarrelli A, Rouphael Y, El-Nakhel C (2022) Nutritive and phytochemical composition of aromatic microgreen herbs and spices belonging to the Apiaceae family. Plants 11(22):3057

Gunjal M, Singh J, Kaur J, Kaur S, Nanda V, Sharma A, Rasane P (2024) Microgreens: cultivation practices, bioactive potential, health benefits, and opportunities for its utilization as value-added food. Food Biosci 62:105133. https://doi.org/10.1016/j.fbio.2024.105133

Gurib-Fakim A (2006) Medicinal plants: traditions of yesterday and drugs of tomorrow. Mol Asp Med 27(1):1–93. https://doi.org/10.1016/j.mam.2005.07.008

Huang H, Jiang X, Xiao Z, Yu L, Pham Q, Sun J, Chen P, Yokoyama W, Yu LL, Luo YS, Wang TTY (2016) Red cabbage microgreens lower circulating low-density lipoprotein (LDL), liver cholesterol, and inflammatory cytokines in mice fed a high-fat diet. J Agric Food Chem 64(48):9161–9171. https://doi.org/10.1021/acs.jafc.6b03805

Islam MZ, Park B-J, Lee Y-T (2019) Effect of salinity stress on bioactive compounds and antioxidant activity of wheat microgreen extract under organic cultivation conditions. Int J Biol Macromol 140:631–636. https://doi.org/10.1016/j.ijbiomac.2019.08.090

Islam Z, Sabiha NE, Salim R (2022) Integrated environment-smart agricultural practices: a strategy towards climate-resilient agriculture. Econ Anal Policy 76:59–72

Jayaraman S, Ramasamy T (2024) Metabolomics and bioactive attributes of fenugreek microgreens: insights into antioxidant, antibacterial and antibiofilm potential. Food Biosci 60:104316. https://doi.org/10.1016/j.fbio.2024.104316

Jiang X, Huang H, Xiao Z, Yu L, Pham Q, Yu L (Lucy), Luo Y (Sunny), Wang TTY (2016). Lipids and cholesterol-lowering activity of red cabbage microgreens. FASEB J 30(S1):431.8. https://doi.org/10.1096/fasebj.30.1_supplement.431.8

Junpatiw A, Sangpituk A (2021) Effects of seed preparation, sowing media, seed sowing rate and harvesting period on the production of chia microgreens. GEOMATE J 17(61):Article 61

Kaimuangpak K, Tamprasit K, Thumanu K, Weerapreeyakul N (2022) Extracellular vesicles derived from microgreens of Raphanus sativus L. var. Caudatus Alef contain bioactive macromolecules and inhibit HCT116 cells proliferation. Sci Rep 12(1):Article 1. https://doi.org/10.1038/s41598-022-19950-7

Kaimuangpak K, Rosalina R, Thumanu K, Weerapreeyakul N (2024) Macromolecules with predominant β-pleated sheet proteins in extracellular vesicles released from Raphanus sativus L. var. Caudatus Alef microgreens induce DNA damage-mediated apoptosis in HCT116 colon cancer cells. Int J Biol Macromol 269:132001. https://doi.org/10.1016/j.ijbiomac.2024.132001

Karuthedath S (2024) Microgreens: A potential source of nourishment. In: Microgreens [PDF]. Unity Women's College. https://unitywomenscollege.ac.in/uploads/Microgreens-2024-10-30-08-58-17.911.pdf

Kavalappa YP, Udayawara Rudresh D, Gopal SS, Haranahalli Shivarudrappa A, Stephen NM, Rangiah K, Ponesakki G (2019) β-carotene isolated from the marine red alga, Gracillaria sp. Potently attenuates the growth of human hepatocellular carcinoma (HepG2) cells by modulating multiple molecular pathways. J Funct Foods 52:165–176. https://doi.org/10.1016/j.jff.2018.11.015

Khanna R, MacDonald JK, Levesque BG (2014) Peppermint oil for the treatment of irritable bowel syndrome: a systematic review and meta-analysis. J Clin Gastroenterol 48(6):505. https://doi.org/10.1097/MCG.0b013e3182a88357

Kou L, Yang T, Liu X, Luo Y (2015) Effects of pre- and postharvest calcium treatments on shelf life and postharvest quality of broccoli microgreens. HortScience 50(12):1801–1808. https://doi.org/10.21273/HORTSCI.50.12.1801

Kyriacou MC, Rouphael Y, Di Gioia F, Kyratzis A, Serio F, Renna M, De Pascale S, Santamaria P (2016) Micro-scale vegetable production and the rise of microgreens. Trends Food Sci Technol 57:103–115. https://doi.org/10.1016/j.tifs.2016.09.005

Lakhiar IA, Yan H, Syed TN, Zhang C, Shaikh SA, Rakibuzzaman M, Vistro RB (2025) Soilless agricultural systems: opportunities, challenges, and applications for enhancing horticultural resilience to climate change and urbanization. Horticulturae 11(6):568

Lee JS, Pill WG, Cobb BB, Olszewski M (2004) Seed treatments to advance greenhouse establishment of beet and chard microgreens. J Hortic Sci Biotechnol. https://www.tandfonline.com/doi/abs/10.1080/14620316.2004.11511806

Li X, Tian S, Wang Y, Liu J, Wang J, Lu Y (2021) Broccoli microgreens juice reduces body weight by enhancing insulin sensitivity and modulating gut microbiota in high-fat diet-induced C57BL/6J obese mice. Eur J Nutr 60(7):3829–3839

Liu J, Li S, O'Keefe S, Hurley K, Rutto L, Eriksen R, Yin Y (2024) Characterization of key aroma compounds in microgreens and mature plants of hydroponic fennel (Foeniculum vulgare Mill.). Food Res Int 197:115229

Liu Y, Fang M, Tu X, Mo X, Zhang L, Yang B, Wang F, Kim Y-B, Huang C, Chen L, Fan S (2024) Dietary polyphenols as anti-aging agents: targeting the hallmarks of aging. Nutrients 16(19):19. https://doi.org/10.3390/nu16193305

Lone JK, Pandey R, Gayacharan. (2024) Microgreens on the rise: expanding our horizons from farm to fork. Heliyon 10(4):e25870. https://doi.org/10.1016/j.heliyon.2024.e25870

Lu Y, Dong W, Yang T, Luo Y, Chen P (2021) Preharvest UVB application increases glucosinolate contents and enhances postharvest quality of broccoli microgreens. Molecules 26(11):3247

Ma S, Tian S, Sun J, Pang X, Hu Q, Li X, Lu Y (2022) Broccoli microgreens have hypoglycemic effect by improving blood lipid and inflammatory factors while modulating gut microbiota in mice with type 2 diabetes. J Food Biochem 46(7):e14145. https://doi.org/10.1111/jfbc.14145

Maina S, Ryu DH, Cho JY, Jung DS, Park JE, Nho CW et al (2021) Exposure to salinity and light spectra regulates glucosinolates, phenolics, and antioxidant capacity of Brassica carinata L. microgreens. Antioxidants 10(8):1183

Marchioni I, Martinelli M, Ascrizzi R, Gabbrielli C, Flamini G, Pistelli L, Pistelli L (2021) Small functional foods: comparative phytochemical and nutritional analyses of five microgreens of the Brassicaceae family. Foods 10(2):427

Martemucci G, Portincasa P, Di Ciaula A, Mariano M, Centonze V, D'Alessandro AG (2022) Oxidative stress, aging, antioxidant supplementation and their impact on human health: an overview. Mech Ageing Dev 206:111707. https://doi.org/10.1016/j.mad.2022.111707

Maru RN (2024) Influence of substrate and LED light Spectrum on growth, yield, and phytochemical content of Ethiopian Kale (Brassica carinata A. braun) microgreens (Doctoral dissertation, JKUAT-COANRE)

Michele Paradiso V, Castellino M, Renna M, Eliana Gattullo C, Calasso M, Terzano R, Allegretta I, Leoni B, Caponio F, Santamaria P (2018) Nutritional characterization and shelf-life of packaged microgreens. Food Funct 9(11):5629–5640. https://doi.org/10.1039/C8FO01182F

Michell KA, Isweiri H, Newman SE, Bunning M, Bellows LL, Dinges MM, Grabos LE, Rao S, Foster MT, Heuberger AL, Prenni JE (2020) Microgreens: consumer sensory perception and acceptance of an emerging functional food crop. J Food Sci 85(4):926–935

Mishra N, Tripathi R, Pandey D, Shah K, Chauhan NS (2025) Wheatgrass (Triticum aestivum): a miraculous microgreen: an overview. J Future Foods 5(3):239–247

Nakakaawa L, Gbala ID, Bargul JL, Cheseto X, Wesonga JM (2025) Therapeutic potential of Brassica carinata microgreens extract in alleviating symptoms of type 2 diabetes in Wistar rats. Food Sci Nutr 13(1):e4635. https://doi.org/10.1002/fsn3.4635

Ni Y, Li J, Panagiotou G (2015) A molecular-level landscape of diet-gut microbiome interactions: toward dietary interventions targeting bacterial genes. MBio 6(6). https://doi.org/10.1128/mbio.01263-15

Niroula A, Khatri S, Timilsina R, Khadka D, Khadka A, Ojha P (2019) Profile of chlorophylls and carotenoids of wheat (Triticum aestivum L.) and barley (Hordeum vulgare L.) microgreens. J Food Sci Technol 56(5):2758–2763. https://doi.org/10.1007/s13197-019-03768-9

Ogal M, Johnston SL, Klein P, Schoop R (2021) Echinacea reduces antibiotic usage in children through respiratory tract infection prevention: A randomized, blinded, controlled clinical trial. Eur J Med Res 26(1):33. https://doi.org/10.1186/s40001-021-00499-6

Palmitessa OD, Renna M, Crupi P, Lovece A, Corbo F, Santamaria P (2020) Yield and quality characteristics of brassica microgreens as affected by the NH4:NO3 molar ratio and strength of the nutrient solution. Foods 9(5):5. https://doi.org/10.3390/foods9050677

Pant P, Pandey S, Dall'Acqua S (2021) The influence of environmental conditions on secondary metabolites in medicinal plants: a literature review. Chem Biodivers 18(11):e2100345. https://doi.org/10.1002/cbdv.202100345

Patel S, Rauf A (2017) Adaptogenic herb ginseng (Panax) as medical food: status quo and future prospects. Biomed Pharmacother 85:120–127. https://doi.org/10.1016/j.biopha.2016.11.112

Patil M, Sharma S, Sridhar K, Anurag RK, Grover K, Dharni K, Mahajan S, Sharma M (2024) Effect of postharvest treatments and storage temperature on the physiological, nutritional, and shelf-life of broccoli (Brassica oleracea) microgreens. Sci Hortic 327:112805

Patras A (2021) Effects of development stage and sodium salts on the antioxidant properties of white cabbage microgreens. Agriculture 11(3):200

Perera WPTD, Navaratne S, Wickramasinghe I (2025) Climate adaptation in built environment: urban agriculture and food security with innovative approaches. In: Climate change adaptation in the built environment: transdisciplinary and innovative learning. Springer Nature Switzerland, Cham, pp 165–187

Phornvillay S, Yodsarn S, Oonsrithong J, Srilaong V, Pongprasert N (2022) A novel technique using advanced oxidation process (UV-C/H2O2) combined with micro-nano bubbles on decontamination, seed viability, and enhancing phytonutrients of roselle microgreens. Horticulturae 8(1):53

Poljšak B, Dahmane R (2012) Free radicals and extrinsic skin aging. Dermatol Res Pract 2012:135206. https://doi.org/10.1155/2012/135206

Poudineh M, Nikzad F, Parvin S, Ghaheri M, Sabbaghi S, Kazemi E, Ghodrati MM, Mohammadyari F, Saeedpour S, Mohammadpour S, Sadat Farizani Gohari N, Heydarasadi F, Abolhasani D, Olangian-Tehrani S, Alinezhad A (2024) Therapeutic effects of vitamins and nutritional supplements on sinusitis: a narrative review. Nutr Metab Insights 17:11786388241263680. https://doi.org/10.1177/11786388241263680

Puccinelli M, Maggini R, Angelini LG, Santin M, Landi M, Tavarini S, Castagna A, Incrocci L (2022) Can light spectrum composition increase growth and nutritional quality of Linum usitatissimum L. sprouts and microgreens? Horticulturae 8(2):98

Rebolledo P, Carrasco G, Moggia C, Gajardo P, Sant'Ana GR, Fuentes-Peñailillo F, Urrestarazu M, Vendruscolo EP (2024) Assessment of vegetable species for microgreen production in unheated greenhouses: yield, nutritional composition, and sensory perception. Plants 13(19):2787

Renna M, Di Gioia F, Leoni B, Mininni C, Santamaria P (2017) Culinary assessment of self-produced microgreens as basic ingredients in sweet and savory dishes. J Culin Sci Technol 15(2):126–142. https://doi.org/10.1080/15428052.2016.1225534

Rizvi A, Sharma M, Saxena S (2023) Microgreens: A next generation nutraceutical for multiple disease management and health promotion. Genet Resour Crop Evol 70(2):311–332. https://doi.org/10.1007/s10722-022-01506-3

Rizvi A, Kumar S, Sharma M, Saxena S (2024) Functional and antioxidant potential of beetroot, mustard and radish microgreens using spectroscopic techniques. Agriculture (Pol'nohospodárstvo) 70:53

Rozali SE, Rahim NRA (2024) Sensory attributes and phytochemical composition in microgreens: implications for acceptability among school students in Selangor, Malaysia. KnE Engineering, pp 32–43

Rusu T, Cowden RJ, Moraru PI, Maxim MA, Ghaley BB (2021) Overview of multiple applications of basil species and cultivars and the effects of production environmental parameters on yields and secondary metabolites in hydroponic systems. Sustainability 13(20):20. https://doi.org/10.3390/su132011332

Sehrish A, Majeed I, Zongo E, Ayub H, Rasul H, Rahim MA, AL-Asmari F (2023) A review on various extraction and detection methods of bio-functional components from microgreens: food applications and health properties. Int J Food Prop 26(2):3082–3105

Senevirathne GI, Gama-Arachchige NS, Karunaratne AM (2019) Germination, harvesting stage, antioxidant activity and consumer acceptance of ten microgreens. Ceylon J Sci 48(91):10–4038

Shah KH, Rathod MC (2024) Comparative study of qualitative phytochemical analysis and inorganic compound of coriander (Coriandrum sativum L.): microgreen and mature green. J Plant Sci Res 40(2):289–295

Sheng Y-N, Luo Y-H, Liu S-B, Xu W-T, Zhang Y, Zhang T, Xue H, Zuo W-B, Li Y-N, Wang C-Y, Jin C-H (2020) Zeaxanthin induces apoptosis via ROS-regulated MAPK and AKT signaling pathway in human gastric cancer cells. Onco Targets Ther 13:10995–11006. https://doi.org/10.2147/OTT.S272514

Signore A, Somma A, Leoni B, Santamaria P (2024) Optimising sowing density for microgreens production in Rapini, Kale and Cress. Horticulturae 10(3):3. https://doi.org/10.3390/horticulturae10030274

Singh A, Singh J, Kaur S, Gunjal M, Kaur J, Nanda V, Ullah R, Ercisli S, Rasane P (2024) Emergence of microgreens as a valuable food, current understanding of their market and consumer perception: a review. Food Chem 23:101527. https://doi.org/10.1016/j.fochx.2024.101527

Sun J, Xiao Z, Lin L-Z, Lester GE, Wang Q, Harnly JM, Chen P (2013) Profiling polyphenols in five brassica species microgreens by UHPLC-PDA-ESI/HRMS(n.). J Agric Food Chem 61(46):10960–10970. https://doi.org/10.1021/jf401802n

Supapvanich S, Sangsuk P, Sripumimas S, Anuchai J (2020) Efficiency of low dose cyanocobalamin immersion on bioactive compounds contents of ready to eat sprouts (sunflower and daikon) and microgreens (red-amaranth) during storage. Postharvest Biol Technol 160:111033. https://doi.org/10.1016/j.postharvbio.2019.111033

Tallei TE, Kepel BJ, Wungouw HIS, Nurkolis F, Adam AA, Fatimawali. (2024) A comprehensive review on the antioxidant activities and health benefits of microgreens: current insights and future perspectives. Int J Food Sci Technol 59(1):58–71. https://doi.org/10.1111/ijfs.16805

Teng Z, Luo Y, Pearlstein DJ, Wheeler RM, Johnson CM, Wang Q, Fonseca JM (2023) Microgreens for home, commercial, and space farming: a comprehensive update of the most recent developments. Annu Rev Food Sci Technol 14:539–562. https://doi.org/10.1146/annurev-food-060721-024636

Thursby E, Juge N (2017) Introduction to the human gut microbiota. Biochem J 474(11):1823–1836. https://doi.org/10.1042/BCJ20160510

Truzzi F, Whittaker A, Roncuzzi C, Saltari A, Levesque MP, Dinelli G (2021) Microgreens: functional food with antiproliferative cancer properties influenced by light. Foods 10(8):8. https://doi.org/10.3390/foods10081690

Tunçtürk R, Yolcu MS, Tunçtürk M, Şelem E, Nohutçu L (2024) Investigation of the biochemical content of some plant microgreens from fhe Asteraceae Family. Yuzuncu Yıl Univ J Agric Sci 34(4):4. https://doi.org/10.29133/yyutbd.1480059

Tzounis X, Rodriguez-Mateos A, Vulevic J, Gibson GR, Kwik-Uribe C, Spencer JP (2011) Prebiotic evaluation of cocoa-derived flavanols in healthy humans by using a randomized, controlled, double-blind, crossover intervention study123. Am J Clin Nutr 93(1):62–72. https://doi.org/10.3945/ajcn.110.000075

Vaou N, Stavropoulou E, Voidarou C, Tsigalou C, Bezirtzoglou E (2021) Towards advances in medicinal plant antimicrobial activity: a review study on challenges and future perspectives. Microorganisms 9(10):10. https://doi.org/10.3390/microorganisms9102041

Varesi A, Chirumbolo S, Campagnoli LIM, Pierella E, Piccini GB, Carrara A, Ricevuti G, Scassellati C, Bonvicini C, Pascale A (2022) The role of antioxidants in the interplay between oxidative stress and senescence. Antioxidants 11(7):7. https://doi.org/10.3390/antiox11071224

Verlinden S (2020) Microgreens. In: Horticultural reviews. John Wiley & Sons, Ltd, pp 85–124. https://doi.org/10.1002/9781119625407.ch3

Vestergaard P (2021) Diabetes and its complications—contemporary treatment and potential side effects of drugs to treat diabetes. Curr Drug Saf 16(1):1–1. https://doi.org/10.2174/157488631601210203093615

Vimalanathan S, Schoop R, Suter A, Hudson J (2017) Prevention of influenza virus induced bacterial superinfection by standardized Echinacea purpurea, via regulation of surface receptor expression in human bronchial epithelial cells. Virus Res 233:51–59. https://doi.org/10.1016/j.virusres.2017.03.006

Weber CF (2017) Broccoli microgreens: A mineral-rich crop that can diversify food systems. Front Nutr 4:7

Wojdyło A, Nowicka P, Tkacz K, Turkiewicz IP (2020) Sprouts vs. microgreens as novel functional foods: variation of nutritional and phytochemical profiles and their in vitro bioactive properties. Molecules 25(20):20. https://doi.org/10.3390/molecules25204648

Xiao Z, Lester GE, Park E, Saftner RA, Luo Y, Wang Q (2015) Evaluation and correlation of sensory attributes and chemical compositions of emerging fresh produce: microgreens. Postharvest Biol Technol 110:140–148. https://doi.org/10.1016/j.postharvbio.2015.07.021

Xiao Z, Codling EE, Luo Y, Nou X, Lester GE, Wang Q (2016) Microgreens of Brassicaceae: mineral composition and content of 30 varieties. J Food Compos Anal 49:87–93

Yordanova M, Miroslavova B (2024) Pre-treatment of slow-germinating Apiaceae seeds for microgreens. Sci Pap Ser B Hortic 68(1)

Zhang Y, Xiao Z, Ager E, Kong L, Tan L (2021) Nutritional quality and health benefits of microgreens, a crop of modern agriculture. J Future Foods 1(1):58–66. https://doi.org/10.1016/j.jfutfo.2021.07.001

Zhou Q, Liang W, Wan J, Wang M (2023) Spinach (Spinacia oleracea) microgreen prevents the formation of advanced glycation end products in model systems and breads. Curr Res Food Sci 6:100490. https://doi.org/10.1016/j.crfs.2023.100490

Zou L, Tan WK, Du Y, Lee HW, Liang X, Lei J, Striegel L, Weber N, Rychlik M, Ong CN (2021) Nutritional metabolites in Brassica rapa subsp. chinensis var. parachinensis (choy sum) at three different growth stages: microgreen, seedling and adult plant. Food Chem 357:129535

Microgreens Farming as Climate-Smart Agriculture: From Soil to Supplement, Next Generation of Nutrient-Dense Foods

10

Rachana Singh, Harshal Singh, Ansh, Shrishty Gupta, Kajal Jaiswal, and Astha Gupta

Abstract

Oilseed-derived microgreens are emerging as high-value functional foods because of their high concentrations of primary metabolites (proteins, essential fatty acids, carbohydrates, minerals) and secondary metabolites (phenolics, glucosinolates, lignans, carotenoids) with numerous antioxidant, anti-inflammatory, cardio-protective, and anti-cancer activities. Compared to mature plants, they exhibit up to 30–40-fold higher micronutrient levels, including vitamins C, E, and K and essential minerals such as iron, zinc, and magnesium. A diverse group of oilseed crops, including sunflower, safflower, mustard, sesame, rapeseed, linseed, niger, groundnut, castor, and soybean, provide distinct nutritional and phytochemical profiles. Oilseed derived microgreen have short growth cycle (7–21 days), low input requirements, and adaptability to controlled environments make them suitable for sustainable production. Multi-OMICS integration (genomics, transcriptomics, proteomics, metabolomics, epigenomics and nutriomics) provides innovative methods to improve flavour, increase post-harvest shelf life, and improve nutrient biosynthesis. In addition to that, association of artificial intelligence with OMICS technology would be a pivotal dynamic to improve the quality and quantity of microgreens and would be a promising resource for climate-resilient smart agriculture. Moreover, oilseed-derived microgreens offer a possible avenue for addressing micronutrient shortages (malnutrition) and improving global health through next-generation functional foods by combining nutritional richness to fulfil the needs of Sustainable Development Goals 2 and 3. In the future, microgreens farming (chemical-free organic green produce) will be a promising small-scale entrepreneurship for

R. Singh · H. Singh · Ansh · S. Gupta · K. Jaiswal · A. Gupta (✉)
Department of Biotechnology, Sharda School of Biosciences and Technology, Sharda University, Greater Noida, Uttar Pradesh, India
e-mail: astha.gupta@sharda.ac.in

V. Gahlaut et al. (eds.), *Smart Crop Development*,
https://doi.org/10.1007/978-981-95-7680-7_10

marginal farmers and active women as an alternative source of income that will fulfil the daily nutrition needs.

Keywords

Microgreens · Oilseed crops · Nutrition · Superfood · Climate-resilient smart agriculture

10.1 Introduction

Microgreens are the young, immature greens and edible seedlings that result from the sprouting of numerous vegetables, herbs, and grains. In addition to the relatives of various plant families such as Brassicaceae, Fabaceae, Asteraceae, Amaranthaceae, Cucurbitaceae, Apiaceae, Amaryllidaceae, and Lamiaceae (Kyriacou et al. 2016). Microgreens are becoming more well-known and drawing interest from consumers as nutritious food options. Microgreens hold enormous promise for drastically lowering health issues. Microgreens are harvested approximately 7 to 21 days after the seedlings appear and have the cotyledon, stalks, and first true leaves. A range of growth media, including soil, sand, a coco peat and vermiculite mixture, and the loose and soil-less germination media, can be used to generate microgreens (Di Gioia et al. 2017). Microgreen manufacturing began in the 1980s when it initially appeared on San Francisco, California, chefs' menus (United States Department of Agriculture 2014). As a result of its increased popularity, microgreens were first cultivated in southern California in the 1990s, and today, they are considered "functional foods" or "superfoods". Numerous publications on the nutritional value of microgreens state that they are rich in antioxidants along with a range of vitamins, minerals, and bioactive substances that are good for your health. The potential of these nutrient-dense plants to avert the onset of numerous chronic inflammatory illnesses may have positive health implications. A possible new food source to meet customer demand for healthier eating could be microgreens (Choe et al. 2018). Microgreens are rich in organic acids, carotenoids, and chlorophylls and low in sugar. They also exhibit antidiabetic and anticholinergic qualities. Microgreens have the potential to significantly improve human health and stave against illnesses linked to oxidative stress (Wojdyło et al. 2020). Because of its short life cycle and limited production space requirements, microgreens are ideal for cultivation under strictly regulated circumstances. Because it allows growers the most control over the growing environment to enhance taste and morphology based on market preferences and the capacity to produce extremely uniform crops year-round, indoor farming is becoming an increasingly popular approach for cultivating green vegetables (Despommier 2013). Because of its positive impacts on human health, microgreens are regarded as a "functional food" (Di Bella et al. 2020).

10.2 Nutritional and Functional Relevance of Microgreens and Role in Sustainable Food System and Human Health

10.2.1 Microgreens in the Context of Global Nutritional Challenges

Many diet-related non-communicable diseases (NCDs), such as underweight, obesity, diabetes, hypertension, cardiovascular illnesses, cancer, etc., are caused by people's imbalanced diets and much lower food intakes in the current hectic lifestyle. Even though urban households in low-income countries like India spend 50–80% of their money on food, they nonetheless experience significant food insecurity (De and Kakar 2021), and dietary deficits in both macro- and micronutrients are prevalent (Gupta et al. 2022). About one-third of Indians eat vegetarian or ovo-vegetarian food, while the majority of Indian states' current diets are deficient in 11 of the 25 important elements (Chaudhary and Krishna 2021). Vegetarian diets may result in inadequate intake of protein, vitamin B12, omega-3 fatty acids, and minerals, including iron and zinc, since plant sources have a lower bioavailability. The agriculture and horticulture sectors must be re-envisioned as the key to decreasing poverty and malnutrition, enabling significant dietary adjustments, and maintaining nutritional security to prevent all forms of malnutrition and reduce the risk of NCDs. Women are mostly in charge of making sure that their family members are eating healthily, and they can improve household nutrition by including microgreens in their regular meals. These problems could be resolved by microgreens since they generate nutrient-rich vegetables from farm to table with minimal acreage and less input, time, effort, and money. As a result, growing microgreens can be a profitable business that helps women in both rural and urban areas maintain their financial security.

10.2.2 Role in Sustainable Food System and Dietary Security

Additionally, eating microgreens will help achieve the Sustainable Development Goals for health and well-being. Since microgreens include a range of phyto-active compounds that are good for human health both on Earth and in space, they are regarded as fresh, ready-to-eat, functional foods rich in nutraceuticals. Microgreens are said to have significantly higher concentrations of bioactive chemicals, phytonutrients, and antioxidants because they only undergo a few days of photosynthesis. Microgreens have been found to have 30–40 times as many nutrients as their mature counterparts (Kyriacou et al. 2016; Mathur and Gupta 2025). Flaxseed microgreens had larger levels of phenol, carotenoids, and chlorophyll than sprouts, while broccoli microgreens had higher levels of antioxidants. Microgreens met 2–12%, 2–13%, and 1–6% of the recommended daily allowance for vitamins and micronutrients. Iron, zinc, selenium, vitamin C, vitamin E, and vitamin A were found to be 28–116%, 28–332%, and 24–72% of the daily recommended consumption, respectively (Ghoora et al. 2020).

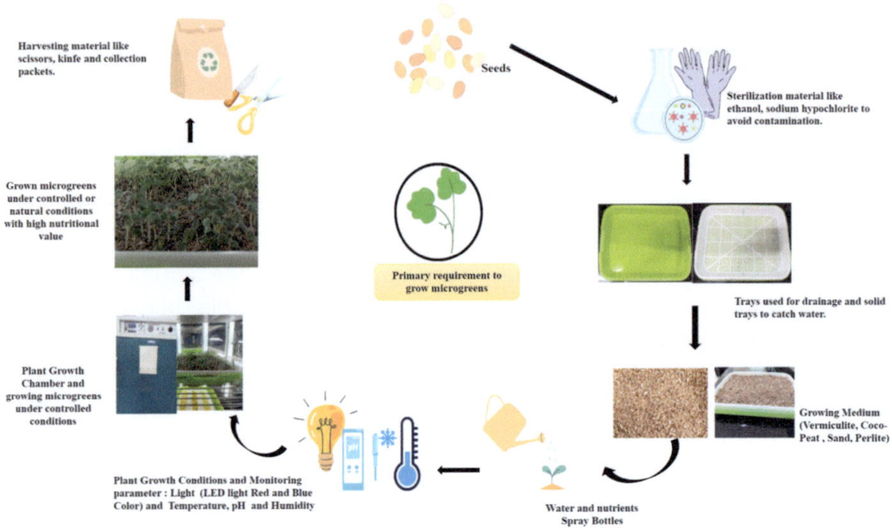

Fig. 10.1 Schematic representation of basic requirements to grow microgreens

10.3 Oilseed Crop with Economic and Nutritional Importance

10.3.1 Edible and Non-edible Oilseed Crops

Edible oilseed crops Edible oil seeds are the most important source of edible oils, and the crops that fall into this category are referred to as edible oilseed crops. For instance, sesame, peanut, niger, sunflower, safflower, soybean, rapeseed, and mustard.

Non-edible oilseed crops Non-edible oil seeds are the most important source of non-edible oil supply, and the crops that fall into this category are referred to as non-edible oil seed crops, for instance, castor, linseed, etc. Category of oilseed crops and their description are mentioned in Fig. 10.1.

10.3.2 Growing Requirement to Grow Microgreen

The right variety selection, seed quality, seed density, substrates, and growing methods are all important components of a successful microgreen production process (Di Gioia and Santamaria 2015). Some basic materials are required to grow microgreens in controlled or natural conditions, as explained in Fig. 10.2. Microgreen growth is significantly influenced by the physical characteristics of the growing media, such as its roughly 85% porosity, 55–70% water-holding capacity, and aeration (20–30% of total volume for roots) (Abad et al. 2001). Throughout the year,

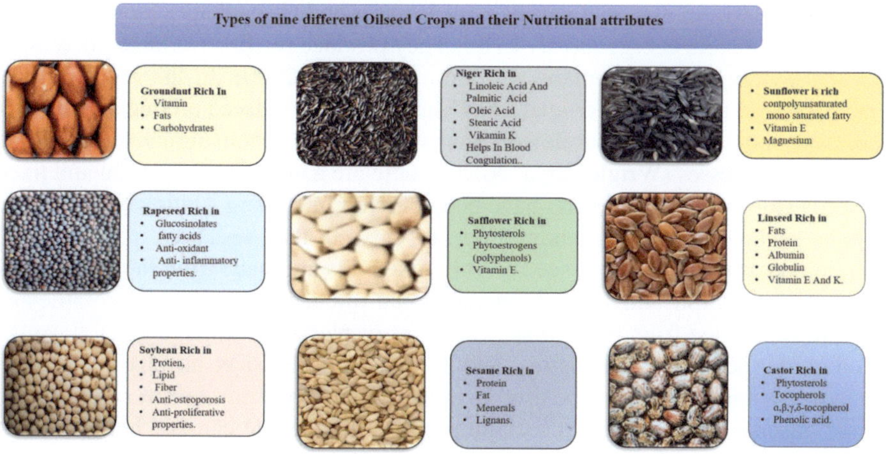

Fig. 10.2 Representation of nine different kinds of oilseed crops and their nutritional attributes

microgreens can be cultivated organically on inexpensive, locally accessible media such as soil and soil-less media (Mir et al. 2017), as well as with contemporary agricultural methods including hydroponics, aquaponics (Kizak and Kapaligoz 2019; Grishin et al. 2021), and vertical farming (AlShrouf 2017). Because of its affordability, accessibility, and biodegradability, cocopeat is a popular soil-less substrate for microgreen development (Islam et al. 2023). For the first three days of their germination, microgreens require a dark environment; to encourage growth, they thereafter require 12–16 hours of daily exposure to sufficient light (Kou et al. 2014; Negri et al. 2019). When growing microgreens in a controlled environment, artificial light sources such as light-emitting diode (LED) (Gao et al. 2021; Truzzi et al. 2021), high-pressure sodium (HPS) lamps, and white fluorescent light were generally preferred.

10.3.3 Economic and Nutritional Importance

Oilseed crops' high nutrient content and ability for biofortification make them a potent remedy for malnutrition when grown as microgreens. Because of their high vitamin, mineral, and antioxidant content, microgreens are a great supplement to diets, particularly in places with limited access to fresh vegetables (Gupta et al. 2025). Oilseed crops are plants grown primarily for the oil content in their seeds or fruits. These oils are used for food, industrial products, and biofuels. There is a need to increase agricultural methods in order to maximize food production for the human population because of the ongoing increase in the human population and the high demand for food, which has led to famine, disease outbreaks, and even death by starvation (Pandey 2018). Humans have been choosing and growing crops to meet their needs for flavour, energy, and health since the dawn of agriculture

(Nagaraj 2009). An essential part of semi-tropical and tropical agriculture, oilseeds are crops that store energy mostly as oil. They produce readily available, incredibly nutrient-dense food for humans and animals (Weiss 2000). In the industrialized world, oilseed crops play a crucial role (Murphy 1991). Oilseed products, including meal, oil, and the hull (for sunflower), play crucial roles in both human nutrition and industrial processes (Wrigley et al. 2015). There are a number of essential and fixed oilseeds from which edible oils have been derived, including sunflower, safflower, canola, soybean, cotton, groundnut, peanut, linseed, flax, sesame, rapeseed, jatropha, jojoba, niger, and castor seeds. While fixed oils are made from the seeds of oil plants and are edible and have a high nutritional value, essential oils are volatile and typically extracted from the non-seed sections of plants and used to make flavours, deodorants, perfumes, antiseptics, and medications (Blasi et al. 2018). Oilseeds include fats and oils that can be used as industrial raw materials and as culinary fats. Some oilseeds and their cakes include proteins that people can eat, but others are beneficial as animal feed. Vitamins, minerals, and carbohydrates are also present in oilseeds. The calorie nutrition and malnutrition of the human and animal populations are significantly reduced by oilseeds and oilseed meals. Additionally, vegetable oils can be used as raw materials for a variety of industrial products, lubricants, surface coatings, and cosmetics. Linseed and castor are inedible, while the first seven are edible sources of oil. Numerous variables, including genotype, season, location, maturity, and growing methods, affect the oilseeds' and their products' composition and quality. The nutritional and other energy-rich compounds found in oilseeds and their products make them extremely beneficial. Their usefulness is occasionally restricted by the presence of anti-nutritional components. For food and feed applications, the oilseeds' and their products' biochemical makeup and quality are crucial. Safflower is well-known for its blooms, which have long been utilized in the food sector for tasting and colouring. They are used in textile industries for the purpose of staining and dyes (Mani et al. 2020).

10.4 Microgreens Derived from Nutrition-Dense Oilseed Crops

Extra nutritional components called bioactive chemicals are present in foods, namely, fruits, vegetables, and grains, and they have the power to alter metabolic processes and have positive health effects (Santos et al. 2019). Because broccoli seedlings have a concentration of bioactive substances that is several times larger than that of mature florets, regular ingestion of these plants may boost the body's defences and reduce the risk of chronic diseases (Moreira-Rodríguez et al. 2017). Oilseed crops are packed with high-quality protein, essential amino acids, unsaturated fatty acids, dietary fibre, minerals, and a wide array of phytochemicals (e.g., phenolics, flavonoids, phytosterols, tocotrienols) that exert antioxidant, anti-inflammatory, antihypertensive, and neuro-protective actions.

Groundnut/peanut oil, with its neutral flavour and high smoke point, is ideal for cooking and deep-frying. It provides protein, B vitamins, and minerals like potassium, calcium, and phosphorus, contributing to heart health, blood sugar control, and digestion. The leftover oil cake is used in animal feed, and roasted peanuts are popular snacks (FAO 2020).

Rapeseed and mustard are important oilseeds in food and industry. Canola oil, a low-erucic-acid rapeseed variety, is low in saturated fats but rich in healthy unsaturated fatty acids, including omega-3 and omega-6, and is used in cooking and processed foods. Mustard oil, valued for its aroma and flavour, is common in Indian cuisine and has antimicrobial and therapeutic uses (Sharma et al. 2022).

Soybean oil, though from a lower-oil-yield crop, is a major global vegetable oil source. It is high in healthy fats, particularly omega-3 and omega-6, and vitamin E and is used in cooking and processed foods. Soy protein is also key in animal feed and plant-based diets (USDA 2021).

Sunflower oil has a mild flavour, high vitamin E content, and good stability, making it suitable for cooking, baking, and salad dressings. Sunflower seeds are eaten as snacks, used in bird feed, and their hulls serve as animal feed or fuel (Ghosh and Chattopadhyay 2019). Linseed/flaxseed oil is rich in alpha-linolenic acid (ALA), offering heart, cholesterol, and digestive benefits, but is unsuitable for high-heat cooking. It also serves industrial purposes like paints and linoleum (Kajla et al. 2015).

Cottonseed oil, a by-product of cotton processing, is used for frying and in margarine and shortening. It is high in unsaturated fats, especially linoleic acid, but must be refined to remove gossypol (Lusas and Jividen 1987). Nutritional composition and bioactive compounds has been described in Tables 10.1 and 10.2 for all nine oilseed crops including sunflower, safflower, mustard, sesame, rapeseed, linseed, niger, groundnut, castor, and soybean (Gupta 2014; Houston 2014; Mir et al. 2017; Ebert 2022; Melrose 2019; Hama 2017; Nandasiri et al. 2020; Thompson et al. 2004; Dutta et al. 1994; Yadav et al. 2018; Sbihi et al. 2018; He and Chen 2013).

Broccoli sprouts and microgreens, like other sprouted seeds, are renowned for having a high concentration of vital nutrients (Turner et al. 2020). Vitamins, minerals, fatty acids, and amino acids are examples of essential nutrients that must be obtained from diet. Normal bodily processes, such as energy production, DNA synthesis, and metabolic pathways, depend on them (Jahangir et al. 2009). Numerous studies have looked into the amount of bioactive substances in microgreens, such as vitamins, minerals, and phytochemicals. Scholars have shown a special interest in examining antioxidants like vitamin C (VC), phytochemicals (such as carotenoids and phenolics), and specific minerals like copper (Cu), zinc (Zn), and selenium (Se) that counteract free radicals and aid in preventing damage brought on by oxidative stress. Ascorbic acid is a strong antioxidant that is necessary for many biological processes, including the production of collagen, wound healing, and immune system control (Akhilender Naidu 2003).

Table 10.1 Representation of nutritional and attributes of oilseed crops

Oilseed crop	Family	Rich in	References
Sunflower (*Helianthus annuus* L.)	Asteraceae	59% polyunsaturated omega-6 linoleic acid 30% monounsaturated omega-9 oleic acid5% palmitic acid, 6% stearic acid, and high in magnesium and vitamin E	Gupta (2014), Houston (2014)
Safflower (*Carthamus tinctorius* L.)	Asteraceae	Phytosterols, phytoestrogens (polyphenols), vitamin E	Mir et al. (2017), Ebert (2022)
Mustard (*Brassica juncea*)	Brassicaceae	Fatty acid (white mustard-oleic acid 22%) and linoleic acid 19.5–2% (brown and black mustard) Linolenic acid 9–15% Antioxidants like 3,4 di-hydroxybenzoic acid, ferulic acid, and sinapic acid Glucosinolates and *omega-3 fatty acids*	Sawicka and Kotiuk (2007), Boscariol Rasera et al. (2019), Melrose (2019)
Sesame (*Sesamum indicum*)	Pedaliaceae	Minerals like iron and calcium, protein content (21.9%) and fat content (61.7%) Lignans, vitamin E, phytosterols, linoleic and linolenic acids	Hama (2017)
Rapeseed (*Brassica napus*)	Brassicaceae	Phenolic compound (sinapic acid), alpha linolenic acid 6–14%, oleic acid 50–60%, saturated fatty acid <7%, monounsaturated fatty acid	Ghazani and Marangoni (2016), Nandasiri et al. (2020)
Linseed (*Linum usitatissimum*)	Linaceae	Saturated fatty acid 9%, monosaturated fatty acid (18%), high percentage of polysaturated fatty acid (73%), antioxidants like vitamin C, lutein, zeaxanthin, vitamin E, vitamin K, protein (albumin, globulin)	Thompson et al. (2004)
Niger (*Guizotia abyssinica*)	Asteraceae	Linoleic acid (18:2), palmitic (16:0), oleic (18:1), stearic acid (18:0)	Dutta et al. (1994)
Groundnut (*Arachis hypogaea*)	Fabaceae	Carbohydrates (21–37%), protein (20.7–25.3%), crude fat (31–46%), crude fibre (1.4–39%). Also rich in protein, lipid, minerals, arginine, phenolic acid, flavonoids, and resveratrol	Yadav et al. (2018), Gagliano et al. (2010)
Castor (*Ricinus communis* L.)	Euphorbiaceae	Monosaturated fatty acid (ricinoleic, palmitic, oleic, linoleic, stearic acid); phytosterols; α-,β-,γ-, and δ-tocopherol; phenolic acid	Omohu and Omale (2017), Sbihi et al. (2018), Ribeiro et al. (2014)
Soybean (*Glycine max*)	Fabaceae	~30–40% protein, ~9% dietary fibre, ~20% lipid, ~8% moisture	He and Chen (2013)

Table 10.2 Bioactive compounds and health-promoting potentials of oilseed crops

Name	Component	Biological constituents	Biological activity	References
Sunflower	Antioxidants, anti-inflammatory, anti-cholesterol, anti-cancer, anti-asthmatic, antidiabetic	Selenium, α-tocopherol, phenolic acids, phytosterols	Improve immunity against cancerous cells, *strong antioxidant protection*; *reduces oxidative stress*; *supports skin health*; *improves HDL/LDL ratio*	Roy et al. (2015), Chopra et al. (2025)
		Glutathione peroxidase	Protects body cell against cancer	Pisoschi and Pop (2015)
		Tocopherol and phytoesterol	Lower cholesterol level	Rani et al. (2017)
Safflower	Anti-inflammatory, anti-cancer, antioxidant effect	Hydroxysafflor yellow A	Inhibits platelet activating factor (PAF), which leads the pro-inflammation	Guo et al. (2019)
		Hydroxysafflor yellow B	Reduces the growth of cancerous cell	Qu et al. (2019)
		Serotonin derivative	Lipid peroxidation and potent radical scavenging action	Khalid et al. (2017)
Mustard	Anti-cancer, antioxidant, and anti-inflammatory	Sulforaphane	*H. pylori* was eliminated from eight out of eleven human stomach tissue xenografts given to immunocompromised mice after 5 days of SFN treatment	Haristoy et al. (2003)
		n-3 Polyunsaturated fatty acid	Preventing cardiovascular diseases.	Oppedisano et al. (2020)
Rapeseed	Anti-cancer, antioxidant, anti-obesity, anti-diabetic	Indole-3-carbinol (I3C)	Chemopreventive phytochemical that works by further hydrolysing rapeseed's indol-3-methyl isothiocyanate	Lee et al. (2019)
		Sterols	Sitosterol may stop chronic inflammation linked to obesity by preventing the release of TNF-α and interleukin-6	Ramprasath and Awad (2015)
Linseed	Anti-cancer, anti-diabetic, anti-inflammatory	Lignans	The lignans found in flaxseed and their metabolites have anti-cancer qualities	Mueed et al. (2023)

(continued)

Table 10.2 (continued)

Name	Component	Biological constituents	Biological activity	References
		ω-3 fatty acids	The potential anti-inflammatory properties of flaxseed and its derivatives make them good for human health. A sensitive indicator of inflammation, C-reactive protein (CRP), was used to assess flaxseeds anti-inflammatory properties	de Abreu Maíra et al. (2018)
Castor	Anti-cytotoxic	Ricinine, epigallocatechin	Cytotoxic properties	Majumder et al. (2019)
Soybean	Anti-osteoporosis, anti-proliferative	Soy-based protein	Helps improve bone health and prevent osteoporosis	Rizzo and Baroni (2018)
		Protease inhibitor	Inhibits the actions of mitogen-activated protein kinase, chymase, trypsin, and chymotrypsin; can suppress protease activities, which are a key factor in cancer	Srikanth and Chen (2016)
Sesame	Antioxidant, anti-tumour, anti-cholesterol	Lignans (sesamin)	Sesamin has antioxidant, and cholesterol-lowering activities	Majdalawieh et al. (2017), Senevirathne et al. (2019)
		Sesamol	Helpful in wound healing	Olasunkanmi et al. (2017), Nagar et al. (2022)
Groundnut	Anti-cholesterol, anti-diabetic	Peanut	Helpful in reducing blood cholesterol and blood pressure	Luu et al. (2015), Bao et al. (2013)
		Peanut	Beneficial in providing glycaemic control	Wien et al. (2014)
Niger	Anti-coagulation	Vitamin K1	Helps in blood clotting metabolism	Dwivedi et al. (2019)

10.4.1 Nutritional Classification of Microgreens

Microgreens are tiny plants with a big reputation. Harvested just after the first true leaves appear, they pack a remarkable punch of flavour, colour, and, most importantly, nutrition (Zhang et al. 2021). Despite their small size, these seedlings are dense with bioactive compounds, vitamins, and minerals, making them a favourite in both gourmet kitchens and health-conscious diets (Fig. 10.3). Below is a clear breakdown of their nutritional classification, covering the major categories that make microgreens such a powerhouse food (Keutgen et al. 2021).

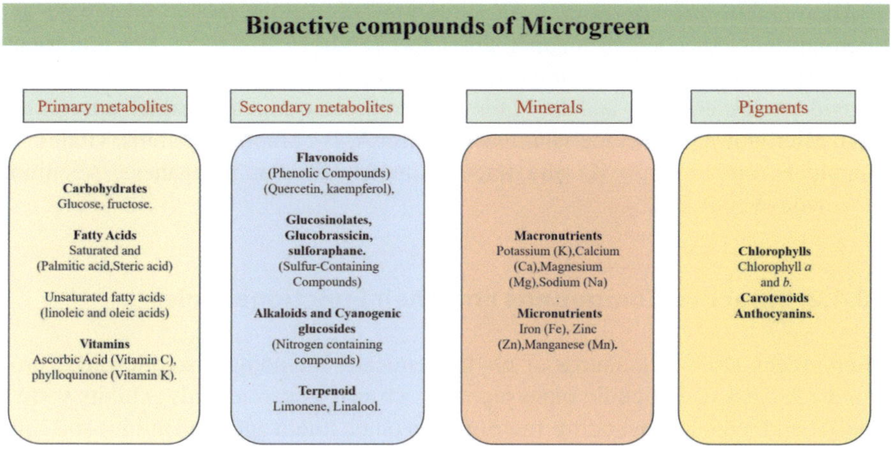

Fig. 10.3 Nutritional classification of microgreens

Carotenoids are plant pigments responsible for yellow, orange, and red colours, while chlorophyll gives plants their green hue. Microgreens contain higher concentrations of these pigments compared to their mature plant counterparts. Carotenoids include beta-carotene, lutein, and zeaxanthin. Beta-carotene acts as a precursor to vitamin A, supporting vision and immune health. Lutein and zeaxanthin are vital for eye protection against oxidative damage. Examples of plants rich in microgreens and carotenoids: red amaranth, red cabbage, and basil (Kyriacou et al. 2021). Chlorophyll, known for its detoxifying potential, also has antioxidant and anti-inflammatory properties. It helps in neutralizing free radicals and may assist in wound healing, for example, wheatgrass, kale, and spinach microgreens (Balik et al. 2025).

Antioxidants help protect the body from oxidative stress caused by free radicals, reducing the risk of chronic diseases (Sharma et al. 2025). Microgreens often have 4–40 times more antioxidant activity than their mature plant forms. Vitamin C, vitamin E and polyphenols of red cabbage, radish, and cilantro strengthens immune system and regulates collagen production, cardiovascular health and protects cell membranes from oxidative damage (De la Fuente et al. 2020).

Secondary metabolites are plant-produced compounds not directly involved in growth but vital for defence and human health benefits. Flavonoids (anti-inflammatory and anti-carcinogenic effects), glucosinolates (found in cruciferous microgreens like broccoli and mustard; they can break down into compounds with potential cancer-protective properties), and phenolic acids (known for antimicrobial and antioxidant activity), for example, broccoli, mustard, and pak choi.

Sunflower, pea shoots, and beet microgreens are a dense source of essential minerals, which are required in small amounts but play large roles in body functions, for example, calcium (bone health, muscle contraction), magnesium (muscle and nerve function, enzyme activity), iron (oxygen transport in the blood), zinc (immune function and wound healing), and potassium (regulates fluid balance and blood pressure) (Di Gioia et al. 2023).

Microgreens provide both macronutrients (needed in larger quantities) and micronutrients (needed in smaller amounts) (Seth et al. 2025). Macronutrients include proteins (essential for muscle repair and enzyme production), carbohydrates (provide quick energy), and fats (some microgreens contain healthy unsaturated fats). Micronutrients include vitamins (vitamin A, B-complex vitamins, vitamin C, vitamin E, and vitamin K) and trace minerals (copper, manganese, selenium) (Sukewijaya et al. 2025).

10.4.2 Bioactive Compounds and Their Functional Roles

Microgreens are a rich source of phytochemicals, including phenolics and carotenoids. A class of lipophilic plant pigments known as carotenoids exhibits yellow, orange, and red hues, including xanthophylls (such lutein and zeaxanthin) and carotenes (like β-carotene and lycopene) (Zhang et al. 2021). Carotenoids have significant physiological functions in the human body and exhibit antioxidant activity (Rodriguez-Amaya 2015). Carotenoids can be found in large quantities in vegetables, particularly those with vibrant colours (Khoo et al. 2011). The most prevalent secondary metabolites of plants are phenolic compounds, which can range in size from tiny molecules like phenolic acids to flavonoids with many rings, as well as to substances that are heavily polymerized, such tannins. Phenolics have demonstrated numerous health advantages for humans and are antioxidants that help plants repair damage from free radicals. In the five Brassica species microgreens, 164 polyphenols, comprising 29 hydroxycinnamic acids, 105 flavanol glycosides, and 30 anthocyanins (Sun et al. 2013). Microgreens are excellent sources of antioxidants because of higher concentrations and more diverse polyphenol profiles than mature *Brassica* plants (Cartea et al. 2010). Detailed information on bioactive compounds is mentioned in Table 10.2.

10.4.3 Micronutrient and Macronutrient Profile: Protein, Lipid, Amino Acid, Carbohydrates, Vitamins, and Minerals

Oilseed crops as microgreen include primary metabolites that are necessary for development and growth, as well as secondary metabolites that are organic compounds and potential signaling molecule which regulates flavour, defence in plants and exhibits numerous health benefits in humans. Several important amino acids are found in different microgreens like lysine, methionine, threonine, tryptophan, leucine, valine, and isoleucine, as well as lipids like phospholipids, triglycerides, and a high concentration of oil (linoleic acid, oleic acid, α-linolenic acid, palmitic, and stearic acids) and organic acids such as fumaric, oxalic, malic, citric, and succinic acid. The secondary metabolites known as glucosinolates are present in the mustard plant, which belongs to the Brassicaceae family. Glucosinolates contain a significant amount of sulphur. They exhibit durability and resistance to elevated temperatures. Cruciferous vegetables are rich in glucosinolates, encompassing families such as Brassicaceae, Capparaceae, and Caricaceae. The vegetables in question are broccoli, brussels sprouts, rocket, cabbage, bok choy, cauliflower, kale, and radish (Connolly et al. 2021; Miao et al. 2021). There are three classifications based on the amino acids and their side chains: indole, aromatic, and aliphatic. Sinalbin is predominantly found in white mustard, ranging from 0.1% to 1.1%, while sinigrin is present in brown mustard at levels of 0.8–0.9%, and in black mustard, it ranges from 0.4% to 0.9% (Sadowska et al. 2023). Because of geographical differences, mustard seeds show different glucosinolate profiles (Frank et al. 2010). Different varieties of microgreen also contain diverse phenolic compounds (at least one aromatic ring with one or more hydroxyl groups: coumarin, lignin, flavanoids, tannins), terpenes (5- carbon isoprene units: monoterpenoids sesquiterenoids, diterpenoids), nitrogen containing compounds (alkaloids, cyanogenic glycosides, non-protein amino acids) and sulphur containing compounds (glucosinolates, glutathione, defensins) that are key regulator and essential in human diet and nutraceuticals field.

10.5 Post-harvest Handling and Value-Added Applications

10.5.1 Shelf-Life Challenges and Preservation Strategies

Depending on the species, microgreens were harvested between 7 and 21 days after the cotyledons and genuine leaves were fully matured. For small-scale production, it is more practical and economical to manually harvest seedlings above the substrate layer using scissors, removing the root part. Particular care should be made to prevent seed integuments and growth media particles from adhering (Di Gioia and Santamaria 2015; Yadav et al. 2025). Harvested microgreens are promptly cleaned and placed in clamshell boxes or polythene bags due to their perishability. The microgreens can be kept for 9 days at 4 °C in a ziplock PPE bag (Arya et al. 2023). Post-harvest handling has an impact on microgreens' nutritional qualities. As a result, microgreens are typically eaten right after harvest. However, utilizing either fresh or dried microgreens powder to prepare a variety of value-added processed items improves the food's nutritional and medicinal qualities in addition to lengthening its shelf life. Even when refrigerated (4 °C), microgreens have a very short shelf life and can only be stored for a maximum of 3–10 days. However, utilizing either fresh or dried microgreens powder to prepare a variety of value-added processed items improves the food's nutritional and medicinal qualities in addition to lengthening its shelf life.

10.5.2 Culinary Application and Valuable Products

Microgreens' fresh, consumable, functional, and nutraceutical qualities might enhance local cuisine (Fig. 10.4). The vivid colour, flavour, and perfume of microgreens make them a popular garnishing item. They can be used as garnishes in salads, soups, smoothies, sandwiches, pizzas, dhoklas, spaghetti, and noodles to increase the nutritional value of the dish, as well as added to breakfast and snack

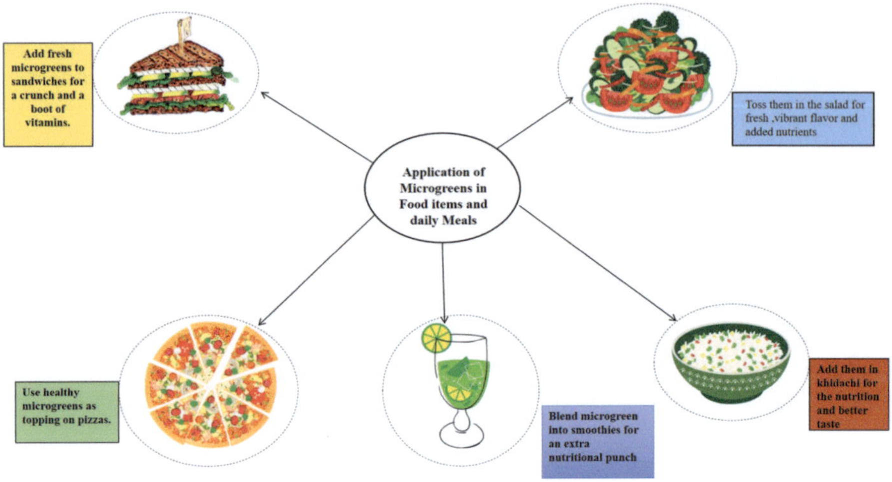

Fig. 10.4 Application of microgreens in food and health sectors

foods such toppings for dosa, idli, roti, bread omelettes, khichdi, and raita (Sreenivasa et al. 2019).

10.5.3 Health-Promoting Potential of Oilseed Crops as Microgreens

Vegetable oils like sunfoil are in high demand as a key component of the human diet with distinctive flavours and textures due to the recent awareness among nutritionists, dietetics, and food chemists about the importance of balancing dietary intake to prevent health issues like obesity and cardiovascular and neurological diseases. Edible oils contain dietary fibre, minerals, vitamins (including vitamin E), and antioxidant and antibacterial properties. Their physical and chemical makeup is mostly determined by their biological origin. Consuming foods high in saturated fatty acids over time can hurt the body's organs, leading to diabetes, cancer, and obesity (Fig. 10.5). Chemical and biological activities can enhance food's useful qualities. Plant-based lipids are thought to be healthier and more advantageous than animal-based fats because they include a higher concentration of polyunsaturated fatty acids, which are vital for organ function, enhance overall health, and lower the risk of serious conditions like cardiovascular disease. Thus, oils from oilseed crops (soybean, sunflower, olive, and rapeseed) are key sources of edible oils with high omega-3 and omega-6 fatty acid levels and other biological components for human consumption in the global market (Van Nieuwenhove et al. 2019).

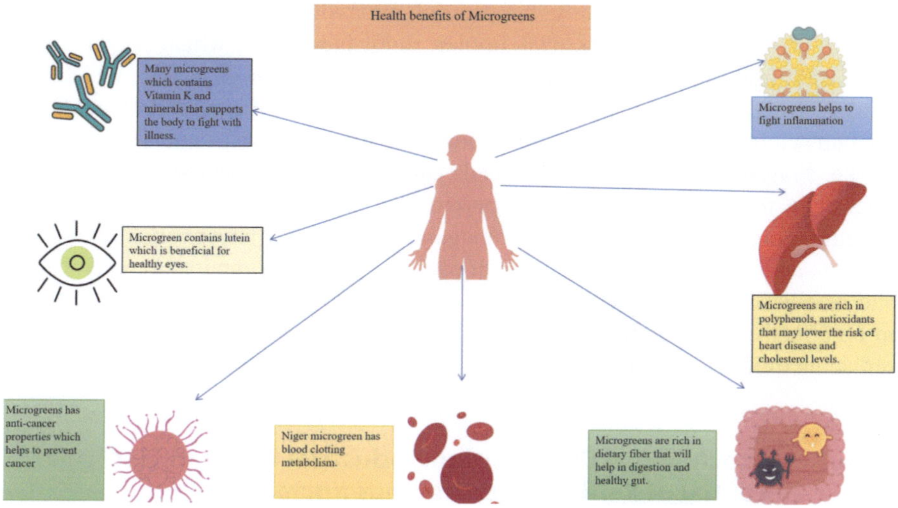

Fig. 10.5 Diagrammatic illustration of microgreens potential and its health benefits

10.6 Microgreens Revolution: Nutritional and Functional Significance of Nine Oilseed Crops

10.6.1 Nutritional and Therapeutic Significance of Sunflower (*Helianthus annuus* L.) as an Oilseed Crop

One of the major oilseed crops cultivated worldwide, sunflowers provide high-quality oil and dietary fibre that are vital to human health (Khan et al. 2015). With almost 70 species known globally, sunflowers are a short-season plant from the genus *Helianthus* and family Asteraceae. The plant's size and resemblance to the sun are the sources of its name, "sunflower". The term also comes from the revolving around the sun. Sunflowers are distinguished by their enormous, round, yellow inflorescence flower head (which bears achenes that mature into seeds), long taproot, hairy stalks, and broad, rough, coarsely serrated leaves that face the sun directly. In the sixteenth century, Spanish explorers brought it to Europe from its original temperate regions in North America, where temperatures often ranged from 20 °C to 25 °C. Sunflowers' flora exhibits their aesthetic and attractive qualities in a variety of sizes and colours, ranging from cream to yellow depending on the cultivar (Vilvert et al. 2018). Linoleic acid, a polyunsaturated fat, and oleic acid, a monounsaturated fat, make up the majority of sunflower oil. The type of plant used and the processing method used during production may be the cause of the variance in the fatty acid content of sunflower oil. The oil is said to have an intriguing flavour and a light golden tint. Minerals and vitamins abound in the oil. Modern technology makes it easier to produce and market sunflower oil, satisfying consumer demand for premium oil. Sunflower oil has 59% linoleic acid (polyunsaturated omega-6), 30% oleic acid (monounsaturated omega-9), 6% stearic acid, and 5% palmitic acid.

Other sunflower oils made through industrial processing and plant breeding include high linoleic, which contains 69% linoleic acid; high oleic, which contains 82% oleic acid; mid-oleic, which contains 65% oleic acid; and high stearic, which contains 72% and 18% oleic acid and stearic acid, respectively (Gupta 2014). Sunflower contains vitamin E, which is helpful in lowering atherosclerosis, artery disease, and stroke. Magnesium-rich sunflower seeds can effectively lessen bronchial asthma, migraines, muscular cramps, and hypertension (Houston 2014). Minerals such as calcium, copper, iron, magnesium, manganese, selenium, phosphorus, potassium, sodium, and zinc are abundant in sunflower seeds, as are vitamins E and B, folate, and niacin. Sunflower seeds' medicinal potential has generally been demonstrated to be effective in treating colds and coughs, acting as a diuretic and expectorant, replacing quinine, and preventing malaria (Islam et al. 2016).

10.6.2 Phytochemical Composition, Industrial Use, and Medicinal Potential of Safflower (*Carthamus tinctorius* L.)

Safflower (*Carthamus tinctorius* L.) is an annual oil-seed crop that originated in the eastern Mediterranean area and is usually grown for its seeds, which are used as a source of pigments, and for its oil, which is used both by food producers and in industry. There are three polyploidy species of safflower, *Carthamus turkestanicus* Popov, *Carthamus lantus* L., and *Carthamus creticus* L. (Downey et al. 1989). Safflower can be grown in arid regions because it is thought to be more resistant to salt and drought than other oil seed crops (Ada 2014). Zeaxanthin, which makes up between 37% and 58% of all carotenoids, is the most common carotenoid constituent in safflower species (Chakradhari et al. 2020). The glycosides generated from quercetin and shannesol, safflower yellow A, hydroxysafflor yellow A, red pigment, apigenin, rutin, myricetin, carthamidin, isocarthamidin, etc. are the primary components of flavonoids, which are secondary metabolites that are present in various plant sections (Asgarpanah and Kazemivash 2013; Ji et al. 2018). According to many studies, hydroxysafflor yellows A and B (HSYA and HSYB) flower extract/compounds cause a variety of inflammatory reactions, such as the growth and inflammatory reactions of human foetal lung fibroblasts (MRC-5 cells) (Liu et al. 2019), the suppression of platelet-activating factor (PAF)-induced proliferation, and an inflammatory response associated with asthma in human bronchial smooth muscle cells (HBSMCs) (Guo et al. 2019). Safflower seed and its flower have various advantages, including cell proliferation and inhibition. Various studies show their effect on brain and liver diseases; it has numerous antioxidant benefits.

10.6.3 Groundnut (*Arachis hypogaea* L.): A Package of Food, Industrial, and Medicinal Benefits

The groundnut, sometimes known as the "king of oilseed", is a type of legume belonging to the Fabaceae family that originated in South America and has since spread around the world. A staple crop cultivated on over 20 million hectares globally, groundnuts rank second in terms of planting area, after rapeseed, in terms of oilseed crops. Approximately 33 million tons of groundnuts are produced worldwide each year. Groundnuts are known to be important for their proteins, fats, carbs, vitamins, minerals, and organic substances. The purine found in groundnuts is referred to as a component that supplements human growth and energy (Woodroof 1983). Most groundnuts grown in the world are used for oil production, groundnut butter, confectionaries such as roasted groundnut, snack products, extenders in meat product formulation, soups, and desserts (Sáiz et al. 2013). The ovary of the groundnut plant turns into a seed, making it a crop that grows underground. Groundnuts have been recognized as a valuable source of protein since the 1800s. Groundnuts contribute a major amount of lipids and carbohydrates that serve as a major energy provider, balancing the basic energy demands of the human body (Tharanathan et al. 1975). In the human diet, vitamins serve a protective function. Generally speaking, vitamins are a class of chemical molecules that human bodies cannot synthesize and should be obtained through diet in order to maintain normal metabolic activity and health (Settaluri et al. 2012). Both fat-soluble and water-soluble vitamins are abundant in the nuts. Thiamine, riboflavin, niacin, pantothenic acid, pyridoxine, and folic acid are water-soluble B-complex vitamins that play a specialized biochemical role by serving as coenzymes that help the enzyme transform a substrate into a product. In the human body, riboflavin, or vitamin B2, is engaged in electron and hydrogen transfer processes as flavin mononucleotide (FMN) and flavin adenine D nucleotide (FAD). In the form of proteins, carbs, lipids, vitamins, minerals, and fibre, groundnuts give the body vital nutrients. Groundnuts are a fantastic and affordable food source. Eating foods high in protein, such as groundnuts, which contain energy-rich nutrients and bioactive chemicals that modulate the human immune system, can help prevent diseases or some immunological-based disorders in humans. In addition to polyunsaturated and monounsaturated fatty acids that can increase levels of HDL cholesterol, a heart-healthy fat, groundnuts are cofactors for a variety of vitamins and enzymes that the body is unable to generate.

10.6.4 Soybean (*Glycine max* L.): A Nutrient-Dense Oilseed with Multifaceted and Health and Economic Value

The soybean (*Glycine max*) crop is regarded as the primary oil source in the world and is significant both commercially and nutritionally (Singh et al. 2019). According to the dry weight of mature raw seeds, soybeans typically contain between 35% and 40% protein, 20% fat, 9% dietary fibre, and about 8% moisture (He and Chen 2013). The location affects the composition, the soybean cultivar, and the planting environment. Foods made from soy are low in saturated fats and high in vitamins, minerals, proteins, and fibre. Many other types of soy products have been made, including tofu, roasted or boiled soybeans, soymilk, soy mayonnaise, miso, soy cheese, soy yoghurt, tempeh, soy sauce, tamari, textured vegetable protein (TVP), or textured soy protein (TSP) (Jayachandran and Xu 2019). It is an excellent source of protein and vegetable oil; thus, it is an economically important crop worldwide (Bezerra et al. 2017). Because of their high isoflavonoids and folic acid content, soybeans are employed in diets all over the world. Because they contain a large amount of plant protein, soybeans and their products are regarded as significant sources of plant protein and a quantity of vital amino acids, and they offer some health advantages to people. According to studies, soybeans may help lower the incidence of osteoporosis and other bone degenerative disorders, as well as breast, prostate, and colon cancers because of their valuable nutritional content. They can decrease menopausal heat flashes (El-Shemy 2011).

10.6.5 Niger (*Guizotia abyssinica* Cass.): A Traditional Oilseed with Modern Nutraceutical and Commercial Relevance

Due to their sticky seed coat, the tiny black seeds don't deteriorate after extended storage. Originating in Ethiopia's highlands, Niger (*Guizotia abyssinica* L. Cass; 2n = 30) is a small oilseed crop that has spread to other regions of the world. It is a member of the Asteraceae family, which also includes other economically significant crops including chrysanthemum, safflower, sunflower, and lettuce. Grown mainly for its edible oil and seeds, the niger crop is robust and upright. Although primarily an oilseed crop, niger can be eaten fried, as a condiment, or dried, powdered, and combined with flour to form sweet cakes, which are customarily presented during Ethiopian coffee ceremonies. Niger oil is a semi-drying, polyunsaturated oil. The main applications for it are in cooking, bird feed, paints, soap, and flower perfume extraction. Linoleic acid (18:2) makes up the majority of the oil content of niger seed, with palmitic (16:0), oleic (18:1), and stearic acid (18:0) being the other main acids that were found to vary from 6% to 11% each (Dutta et al. 1994). Niger is renowned for its dietary fats and linoleic acid, which prevent cardiovascular conditions like high blood pressure, coronary heart disease, and atherosclerosis. It also produces linoleic acid derivatives, which are a precursor to some metabolic regulatory compounds and a structural component of the plasma membrane. Additionally, niger oil is utilized medicinally. It can be utilized in pharmaceutical applications in place of sesame oil (Mohseni et al. 2020).

10.6.6 Mustard (*Brassica* spp.): From Culinary Spice to Multifunctional Resource, Nutritional and Pharmacological Significance of Mustard Oilseed

A member of the Brassicaceae family, mustard is prized for its strong, peppery dried seeds. Black mustard, *Brassica nigra* (L.) W. D. J. Koch; brown mustard, *Brassica juncea* (L.) Czerniak, Brassica rugosa Hort., or *Sinapis juncea* L.; and white mustard, *Brassica hirta* Moench, are a few of the well-known mustard species (Rahman et al. 2018). In sandy loam soils with little rainfall, mustard thrives. Although it may be grown in tropical and subtropical areas, it is typically grown in temperate settings. It is widely farmed in Asia, North Africa, and Europe and is regarded as one of the earliest domesticated crops (Divakaran and Babu 2016). The food business frequently uses mustard plants. While black and brown mustard are typically utilized for their scent, white mustard is frequently used to flavour dishes. Traditional healers also utilize some mustard plants, like *Brassica alba* and *Brassica juncea*, as herbal remedies to treat colds, arthritis, cough, sore throats, muscle aches, and diabetes (Rahman et al. 2018). Omega-3 fatty acids and glucosinolates (GSLs) are two of the bioactive substances found in mustard seeds (Melrose 2019). Beta-thioglucose, thiohydroximate-O-sulfonate, and a variable aglycone side chain made from an α-amino acid make up the three compartments that make up GSLs (Ladak et al. 2021). With a high fat content of 23–47%, mustard seeds are primarily composed of erucic acid (26.5–36.5%). The percentage of other fatty acids varies by species; for example, white mustard oil contains a large amount of oleic acid (22%); linoleic acid is most prevalent in brown and black mustard oil (19.5–22%). Significant amounts of linolenic acid (9–15%) are also present in the seeds of all mustard species (Sawicka and Kotiuk 2007).

10.6.7 Castor (*Ricinus communis* L.): A Non-edible Oilseed Crop with Diverse Medicinal, Cosmetic, and Industrial Applications

Castor is the popular name for the annual oilseed crop *Ricinus communis*. Although it is not a real bean in nature, it is occasionally referred to as castor bean. The castor plant, a member of the Euphorbiaceae spurge family, can grow in a variety of climates (Anjani et al. 2014). One of the oldest and most important non-edible oilseed crops is castor, or *Ricinus communis* L. It is a significant member of the medicinal plant family Euphorbiaceae. Flavonoids, phenolic compounds, fatty acids, amino acids, terpenoids, phytosterol, and other chemical components are all present in castor. These compounds have been shown to have anti-inflammatory, antibacterial, antioxidant, hepatoprotective, insecticidal, anticonception, antidiabetic, antifertility, and wound-healing qualities. Castor oil is utilized in the food business as a mould inhibitor, flavouring, and food additive (Wilson et al. 1998). In the southeast region of Nigeria, the large, white seeds of castor are a major source of the food condiment known as ogiri. Ricinus communis seeds are traditionally fermented to produce ogiri, a culinary condiment. It is said that the condiment helps with vision. One natural product that helps with several illnesses is castor oil. Because of the active chemicals it contains, it plays a key role in the creation of numerous pharmaceutical and cosmetic goods. When it comes to treating skin conditions like sunburn, ringworm, stretch marks, wrinkles, and acne, castor oil works incredibly well. Additionally, it guards against warts, athlete's foot, boils, and persistent itching. The oil works well as a wound disinfectant and skin moisturizer. When combined with coconut or almond oil, castor oil promotes hair growth and thickens eyelashes and eyebrows. Faster hair growth results from the oil's increased blood flow to the follicles. Menstrual disorders, migraines, skin abrasions, rheumatism, and inflammation can all be treated with castor oil, a good addition and potent laxative.

Additionally, the oil contains omega-6 necessary fatty acids, which are essential for healthy hair and also dark hair and bald spots.

10.6.8 Linseed (*Linum usitatissimum* L.): A Rich Source of Omega-3 Fatty Acid, Essential Nutrients, and Health-Promoting Bioactive

Flaxseed, also called linseed (*Linum usitatissimum*), is one of the oldest oilseed crops in the world. It comes from a Latin word that means "extremely useful" (Ganorkar and Jain 2013). It has been grown for many years and is used as both food and fibre. Flaxseed is grown in many parts of the world for its fibre and oil, nutritional content, and medicinal uses. Despite its Egyptian origins, flaxseed is currently farmed in India, Holland, Russia, and Britain for its oil and fibre (Soni et al. 2016). Thiamine, magnesium, phosphorus, dietary fibre, protein, and B vitamins are all abundant in flaxseeds. It is rich in linoleic acid, a crucial omega-6 fatty acid. Flax fibres are a good source of polyunsaturated fatty acids, including α-linolenic acid, an important omega-3 fatty acid. According to Tourre and Xueming (2010), flaxseed contains the highest quantity of omega-3 fatty acids. It contains a low percentage of saturated fatty acids (9%), a moderate amount of monounsaturated fatty acids (18%), and a high percentage of polyunsaturated fatty acids (73%). Antioxidants such as vitamin C, lutein and zeaxanthin, vitamin E, and vitamin K are also found in flax. According to research, flax lowers the incidence of many cancers, alters autoimmune illnesses, and lowers the risk of heart disease (Thompson et al. 2004). Albumin and globulin are the two main proteins found in flax. About 26.6% of total protein is made up of albumin, whereas 73.4% is made up of globulin fraction.

10.6.9 Bioactive Constituents and Nutritional Importance of Rapeseed (*Brassica napus* L.)

The seeds of brassica plants, such as *Brassica napus* and *Brassica rapa*, are sources of rapeseed oil. The high oil content of the seeds of the *Brassica* plant makes it a major cash crop (Chew 2020). Over 1.9 billion persons were overweight between 1975 and 2016, which can result in a number of illnesses such as obesity, heart disease, high blood pressure, and hyperlipidemia (Mathews et al. 2023). Therefore, there is a pressing demand for nutrient-dense, easily digested foods. Rapeseed oil is good for human health since it is rich in bioactive compounds and unsaturated fatty acid. Although most rapeseed have low levels of erucic acid and glucosinolate, these substances can be categorized as antioxidants. The phytochemical compound's nutrients are either lipid or water soluble. Lipids are therefore very important for health. Phenolic acids, phytosterols, diglycerides, flavones, vitamin E, and flavonols are the bioactive substances found in rapeseed. Since they must be obtained through diet, linoleic acid and α-linolenic acid are both considered necessary fatty acids for humans. Moreover, the human body is not able to synthesized some essential fatty acid (omega-3 fatty acid: essential polyunsaturated fats) due to absence of certain enzymes and their cofactors therefore depends on diet and nutritious food which is important for proper functioning of brain, eye and heart (Jain and Lingwan 2026).

10.6.10 Biochemical Richness and Health Protective Role of Sesame (*Sesamum indicum* L.)

The Pedaliaceae family includes the genus *Sesame*, which is a subgenus of Sesamum. Sesame can be categorized as white, black, or yellow based on the colour of the germplasm; black and white sesame are the most prevalent and extensively cultivated dominant species. White sesame has a high oil content, is of good quality, and has the widest planting area and distribution, whereas black sesame has a strong growth ability, lodging resistance, and drought tolerance. The majority of the plants in other variegated varieties, like yellow sesame, are branching. Generally speaking, when the germplasm's colour deepens, the oil content gradually drops (Dar et al. 2019). China's four main oil crops are peanuts, soybeans, rape, and sesame (*Sesamum indicum* L.), which are among the oldest oil crops used and produced by humans (Zech-Matterne et al. 2015). Sesame has been farmed for a long time and was first found in ancient Pakistani ruins (Bedigian 2010). Sesame's mild flavour and strong aroma make it a popular and widely produced crop. Sesame seeds are frequently used in people's daily lives to produce a range of dishes, including sesame paste and oil, as well as to garnish other cuisines. When sesame seeds were included to the list of food and medicinal ingredients released by the previous Chinese Ministry of Health in 2002, their scientific nutritional value status was established legally. Fat, protein, vitamins, minerals, and dietary fibre are all abundant in sesame seeds. Traditional oil manufacturing techniques yield sesame oil, which is high in fat-soluble vitamins, amino acids, and unsaturated fatty acids. According to studies, sesame seeds are high in minerals like iron and calcium and have a protein content of 21.9% and a fat content of 61.7% (Hama 2017). Rich in nutrients, sesame seeds are known as the "crown of eight grains" and a "all-purpose nutrient bank" (Rout et al. 2018). One of the traditional products of the basic processing of sesame seeds is sesame oil, a fragrant oil that can be used as an edible oil. In addition to having significant concentrations of physiologically active compounds such lignans, natural vitamin E, and phytosterols, sesame oil is also rich in linoleic and linolenic acids. The oil extracted by cold pressing sesame seeds has a high quality and nutritional value. Linoleic acid (46.9%) and oleic acid (37.4%) are the two main unsaturated fatty acids found in sesame oil. Since the body is unable to produce certain fatty acids, they must be received from diet. This makes them essential fatty acids. The predominant form of vitamin E found in sesame oil is gamma-tocopherol (90.5%) (Gharby et al. 2017). Lignans are also found in sesame seeds which have antioxidant properties. Sesame lignans help improve liver function and control blood lipid levels. Sesamin has good physiological activity and is one of the most prevalent lignans found in sesame seeds. It has been shown in studies to have high antioxidant qualities, decrease cholesterol, regulate lipid metabolism, stabilize blood pressure, and have anti-tumour actions (Majdalawieh et al. 2017). Sesamin has strong antitumor effects, according to several research. It's anti-proliferative, pro-apoptotic, anti-inflammatory, anti-metastatic, anti-angiogenic, and pro-autophagocytic properties are primarily responsible for its anti-cancer actions. The precise signaling events that sesamin triggers in cancer cells are yet unknown, although the anti-cancer actions of sesamin are mediated in large part by the STAT3, JNK, ERK1/2, p38 MAPK, PI3K/AKT, caspase-3, and p53 signaling pathways.

10.7 Integration of OMICS Technologies with Artificial Intelligence to Improve Nutritional and Shelf-Life Traits

With a wide range of applications, OMICS is a relatively new technique that addresses the universal detection of genes (genomics), mRNAs (transcriptomics), proteins (proteomics), and metabolites (metabolomics) in a particular biological sample (Misra et al. 2019; Gupta et al. 2023). A potent method for deciphering intricate biological processes that control nutritional quality and shelf life features in crops is the integration of multi-omics technologies, including transcriptomic, proteomics, metabolomics, epigenomics, and genomics. This all-encompassing approach makes it possible to precisely identify the important genes, proteins, and metabolites that are involved in post-harvest deterioration as well as nutrient production, accumulation, and degradation. Enhancing pre-harvest nutritional quality and post-harvest shelf-life features in crops through the integration of omics technology has enormous potential to improve food security and health benefits by lowering nutrient loss and spoilage. Integration of advanced OMICS technologies (genomics, transcriptomics, proteomics, metabolomics and nutriomics) along with re-sequencing of complete genome of chickpea would be fruitful to explore the data for several studies to improve the crop and its potential traits (enhanced nutritional quality and shelf-life traits), includes linkage disequilibrium, domestication patterns, genetic diversity, evolutionary analysis, and population structure, transforming the orphan crop chickpea into a promising genetic resource as a result of the technological developments (Varshney et al. 2019).

10.7.1 Nutritional Trait Management in the Pre-harvest Stage

The identification of candidate genes governing the amount of micronutrients (such as vitamins and minerals) and bioactive substances (such as antioxidants and phenolics) is made easier by genomic methods like genome-wide association studies (GWAS) and QTL mapping. Differential gene expression patterns linked to food metabolism and stress tolerance during crop development are revealed by transcriptomic analysis employing RNA-seq (Cao et al. 2025). For instance, genes controlling the manufacture of glucosinolate and the metabolism of fatty acids have been found in microgreens and oilseed crops, offering possibilities for the breeding of nutrient-dense cultivars. Protein modifications and enzyme activity involved in biosynthetic pathways are further clarified by proteomics. Metabolomics profiling identifies the build-up of vital metabolites that contribute to nutritional value, such as lipids, flavonoids, and amino acids. Histone alterations and DNA methylation that affect gene expression for quality attributes are discovered by epigenomic research (Ibragic et al. 2025).

10.7.2 Post-harvest Shelf Life Enhancement

OMICS analyses clarify molecular alterations that occur during storage, such as pathogen contacts, cell wall deterioration, and oxidative stress response. Stress-responsive genes and proteins that preserve cellular integrity and postpone senescence are identified by transcriptomic and proteomics data. Changes in primary and secondary metabolites linked to flavour, texture, and nutritional deterioration are revealed by metabolomics (Firdous et al. 2025). For example, OMICS research has identified enzymes associated with lignin build-up and chlorophyll degradation in leafy greens and microgreens that impact shelf life. The creation of molecular markers for choosing genotypes with improved post-harvest durability is made possible by the integration of these data. Furthermore, to maintain nutritional and sensory quality, metabolite indicators can direct the creation of ideal storage conditions.

A comprehensive understanding of the regulatory networks governing nutritional characteristics and shelf life is made possible by combining datasets from various omics platforms with cutting-edge bioinformatics and systems biology methods. Precision breeding and genome editing (e.g., CRISPR-Cas9) for targeted improvement are accelerated by this integrative approach (Sasaki et al. 2025). It also aids in the development of innovative post-harvest technologies and treatments based on the molecular profiles of particular crops.

10.7.3 Comparative Analysis of Conventional and Modern Agriculture

Conventional agriculture relies heavily on traditional breeding techniques, manual labour, and the application of synthetic fertilizers to enhance crop yields. For centuries, this approach has served as the backbone of food production in many countries, ensuring food security and livelihoods for millions of smallholder farmers. However, over time, heavy reliance on chemical inputs has led to several environmental and ecological concerns. Soil degradation, nutrient depletion, and water contamination due to fertilizer runoff have become major challenges associated with conventional farming (Khan 2020). Furthermore, excessive use of pesticides has been linked to declining biodiversity and the emergence of resistant pests, which further exacerbate the vulnerabilities of farming systems. Despite these drawbacks, conventional methods remain popular because they are cost-effective and easily accessible and require relatively little specialized training. In regions where capital investment in technology is not feasible, conventional agriculture continues to be the most practical choice for meeting immediate food demands. It provides employment opportunities, supports rural economies, and helps maintain traditional farming knowledge, which is an important cultural asset (Khan 2020).

In contrast, modern agriculture has evolved to address many of the limitations inherent in conventional practices. By incorporating advanced technologies such as drones, sensors, artificial intelligence (AI), and machine learning, modern farming aims to optimize planting, irrigation, and pest management strategies (Smith 2019; Brown and Wang 2021). The adoption of precision agriculture techniques allows farmers to apply inputs only where and when they are needed, thereby reducing waste and improving efficiency. For instance, drones can be used to monitor crop health in real time, identifying areas of stress or nutrient deficiency, while AI-based systems can predict weather changes and guide irrigation schedules. These innovations not only enhance productivity but also contribute to environmental sustainability by minimizing the ecological footprint of farming. Although the initial cost of implementing such technologies can be high, the long-term benefits often outweigh these challenges. Farmers adopting modern practices report higher yields, better soil health, and reduced dependency on harmful chemicals, thereby making agriculture more sustainable and resilient in the face of climate change (Patel and Sharma 2022).

Both systems present distinct advantages and challenges, and their adoption often depends on the socioeconomic and geographical context in which they are practiced. Conventional agriculture, for example, is particularly suited for small-scale farmers in developing regions who may lack access to modern infrastructure, capital investment, or digital literacy. In such cases, the accessibility of fertilizers and traditional techniques ensures that farming can still be productive, albeit with limitations related to long-term sustainability (Khan 2020). On the other hand, modern agriculture offers scalability and sophisticated monitoring capabilities that are increasingly essential to meet the rising global food demand driven by population growth and urbanization (Smith 2019; Brown and Wang 2021). Importantly, these modern methods allow for better adaptation to changing climatic conditions by providing predictive insights and data-driven decision-making tools that can mitigate risks of crop failure.

Recent research highlights the potential of hybrid approaches that integrate the strengths of both conventional and modern systems. For example, farmers may continue to rely on organic manure and local seed varieties while also adopting data-driven irrigation scheduling or drone-based crop monitoring. This blending of methods ensures resilience, reduces the environmental impact, and maintains affordability for farmers who cannot fully transition to high-tech agriculture (Patel and Sharma 2022). Moreover, such approaches can safeguard cultural traditions and biodiversity while also introducing scientific innovation to improve productivity. Hybrid systems can therefore play a key role in ensuring food security and sustainability in the coming decades, especially in regions facing resource constraints or high vulnerability to climate change. By aligning traditional wisdom with technological advancements, agriculture can achieve a balance between productivity, sustainability, and inclusivity, thereby addressing the needs of both present and future generations.

10.7.4 Development of Climate-Resilient Nutritionally Rich Microgreens Through Advancements of Deep Learning in Smart Agriculture

Agriculture continues to be a cornerstone of global food production and economic stability. With the ongoing rise in population, the sector is under pressure to increase productivity while reducing its environmental footprint. To address these challenges, smart farming practices have emerged, leveraging artificial intelligence (AI), the Internet of Things (IoT), and, more recently, deep learning technologies. Deep learning, particularly through Convolutional Neural Networks (CNNs) and Recurrent Neural Networks (RNNs), has shown exceptional promise in automating crop monitoring, disease diagnosis, and yield prediction (Kamilaris and Prenafeta-Boldú 2018).

CNNs have become the leading tool in image-based applications such as plant disease detection, fruit counting, and crop classification. Their layered structure allows efficient extraction of complex visual features from leaves and aerial imagery (Mohanty et al. 2016). In contrast, RNNs, especially Long Short-Term Memory (LSTM) models, excel in processing sequential data, making them highly suitable for forecasting weather trends and modelling crop growth and yield prediction over time (Joshi et al. 2025). Despite their success, a persistent issue remains: CNNs are highly accurate at classifying visible symptoms but often struggle with early disease detection. Hybrid frameworks, combining CNN with Support Vector Machines (SVM), have been proposed to overcome this limitation by enhancing precision in identifying early-stage infections (Singh et al. 2025).

Several studies demonstrate that CNN-based models can distinguish between healthy and infected leaves with accuracies above 95%. For instance, AlexNet and VGG-16 architectures have been applied to banana and sunflower crops, successfully detecting fungal diseases such as Sigatoka and Alternaria leaf rot (Mohanty et al. 2016). Identifying crop types and weeds is crucial for yield optimization. Residual networks (ResNet) trained on UAV imagery have achieved up to 94% accuracy in separating weeds from sugar beet fields (Singh et al. 2016). Similarly, satellite-based CNN models allow farmers to classify maize, soybeans, and wheat across large-scale farmland (Kamilaris and Prenafeta-Boldú 2018). Furthermore, CNN-based fruit-counting systems using UAV images enable accurate yield forecasting, reducing manual labour (Mohanty et al. 2016). Also, a large proportion of research relies on the PlantVillage dataset, which contains over 54,000 labeled images of healthy and diseased plant leaves (Kamilaris and Prenafeta-Boldú 2018). Deep learning is transforming smart agriculture by providing tools for automated plant disease detection, weed management, water stress monitoring, weather forecasting, and yield estimation. CNNs dominate in handling image-based agricultural data, while RNNs prove invaluable in sequential forecasting tasks. A promising direction lies in hybrid models, which combine the strengths of CNNs with classical classifiers to enhance early detection capabilities. With advances in IoT integration and UAV-based data collection, deep learning will continue to play a central role in ensuring sustainable and resilient farming systems.

10.8 Future Perspectives

Oilseed-derived microgreens represent a promising class of next-generation functional foods due to their exceptional nutrient density, rapid growth cycle, and adaptability to sustainable cultivation systems. Rich in both primary and secondary metabolites, they exhibit potent antioxidant, anti-inflammatory, cardio-protective, and anti-cancer activities while offering higher levels of key micronutrients such as vitamins C, E, and K and minerals compared to mature plants. The diverse phytochemical profiles of such crops including sunflower, safflower, mustard, sesame, rapeseed, linseed, niger, groundnut, castor, and soybean further enhance their nutritional and therapeutic potential. The integration of multi-omics approaches enables targeted improvement of flavour, nutrient biosynthesis, and post-harvest stability, positioning these crops as ideal candidates for biofortification and personalized nutrition based on health needs. By contributing to nutrient security and reducing the global burden of diet-related diseases, oilseed-derived microgreens align strongly with United Nations Sustainable Development Goal 3 (Good Health and Well-Being). Their cultivation and consumption thus provide an innovative, sustainable, and health-promoting strategy for advancing global nutrition and wellness. There are several nutrition-related products available in the market related to microgreens, which are very helpful in terms of nutrients. Moreover, value-added products may be derived from oilseeds microgreens, such as bakery products (biscuits and breads), and can be consumed as fresh microgreens (salad, pizzas), which would be a promising resource to fulfil the basic desirable needs of nutrients with foods. Moreover, microgreens farming promises small-scale entrepreneurship for marginal farmers under adverse climatic conditions. However, kitchen gardening and rooftop gardening are a promising approach to produce chemical-free green economically. In addition to that, active and enthusiastic women may initiate a microgreen farming start-up as an income source and an alternative approach for the development of smart crop agriculture.

Declarations The authors declare no conflict of interest.

References

Abad M, Noguera P, Bures S (2001) National inventory of organic wastes for use as growing media for ornamental potted plant production: case study in Spain. Bioresource Technol 77(2):197–200

de Abreu Maíra DC, Letícia RP, da Costa LR, da Camara Boueri BF, Carolina RP, Pereira ADA, Ribeiro DC, da Silva EM, da Costa CAS, Gilson TB (2018) Flaxseed (Linum usitatissimum) flour contributes to bone health in adult male rats. Nutrition 49:48–50

Ada R (2014) Dimension, geometric, agricultural and quality characteristics of safflower seeds. Turk J Field Crops 19(1):7–12

Akhilender Naidu K (2003) Vitamin C in human health and disease is still a mystery? An overview. Nutr J 2:2–7

AlShrouf A (2017) Hydroponics, aeroponic and aquaponic as compared with conventional farming. Am Sci Res J Eng Technol Sci 27(1):247–255

Anjani K, Raoof MA, Desai AG (2014) Evaluation of world castor (Ricinus communis L.) germplasm for resistance to Fusarium wilt (Fusarium oxysporum f. sp. ricini). Eur J Plant Pathol 139(3):567–578

Arya KS, Kutty MS, Pradeepkumar T (2023) Microgreens of tropical edible-seed species, an economical source of phytonutrients-insights into nutrient content, growth environment and shelf life. Future Foods 8:100262

Asgarpanah J, Kazemivash N (2013) Phytochemistry, pharmacology and medicinal properties of Carthamus tinctorius L. Chin J Integr Med 19(2):153–159

Balik S, Elgudayem F, Dasgan HY, Kafkas NE, Gruda NS (2025) Nutritional quality profiles of six microgreens. Sci Rep 15(1):6213

Bao Y, Han J, Hu FB, Giovannucci EL, Stampfer MJ, Willett WC, Fuchs CS (2013) Association of nut consumption with total and cause-specific mortality. N Engl J Med 369(21):2001–2011

Bedigian D (2010) Characterization of sesame (Sesamum indicum L.) germplasm: a critique. Genet Resour Crop Evol 57:641–647

Bezerra ARG, Sediyama T, da Silva FL, Borém A, da Silva AF, dos Santos Silva FC (2017) Agronomical aspects of the development of cultivars. In: Soybean breeding. Cham: Springer International Publishing, pp 395–411

Blasi F, Rocchetti G, Montesano D, Lucini L, Chiodelli G, Ghisoni S, Baccolo G, Simonetti MS, Cossignani L (2018) Changes in extra-virgin olive oil added with Lycium barbarum L. carotenoids during frying: chemical analyses and metabolomic approach. Food Res Int 105:507–516

Brown L, Wang Y (2021) Precision agriculture using drones: a review. Comput Electron Agric 185:106150

Cao Y, Hao F, Li J, Zhang B, Li Z, Liu T, Gao Y, Niu X, Liu X, Zhang H, Yang L (2025) Integrated transcriptome and metabolome analyses reveal complex oxidative damage mechanisms in rice seedling roots under different carbonate stresses. Antioxidants 14(6):658

Cartea ME, Francisco M, Soengas P, Velasco P (2010) Phenolic compounds in Brassica vegetables. Molecules 16(1):251–280

Chakradhari S, Perkons I, Mišina I, Sipeniece E, Radziejewska-Kubzdela E, Grygier A, Rudzińska M, Patel KS, Radzimirska-Graczyk M, Górnaś P (2020) Profiling of the bioactive components of safflower seeds and seed oil: cultivated (Carthamus tinctorius L.) vs. wild (Carthamus oxyacantha M. Bieb.). Eur Food Res Technol 246(3):449–459

Chaudhary A, Krishna V (2021) Region-specific nutritious, environmentally friendly, and affordable diets in India. One Earth 4(4):531–544

Chew SC (2020) Cold-pressed rapeseed (*Brassica napus*) oil: chemistry and functionality. Food Res Int 131:108997

Choe U, Yu LL, Wang TT (2018) The science behind microgreens as an exciting new food for the 21st century. J Agric Food Chem 66(44):11519–11530

Chopra P, Singh G, Kumar A, Bisht A (2025) Impact of foliar-applied growth regulators on productivity and profitability of linseed (Linum usitatissimum L.). Biol Forum 17(5):124–128

Connolly EL, Sim M, Travica N, Marx W, Beasy G, Lynch GS, Bondonno CP, Lewis JR, Hodgson JM, Blekkenhorst LC (2021) Glucosinolates from cruciferous vegetables and their potential role in chronic disease: investigating the preclinical and clinical evidence. Front Pharmacol 12:767975

Dar AA, Kancharla PK, Chandra K, Sodhi YS, Arumugam N (2019) Assessment of variability in lignan and fatty acid content in the germplasm of Sesamum indicum L. J Food Sci Technol 56:976–986

De K, Kakar V (2021) Effects of monetary policy on food inequality in India. The J Dev Stud 57(11):1852–1870

De la Fuente M, Sánchez C, Vallejo C, Díaz-Del Cerro E, Arnalich F, Hernanz Á (2020) Vitamin C and vitamin C plus E improve the immune function in the elderly. Exp Gerontol 142:111118

Despommier D (2013) Farming up the city: the rise of urban vertical farms. Trends Biotechnol 31(7):388–389

Di Bella MC, Niklas A, Toscano S, Picchi V, Romano D, Lo Scalzo R, Branca F (2020) Morphometric characteristics, polyphenols and ascorbic acid variation in Brassica oleracea L. novel foods: sprouts, microgreens and baby leaves. Agronomy 10(6):782

Di Gioia F, Santamaria P (2015) Microgreens. Nuovi alimenti freschi e funzionali per esplorare tutto il valore della biodiversità. Novel fresh and functional food to explore all the value of biodiversity. Nuevos alimentos frescos y funcional para explorar todo el valor de la biodiversidad

Di Gioia F, Renna M, Santamaria P (2017) Sprouts, microgreens and "baby leaf" vegetables. In: Minimally processed refrigerated fruits and vegetables, pp 403–432

Di Gioia F, Hong JC, Pisani C, Petropoulos SA, Bai J, Rosskopf EN (2023) Yield performance, mineral profile, andnitrate content in a selection of seventeen microgreen species. Front Plant Sci 14:1220691

Divakaran M, Babu KN (2016) Encyclopedia of food and health. Science Direct, pp 9–19

Downey RK, Röbbelen G, Ashri A (eds) (1989) Oil crops of the world: their breeding and utilization. McGraw-Hill

Dutta PC, Helmersson S, Kebedu E, Alemaw G, Appelqvist LÅ (1994) Variation in lipid composition of niger seed (Guizotia abyssinica Cass.) samples collected from different regions in Ethiopia. J Am Oil Chem Soc 71(8):839–843

Dwivedi A, Bhatt DK, Jatav AK (2019) Niger seed as a nourishing and therapeutic agent–a review. J Emerg Technol Innov Res 6(5)

Ebert AW (2022) Sprouts and microgreens—Novel food sources for healthy diets. Plants 11(4):571

El-Shemy H (ed) (2011) Soybean and health. BoD–Books on Demand

FAO (2020) Groundnut: post-harvest operations. Food and Agriculture Organization of the United Nations. Available at: http://www.fao.org/3/AC482E/ac482e00.htm. Accessed 10 Aug 2025

Firdous N, Farooq U, Batool A, Aslam A, Erum A, Ahmad S, Sibt-e-Abbas M, Usman M, Khalil S (2025) Postharvest metabolomics: prolonging shelf life and quality of tomatoes. In: Omics approaches for tomato yield and quality trait improvement. Springer Nature Singapore, Singapore, pp 147–176

Frank N, Dubois M, Goldmann T, Tarres A, Schuster E, Robert F (2010) Semiquantitative analysis of 3-butenyl isothiocyanate to monitor an off-flavor in mustard seeds and glycosinolates screening for origin identification. J Agric Food Chem 58(6):3700–3707

Gagliano N, Aldini G, Colombo G, Rossi R, Colombo R, Gioia M, Milzani A, Dalle-Donne I (2010) The potential of resveratrol against human gliomas. Anti-cancer Drugs 21(2):140–150

Ganorkar PM, Jain RK (2013) Flaxseed--a nutritional punch. Int Food Res J 20(2)

Gao M, He R, Shi R, Zhang Y, Song S, Su W, Liu H (2021) Differential effects of low light intensity on broccoli microgreens growth and phytochemicals. Agronomy 11(3):537

Gharby S, Harhar H, Bouzoubaa Z, Asdadi A, El Yadini A, Charrouf Z (2017) Chemical characterization and oxidative stability of seeds and oil of sesame grown in Morocco. J Saudi Soc Agric Sci 16(2):105–111

Ghazani SM, Marangoni AG (2016) Healthy fats and oils

Ghoora MD, Babu DR, Srividya N (2020) Nutrient composition, oxalate content and nutritional ranking of ten culinary microgreens. J Food Compos Anal 91:103495

Ghosh PK, Chattopadhyay K (2019) Sunflower: production, processing, and uses. Oilseeds Oilseed Prod 2(1):45–60

Grishin A, Grishin A, Semenova N, Grishin V, Knyazeva I, Dorochov A (2021) The effect of dissolved oxygen on microgreen productivity. In: BIO Web of Conferences, vol 30. EDP Sciences, p 05002

Guo X, Zheng M, Pan R, Zang B, Gao J, Ma H, Jin M (2019) Hydroxysafflor yellow A (HSYA) targets the platelet-activating factor (PAF) receptor and inhibits human bronchial smooth muscle activation induced by PAF. Food Funct 10(8):4661–4673

Gupta MK (2014) Sunflower oil and its applications. Lipid Technol 26(11–12):260–263

Gupta S, Seth P, Abraham M, Pingali P (2022) COVID-19 and women's nutrition security: panel data evidence from rural India. Econ Polit 39(1):157–184

Gupta A, Sharma T, Singh SP, Bhardwaj A, Srivastava D, Kumar R (2023) Prospects of microgreens as budding living functional food: breeding and biofortification through OMICS and other approaches for nutritional security. Front Genet 14:1053810

Gupta A, Sharma T, Bhardwaj V, Shubhamkumar Manojkumar S, Sanjaykumar DY, Singh DV, Dhulipalla M (2025) Microgreens as budding medicine and bountiful source of nutrition. In: Recent trends and applications of leguminous microgreens as functional foods. Springer Nature Switzerland, Cham, pp 579–604

Hama JR (2017) Comparison of fatty acid profile changes between unroasted and roasted brown sesame (Sesamum indicum L.) seeds oil. Int J Food Prop 20(5):957–967

Haristoy X, Angioi-Duprez K, Duprez A, Lozniewski A (2003) Efficacy of sulforaphane in eradicating Helicobacter pylori in human gastric xenografts implanted in nude mice. Antimicrob Agents Chemother 47(12):3982–3984

He FJ, Chen JQ (2013) Consumption of soybean, soy foods, soy isoflavones and breast cancer incidence: differences between Chinese women and women in Western countries and possible mechanisms. Food Sci Hum Wellness 2(3–4):146–161

Houston M (2014) The role of nutrition and nutraceutical supplements in the treatment of hypertension. World J Cardiol 6(2):38

Ibragic S, Dahija S, Karalija E (2025) The good, the bad, and the epigenetic: stress-induced metabolite regulation and transgenerational effects. Epigenomes 9(2):10

Islam RT, Hossain MM, Majumder K, Tipu AH (2016) In vitro phytochemical investigation of Helianthus annuus seeds. Bangladesh Pharm J 19(1):100–105

Islam MZ, Park BJ, Lee YT (2023) Effects of seed pre-soaking on bioactive phytochemical levels of wheat and barley microgreens grown under hydroponics versus organic soil conditions. Ital J Agron 18(1):2183

Jahangir M, Kim HK, Choi YH, Verpoorte R (2009) Health-affecting compounds in Brassicaceae. Compr Rev Food Sci Food Saf 8(2):31–43

Jain P, Lingwan M (2026) No more fishes: Plants as a bio-factory for omega-3 fatty acid

Jayachandran M, Xu B (2019) An insight into the health benefits of fermented soy products. Food Chem 271:362–371

Ji Y, Guo S, Wang B, Yu M (2018) Extraction and determination of flavonoids in Carthamus tinctorius. Open Chem 16(1):1129–1133

Joshi A, Pradhan B, Chakraborty S, Varatharajoo R, Alamri A, Gite S, Lee CW (2025) An explainable Bi-LSTM model for winter wheat yield prediction. Front Plant Sci 15:1491493

Kajla P, Sharma A, Sood DR (2015) Flaxseed—a potential functional food source. J Food Sci Technol 52(4):1857–1871. https://doi.org/10.1007/s13197-014-1293-y

Kamilaris A, Prenafeta-Boldú FX (2018) Deep learning in agriculture: a survey. Comput Electron Agric 147:70–90

Keutgen N, Hausknecht M, Tomaszewska-Sowa M, Keutgen AJ (2021) Nutritional and sensory quality of two types of cress microgreens depending on the mineral nutrition. Agronomy 11(6):1110

Khalid N, Khan RS, Hussain MI, Farooq M, Ahmad A, Ahmed I (2017) A comprehensive characterisation of safflower oil for its potential applications as a bioactive food ingredient-A review. Trends Food Sci Technol 66:176–186

Khan R (2020) Conventional farming practices and the role of fertilizers in crop yield. Plant Soil 449(1):123–139

Khan S, Chaudhary S, Pandey A, Khan MK, Thomas G (2015) Sunflower oil: efficient oil source for human consumption. Emerg Life Sci Res 1:1–3

Khoo HE, Prasad KN, Kong KW, Jiang Y, Ismail A (2011) Carotenoids and their isomers: color pigments in fruits and vegetables. Molecules 16(2):1710–1738

Kizak V, Kapaligoz S (2019) Water quality changes and goldfish growth (Carassius auratus) in microgreen aquaponic and recirculating systems. Fresenius Environ Bull 28(9):6460–6466

Kou L, Yang T, Luo Y, Liu X, Huang L, Codling E (2014) Pre-harvest calcium application increases biomass and delays senescence of broccoli microgreens. Postharvest Biol Technol 87:70–78

Kyriacou MC, Rouphael Y, Di Gioia F, Kyratzis A, Serio F, Renna M, De Pascale S, Santamaria P (2016) Micro-scale vegetable production and the rise of microgreens. Trends Food Sci Technol 57:103–115

Kyriacou MC, El-Nakhel C, Soteriou GA, Graziani G, Kyratzis A, Antoniou C, Ritieni A, De Pascale S, Rouphael Y (2021) Preharvest nutrient deprivation reconfigures nitrate, mineral, and phytochemical content of microgreens. Foods 10(6):1333

Ladak Z, Khairy M, Armstrong EA, Yager JY (2021) Glucosinolates: paradoxically beneficial in fighting both brain cell death and cancer. In: Nutraceuticals in brain health and beyond. Academic Press, pp 155–167

Lee YR, Chen M, Lee JD, Zhang J, Lin SY, Fu TM, Chen H, Ishikawa T, Chiang SY, Katon J, Zhang Y (2019) Reactivation of PTEN tumor suppressor for cancer treatment through inhibition of a MYC-WWP1 inhibitory pathway. Science 364(6441):eaau0159

Liu S, Wang Y, Wen H, Sun X, Wang Y (2019) Hydroxysafflor Yellow A inhibits TNF-α-induced inflammation of human fetal lung fibroblasts via NF-κB signaling pathway. Evid-Based Complement Alternat Med 2019(1):4050327

Lusas EW, Jividen GM (1987) Cottonseed oil. In: Bailey's industrial oil and fat products, vol 4, pp 257–337

Luu HN, Blot WJ, Xiang YB, Cai H, Hargreaves MK, Li H, Yang G, Signorello L, Gao YT, Zheng W, Shu XO (2015) Prospective evaluation of the association of nut/peanut consumption with total and cause-specific mortality. JAMA Intern Med 175(5):755–766

Majdalawieh AF, Massri M, Nasrallah GK (2017) A comprehensive review on the anti-cancer properties and mechanisms of action of sesamin, a lignan in sesame seeds (Sesamum indicum). Eur J Pharmacol 815:512–521

Majumder M, Debnath S, Gajbhiye RL, Saikia R, Gogoi B, Samanta SK, Das DK, Biswas K, Jaisankar P, Mukhopadhyay R (2019) Ricinus communis L. fruit extract inhibits migration/invasion, induces apoptosis in breast cancer cells and arrests tumor progression in vivo. Sci Rep 9(1):14493

Mani V, Lee SK, Yeo Y, Hahn BS (2020) A metabolic perspective and opportunities in pharmacologically important safflower. Metabolites 10(6):253

Mathews R, Shete V, Chu Y (2023) The effect of cereal B-glucan on body weight and adiposity: a review of efficacy and mechanism of action. Crit Rev Food Sci Nutr 63(19):3838–3850

Mathur P, Gupta A (eds) (2025) Recent trends and applications of leguminous microgreens as functional foods. Springer Nature Switzerland

Melrose J (2019) The glucosinolates: a sulphur glucoside family of mustard anti-tumour and antimicrobial phytochemicals of potential therapeutic application. Biomedicines 7(3):62

Miao H, Zeng W, Wang J, Zhang F, Sun B, Wang Q (2021) Improvement of glucosinolates by metabolic engineering in Brassica crops. Abiotech 2(3):314–329

Mir SA, Shah MA, Mir MM (2017) Microgreens: production, shelf life, and bioactive components. Crit Rev Food Sci Nutr 57(12):2730–2736

Misra BB, Langefeld C, Olivier M, Cox LA (2019) Integrated omics: tools, advances and future approaches. J Mol Endocrinol 62(1):R21–R45

Mohanty A, Rambhatla SS, Sahay RR (2016) Deep gesture: static hand gesture recognition using CNN. In: Proceedings of International Conference on Computer Vision and Image Processing: CVIP 2016, Volume 2. Springer Singapore, Singapore, pp 449–461

Mohseni NM, Mirzaei H, Moghimi M (2020) Optimized extraction and quality evaluation of Niger seed oil via microwave-pulsed electric field pretreatments. Food Sci Nutr 8(3):1383–1393

Moreira-Rodríguez M, Nair V, Benavides J, Cisneros-Zevallos L, Jacobo-Velázquez DA (2017) UVA, UVB light, and methyl jasmonate, alone or combined, redirect the biosynthesis of glucosinolates, phenolics, carotenoids, and chlorophylls in broccoli sprouts. Int J Mol Sci 18(11):2330

Mueed A, Deng Z, Korma SA, Shibli S, Jahangir M (2023) Anticancer potential of flaxseed lignans, their metabolites and synthetic counterparts in relation with molecular targets: current challenges and future perspectives. Food Funct 14(5):2286–2303

Murphy DJ (1991) Designer oilseed crops. Genetic engineering of new oilseed crops for edible and non-edible applications

Nagar P, Agrawal MAK, Agrawal K (2022) Sesame (Sesamum indicum L.) seed as a functional food: a review. Pharma Innov J 11:893–896

Nagaraj G (2009) Oilseeds: properties, processing, products and procedures. New India Publishing

Nandasiri R, Eskin NM, Eck P, Thiyam-Höllander U (2020) Application of green technology on extraction of phenolic compounds in oilseeds (Canola). Cold Pressed Oils, pp 81–96

Negri M, Bulgari R, Santoro P, Ferrante A (2019) Evaluation of different growing substrates for microgreens production. In: III International Symposium on Growing Media, Composting and Substrate Analysis, vol 1305, pp 109–114

Olasunkanmi GS, Omolayo FT, Olusegun OT (2017) Fatty acid profile, physico-chemical and functional properties of oil and protein isolate simultaneously extracted from sesame (Sesamum indicum) seed. Ann: Food Sci Technol 18(1)

Omohu OJ, Omale AC (2017) Physicochemical properties and fatty acid composition of castor bean Ricinus communis L. seed oil. Am J Appl Ind Chem 3(1):1–4

Oppedisano F, Macrì R, Gliozzi M, Musolino V, Carresi C, Maiuolo J, Bosco F, Nucera S, Caterina Zito M, Guarnieri L, Scarano F (2020) The anti-inflammatory and antioxidant properties of n-3 PUFAs: their role in cardiovascular protection. Biomedicines 8(9):306

Pandey G (2018) Challenges and future prospects of agri-nanotechnology for sustainable agriculture in India. Environ Technol Innov 11:299–307

Patel A, Sharma D (2022) Comparative analysis of conventional and modern agricultural methods. Int J Agric Res 17(4):233–245

Pisoschi AM, Pop A (2015) The role of antioxidants in the chemistry of oxidative stress: a review. Eur J Med Chem 97:55–74

Qu C, Zhu W, Dong K, Pan Z, Chen Y, Chen X, Liu X, Xu W, Lin H, Zheng Q, Li D (2019) Inhibitory effect of hydroxysafflor yellow B on the proliferation of human breast cancer MCF-7 cells. Recent Pat Anti-Cancer Drug Discov 14(2):187–197

Rahman M, Khatun A, Liu L, Barkla BJ (2018) Brassicaceae mustards: traditional and agronomic uses in Australia and New Zealand. Molecules 23(1):231

Ramprasath VR, Awad AB (2015) Role of phytosterols in cancer prevention and treatment. J AOAC Int 98(3):735–738

Rani R, Sheoran R, Sharma B (2017) Perspectives of breeding for altering sunflower oil quality to obtain novel oils. Int J Curr Microbiol App Sci 6(8):949–962

Rasera GB, Hilkner MH, de Alencar SM, de Castro RJS (2019) Biologically active compounds from white and black mustard grains: an optimization study for recovery and identification of phenolic antioxidants. Ind Crops Prod 135:294–300

Ribeiro PR, Fernandez LG, de Castro RD, Ligterink W, Hilhorst HW (2014) Physiological and biochemical responses of Ricinus communis seedlings to different temperatures: a metabolomics approach. BMC Plant Biol 14:1–14

Rizzo G, Baroni L (2018) Soy, soy foods and their role in vegetarian diets. Nutrients 10(1):43

Rodriguez-Amaya DB (2015) Carotenes and xanthophylls as antioxidants. In: Handbook of antioxidants for food preservation. Woodhead Publishing, pp 17–50

Rout K, Yadav BG, Yadava SK, Mukhopadhyay A, Gupta V, Pental D, Pradhan AK (2018) QTL landscape for oil content in Brassica juncea: analysis in multiple bi-parental populations in high and "0" erucic background. Front Plant Sci 9:1448

Roy RD, Hossan MS, Rahmatullah M (2015) A review of anticancer potential of elephantopus scaber and its phytoconstituents. World J Pharm Pharma Sci 4(10):86–94

Sadowska U, Jewiarz K, Kopak M, Dziadek K, Francik R, Kopeć A (2023) Proximate analysis and antioxidant properties of young plants of Sinapis alba L. depend on the time of harvest and variety. Appl Sci 13(13):7980

Sáiz J, Montealegre C, Marina ML, García-Ruiz C (2013) Peanut allergens: an overview. Crit Rev Food Sci Nutr 53(7):722–737

Santos DI, Saraiva JMA, Vicente AA, Moldão-Martins M (2019) Methods for determining bioavailability and bioaccessibility of bioactive compounds and nutrients. In: Innovative thermal and non-thermal processing, bioaccessibility and bioavailability of nutrients and bioactive compounds. Woodhead Publishing, pp 23–54

Sasaki K, Urano K, Mimida N, Nonaka S, Ezura H, Imai R (2025) A long shelf-life melon created via CRISPR/Cas9 RNP-based in planta genome editing. Front Genome Ed 7:1623097

Sawicka B, Kotiuk E (2007) Mustard species as multi-functional plants

Sbihi HM, Nehdi IA, Mokbli S, Romdhani-Younes M, Al-Resayes SI (2018) Hexane and ethanol extracted seed oils and leaf essential compositions from two castor plant (Ricinus communis L.) varieties. Ind Crops Prod 122:174–1181

Senevirathne GI, Gama-Arachchige NS, Karunaratne AM (2019) Germination, harvesting stage, antioxidant activity and consumer acceptance of ten microgreens. Ceylon J Sci 48(1)

Seth T, Mishra GP, Chattopadhyay A, Deb Roy P, Devi M, Sahu A, Sarangi SK, Mhatre CS, Lyngdoh YA, Chandra V, Dikshit HK (2025) Microgreens: functional food for nutrition and dietary diversification. Plants 14(4):526

Seth T, Mishra GP, Chattopadhyay A, Deb Roy P, Devi M, Sahu A, Sarangi SK, Mhatre CS, Lyngdoh YA, Chandra V, Dikshit HK, Nair RM (2025) Microgreens: functional food for nutrition and dietary diversification. Plants 14(4):526

Settaluri VS, Kandala CVK, Puppala N, Sundaram J (2012) Peanuts and their nutritional aspects—a review. Food Nutr Sci 03:1644

Sharma M, Singh S, Gupta N (2022) Nutritional and functional properties of rapeseed and mustard oil. J Oilseed Res 39(2):89–98

Sharma T, Singh KP, Chudali S, Limbu BH, Gupta A (2025) Phytochemical and nutritional properties of leguminous microgreens and its application in food industry. In: Recent trends and applications of leguminous microgreens as functional foods. Springer Nature Switzerland, Cham, pp 605–619

Singh A, Ganapathysubramanian B, Singh AK, Sarkar S (2016) Machine learning for high-throughput stress phenotyping in plants. Trends Plant Sci 21(2):110–124

Singh G, Ratnaparkhe M, Kumar A (2019) Comparative analysis of transposable elements from Glycine max, Cajanus cajan and Phaseolus vulgaris. J Exp Biol Agric Sci 7(2):167–177

Singh K, Joshi A, Kumari P (2025) Hybrid CNN-SVM framework for rice leaf disease detection to promote sustainable agriculture. In: 2025 3rd International Conference on Disruptive Technologies (ICDT), pp 1292–1295

Smith J (2019) Advances in modern agriculture: AI and machine learning applications. J Agric Sci 157(3):45–60

Soni RP, Katoch M, Kumar A, Verma P (2016) Flaxseed—Composition and its health benefits. Res Environ Life Sci 9:310–316

Sreenivasa RJ, Sri TSD, Shivudu G, Kumari V (2019) Microgreens: a new initiative and harmonizing approach for promoting livelihood and nutritional security-India. Int J Food Biosci 2(1):01–05

Srikanth S, Chen Z (2016) Plant protease inhibitors in therapeutics-focus on cancer therapy. Front Pharmacol 7:470

Sukewijaya IM, Dwiyani R, Bimantara PO (2025) Optimization of growing media to support microgreens growth and nutritional profile. Agro Bali: Agric J 8(1):102–113

Sun J, Xiao Z, Lin LZ, Lester GE, Wang Q, Harnly JM, Chen P (2013) Profiling polyphenols in five Brassica species microgreens by UHPLC-PDA-ESI/HRMS n. J Agric Food Chem 61(46):10960–10970

Tharanathan RN, Wankhede DB, Rao MRRR (1975) Carbohydrate composition of groundnuts (Arachis hypogea). J Sci Food Agric 26(6):749–754

Thompson LU, Chen J, Hui E, Mann J, Ip T (2004) Interactive effects of Flax seeds and tamoxifen on human breast cancer. In: Proc. 60th Flax Institute, pp 86–90

Touré A, Xueming X (2010) Flaxseed lignans: source, biosynthesis, metabolism, antioxidant activity, bio-active components, and health benefits. Compr Rev Food Sci Food Saf 9(3):261–269

Truzzi F, Whittaker A, Roncuzzi C, Saltari A, Levesque MP, Dinelli G (2021) Microgreens: functional food with antiproliferative cancer properties influenced by light. Foods 10(8):1690

Turner ER, Luo Y, Buchanan RL (2020) Microgreen nutrition, food safety, and shelf life: a review. J Food Sci 85(4):870–882

USDA (2021) Soybeans and oil crops. United States Department of Agriculture. Available at: https://www.ers.usda.gov/topics/crops/soybeans-oil-crops

United States Department of Agriculture (2014) Specialty greens pack a nutritionalpunch, AgResearch Magazine. Available at: http://agresearchmag.ars.usda.gov/2014/jan/greens

Van Nieuwenhove CP, Moyano A, Castro-Gómez P, Fontecha J, Sáez G, Zárate G, Pizarro PL (2019) Comparative study of pomegranate and jacaranda seeds as functional components for the conjugated linolenic acid enrichment of yogurt. LWT 111:401–407

Varshney RK, Thudi M, Roorkiwal M, He W, Upadhyaya HD, Yang W, Bajaj P, Cubry P, Rathore A, Jian J, Doddamani D (2019) Resequencing of 429 chickpea accessions from 45 countries provides insights into genome diversity, domestication and agronomic traits. Nat Genet 51(5):857–864

Vilvert E, Lana M, Zander P, Sieber S (2018) Multi-model approach for assessing the sunflower food value chain in Tanzania. Agric Syst 159:103–110

Weiss EA (2000) Sesame. In: Oilseed crops. Blackwell Science, London

Wien M, Oda K, Sabaté J (2014) A randomized controlled trial to evaluate the effect of incorporating peanuts into an American Diabetes Association meal plan on the nutrient profile of the total diet and cardiometabolic parameters of adults with type 2 diabetes. Nutr J 13:1–9

Wilson R, Van Schie BJ, Howes D (1998) Overview of the preparation, use and biological studies on polyglycerol polyricinoleate (PGPR). Food Chem Toxicol 36(9–10):711–718

Wojdyło A, Nowicka P, Tkacz K, Turkiewicz IP (2020) Sprouts vs. microgreens as novel functional foods: variation of nutritional and phytochemical profiles and their in vitro bioactive properties. Molecules 25(20):4648

Woodroof JG (1983) Historical background in peanuts: production, processing, products. AVI Westport CT, p 181

Wrigley CW, Corke H, Seetharaman K, Faubion J (eds) (2015) Encyclopedia of food grains. Academic Press

Yadav PB, Edukondalu L, Patel S, Rao DB (2018) Proximate composition of peanut milk prepared by different methods. Int J Curr Microbiol Appl Sci 7(10):2388–2391

Yadav P, Shukla R, Gupta A (2025) Enhancing shelf life and nutritional quality of leguminous microgreens: insights into the application of nanotechnology, hydroponics, and genetic engineering approaches. In: Recent trends and applications of leguminous microgreens as functional foods. Springer Nature Switzerland, Cham, pp 421–448

Zech-Matterne V, Tengberg M, Van Andringa W (2015) Sesamum indicum L.(sesame) in 2nd century BC Pompeii, southwest Italy, and a review of early sesame finds in Asia and Europe. Veget Hist Archaeobot 24:673–681

Zhang Y, Xiao Z, Ager E, Kong L, Tan L (2021) Nutritional quality and health benefits of microgreens, a crop of modern agriculture. J Future Foods 1(1):58–66

Genomic Advances in Millets for Climate-Resilient Agriculture and Nutritional Security

11

Ayushi Kukreti, Anita Choudhary, and Vandana Jaiswal

Abstract

Climate change is leading to higher temperatures, altered rainfall patterns, and an increase in extreme weather events like droughts, floods, and heatwaves. These changes negatively affect global agriculture, leading to lower crop yields and threatening food security. Millets are climate-resilient crops that have ability to tolerate environmental stress like heat, salinity, poor soil condition, and drought stress. They stand with very less inputs, which makes them necessary for sustainable agriculture under harsh conditions. Millets are considered to be more nutritious than the majority of main cereals, as they provide high levels of dietary fibre, essential amino acids, vitamins, storage proteins, and various health-promoting bioactive compounds. This chapter includes the effects of climate change on major and minor millets with impact of heat, waterlogging, elevated CO_2, salinity, and soil changes. It also focuses on the genetic and genomic resources available for millet improvement, along with recent advances in transgenic and genomic editing methods such as CRISPR/Cas9 to increase climate resilience. The kranz anatomy of the C4 plant makes them more valuable than the C3 plant along with its nutritional benefits. Millets are rich sources of protein, calcium, and viable minerals as compared to cereals which make them nutri-cereals. Recent advancement in genomics have made millets more

A. Kukreti
Biotechnology, Division, CSIR-Institute of Himalayan Bioresource Technology, Palampur, Himachal Pradesh, India

A. Choudhary · V. Jaiswal (✉)
Biotechnology, Division, CSIR-Institute of Himalayan Bioresource Technology, Palampur, Himachal Pradesh, India

Academy of Scientific and Innovative Research (AcSIR), Ghaziabad, India
e-mail: vandana.jaiswal@csir.res.in

V. Gahlaut et al. (eds.), *Smart Crop Development*,
https://doi.org/10.1007/978-981-95-7680-7_11

promising crop not only for their nutrition point but also for climatic stress tolerance which make millets a future crop.

Keywords

Millets · Climate resilience · GWAS · Drought · Tolerance · Elevated CO_2

11.1 Introduction

Climate change is causing rising temperatures, changing rainfall patterns, and more extreme weather events such as droughts, floods, and heatwaves. These changes negatively affect global agriculture, leading to lower crop yields and threatening food security, especially in regions where farming depends on rainfall, like South Asia and sub-Saharan Africa (Ray et al. 2019). Studies show that food production is already declining, which poses a serious challenge as the global population is expected to exceed 9 billion by 2050. This could leave 2–3 billion people facing hunger and malnutrition (Wheeler and Von Braun 2013). Crops like rice and wheat require large amounts of external resources, making them less sustainable under future climate conditions (Bhatt et al. 2016). To address this, there is a growing need to promote crops that can withstand environmental stress while requiring fewer resources.

Millets are climate-friendly crops that can survive in tough conditions such as drought, high temperatures, and even saline soils. Additionally, they grow well in environments with elevated levels of carbon dioxide. Compared to other cereals, millets leave behind a smaller carbon footprint, help prevent soil erosion, and keep the soil healthy (Maitra et al. 2024). From the perspective of nutrition, millets are very rich. They provide fibre, protein, essential amino acids, vitamins, and beneficial natural compounds (Amadou et al. 2013; Babypriyanka et al. 2025). These grains have been part of human diets for more than 5000 years, especially in the Indian subcontinent (Tripathi and Vyas 2023). The word "millet" comes from the Latin word *Millium*, meaning "tiny seeds". Belonging to the grass family (Poaceae), as per Food and Agriculture Organization of the United Nations (FAO) classification, millets are grouped into two types: major millets (such as sorghum, pearl millet, and finger millet) and minor millets (like foxtail, kodo, barnyard, little, and proso millet) (Vetriventhan et al. 2020). Their description, including scientific and common name, ploidy, and estimated genome size are presented in Fig. 11.1 and Table 11.1. Recently, a chromosome-level reference genome study of finger millet estimated a genome size of 1.11 Gb (Devos et al. 2023). Further, estimated draft or reference genome sizes of different millets are identified as follows: pearl millet, 1.79 Gb; *Sorghum*, ~730 Mb; barnyard millet, 1.27 Gb; proso millet, ~887.8 Mb/~923 Gb; and foxtail millet, ~400 Mb/~423 Mb (Paterson et al. 2009; Varshney et al. 2017; Bannetzen et al. 2012; Zhang et al. 2012; Guo et al. 2017; Zou et al. 2019; Shi

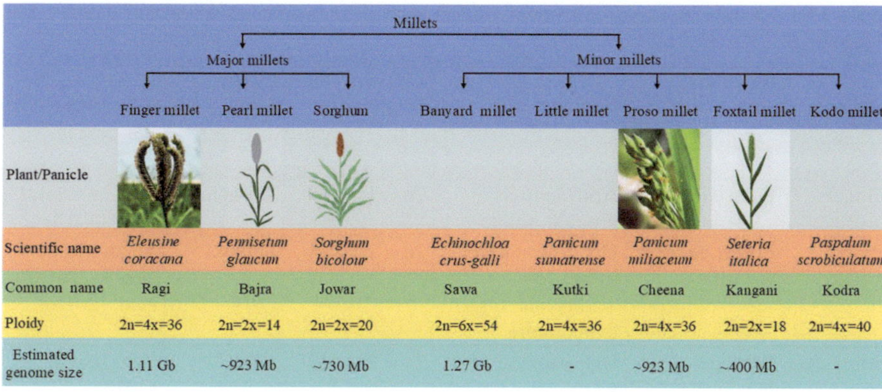

Fig. 11.1 Major and minor millet description, including scientific and common name, ploidy, and estimated genome size

et al. 2019). According to the data of FAO state 2019, worldwide millet is grown on about 71.8 million hectares of land producing 86.3 million metric tons. India alone grows it on 13.8 million hectares and produced 17.3 million tons (Sachan et al. 2023). This makes India the largest producer, accounting for 39% of the world's production, followed by Nigeria (11%) and China (9%) (Meena et al. 2021). Even though they are often called "minor crops" because they occupy less area, millets are very efficient (Yenagi et al. 2010). They require much less water than rice or wheat to grow (Bhatt et al. 2016).

Millets are not only resilient but also highly nutritious. They are packed with complex carbohydrates with a low glycaemic index, which release energy slowly, dietary fibre, and essential minerals such as calcium, magnesium, potassium, manganese, and zinc. They also provide B vitamins and bioactive compounds (Laishram et al. 2023). Since millets are naturally gluten-free, they are a great alternative to wheat and rice, especially for people with celiac disease. Because of their health and environmental benefits, millets are often called "nutri-cereals" or "miracle grains" (Sharma et al. 2021). They supply important macronutrients, micronutrients, and antioxidants (Banerjee et al. 2020; Srivastav et al. 2021). Their nutrition profile includes protein, energy, minerals, and vitamins like thiamine (B1), riboflavin (B2), niacin (B3), and folic acid (B9) (Hassan et al. 2021). On average, millets contain about 65–75% carbohydrates (in the form of starch) and 18–20% dietary fibre. Their starch is made up of 16–28% amylose and 72–84% amylopectin (Mukherjee et al. 2025). Millets are also rich in natural plant compounds such as polyphenols (0.2–0.3%) and phytochemicals, which are known to boost health, and 0.48% phytic acid, which influences the bioavailability of several key nutrients (Sheethal et al. 2022). A detailed breakdown of millet nutrients is provided in Table 11.2.

Table 11.1 Details of major and minor millets with their habitat, chromosome number, and ploidy level

Millet	Botanical name	Habitat	Chromosome number	Ploidy level	References
Major millet					
Finger millet	*Eleusine coracana*	Arid, semi-arid, drought, saline	2n = 4x = 36	Tetraploid	Antony Ceasar et al. (2018)
Pearl millet	*Pennisetum glaucum*	Dry, low nutrient, saline, low acidity, and high temperature	2n = 2x = 14	Diploid	Singh et al. (1997), Uppal et al. (2015)
Sorghum	*Sorghum bicolour*	Dry, parched, and warm areas	2x = 2x = 20	Diploid	Kumar et al. (2013)
Minor millet					
Barnyard millet	*Echinochloa esculenta*	Low-temperature environments, inappropriate for unproductive barren, arid, or oversaturated soils	2n = 6x = 54	Hexaploid	Sood et al. (2020)
Little millet	*Panicum sumatrense*	Tropical regions, water logged, temperate regions, drought area	2n = 4x = 36	Tetraploid	Shweta et al. (2018)
Proso millet	*Panicum miliaceum*	Low moisture content, dry conditions	2n = 4x = 36	Tetraploid	Zhang et al. (2019)
Foxtail millet	*Setaria italica*	Drought sensitive	2n = 2x = 18	Diploid	Hariprasanna (2017)
Kodo millet	*Paspalum scrobiculatum*	Wetland areas are found in tropical and subtropical regions, with soil ranging from highly fertile to poor quality	2n = 4x = 40	Tetraploid	Selvi and Nirmalakumari (2023)

11.2 Climate Resilience: Millets and Its Traits Contribution

Climate change is a well-established global challenge, evidenced by glacier retreat, shifting rainfall patterns, and increasing air temperatures (Hadimani and Malleshi 1993; Zemp et al. 2025). While much emphasis has been placed on staple crops like rice and wheat, their high resource demands and vulnerability to climatic stress raise concerns about their long-term sustainability. This underscores the urgent need to identify and promote climate-smart alternatives. Millets are often positioned as

Table 11.2 Details of different nutritional content present in various type of millets

Millet	Nutritional content	References
Sorghum	High in protein, fibre, thiamine, riboflavin, folic acid, carotene, and numerous minerals such as potassium, phosphorus, calcium, iron, zinc, and sodium	Gerrano et al. (2016)
Pearl millet	Abundant in protein, lipids, and dietary fibre, significant source of calcium, unsaturated fat, and niacin; contains folate, magnesium, iron, copper, zinc, and vitamins E and B complex	Satyavathi et al. (2021), Reddy et al. (2021)
Finger millet	Abundant source of calcium, iron, large amounts of antioxidants, sulphur, amino acids, tannins	Maharanjan et al. (2021), Ramachandran et al. (1977)
Foxtail millet	Rich in carbohydrate, protein, copper, iron, crude fibre, antioxidant	Hariprasanna (2017), Abedin et al. (2022)
Kodo millet	Abundant vitamin C complex, mineral elements	Deshpande et al. (2015)
Barnyard millet	Crude fibre and iron, GABA (present in grains), beta-glycan, low level of lipid	Sharma et al. (2016)
Proso millet	Rich in vitamins, amino acids, starch, phenolic compounds, lipoprotein	Karkannavar et al. (2021)
Little millet	Rich in iron, antioxidant, dietary fibre	Shweta et al. (2018)

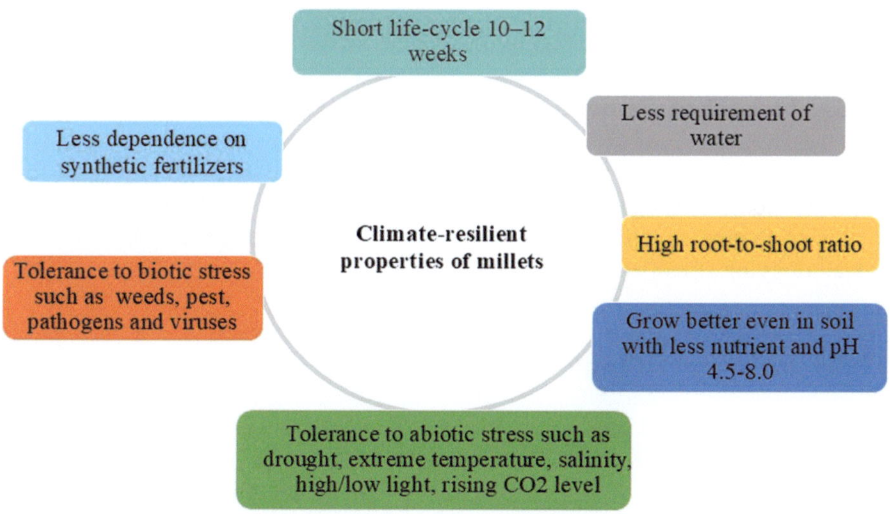

Fig. 11.2 Unique properties of millets as climate-resilient crops

such alternatives due to their short life cycle (12–14 weeks in contrast to rice and wheat which require 20–24 weeks) and adaptive traits, including reduced height, small leaf area, thicker cell walls, and robust root systems (Li and Brutnell 2011). These features enable them to survive in stressful environments (Fig. 11.2). However, while their resilience is widely acknowledged, the extent to which these traits can consistently secure yields under severe climate variability remains an area requiring deeper investigation.

As C4 plants with Kranz anatomy, millets exhibit higher photosynthetic efficiency, improved carbon fixation, and enhanced tolerance to elevated CO_2 compared to C3 crops (Aubry et al. 2011). Their water requirement (200–250 mm) is substantially lower than rice and sugarcane and their nitrogen-use efficiency (NUE) and water-use efficiency (WUE) are 1.3 to 4 times more than those of C3 cereals (Sage and Zhu 2011). For instance, foxtail millet produces 1 g of dry biomass using just 257 g of water, while maize and wheat require nearly double amounts of 470 g and 510 g, respectively (Li and Brutnell 2011). While these physiological advantages are compelling, more empirical field studies are needed across different agro-ecological zones to validate their performance under real-world stress scenarios. Adaptive responses in millets also include phenological shifts (e.g. pearl millet altering flowering with rainfall variation) and morphological adjustments (e.g. increased root length in little millet during drought) (Bidinger et al. 2007; Ajithkumar and Panneerelvam 2014). Additionally, their biochemical resilience is evident in elevated antioxidant levels and stress-responsive enzyme activity. However, the translation of these physiological and biochemical traits into stable yield gains under climate extremes remains insufficiently explored. Environmentally, millets have a considerably smaller carbon footprint than major cereals, positioning them as a sustainable option in low-input agriculture (Prasad and Staggenborg 2009; Jain et al. 2016).

11.3 Impact of Climate Change on Millets

The climate is altering rapidly, causing multiple abiotic stresses that impact food crops in different ways, directly influencing the global food supply. The main effects of climate change on millet production include drought stress, heat stress, flooding or waterlogging stress, and lodging (Numan et al. 2021).

11.3.1 Impact of Drought Stress on Millets

Millets have developed several mechanisms that help them survive under drought and other abiotic stresses (Duque et al. 2013; Mickelbart et al. 2015). Their strategies include physiological adaptations such as stomatal regulation, osmotic adjustment, and maintaining the stability of cell membranes (Kusaka et al. 2005). These mechanisms are often grouped into four categories (Numan et al. 2021):

(i) Drought recovery: plants resume growth and yield once water becomes available again.
(ii) Drought tolerance: plants continue producing biomass even under low water conditions.
(iii) Drought escape: plants complete their life cycle before severe drought begins.

(iv) Drought avoidance: plants avoid tissue dehydration by conserving water.

Experimental evidence shows that drought reduces plant height, biomass, and grain number in finger millet (Khatoon et al. 2016). In barnyard and finger millet, drought stress lowered chlorophyll content, photosynthesis, and relative water content (RWC) while increasing proline accumulation, with finger millet showing greater tolerance (Tiwari et al. 2022). Drought also induced oxidative stress symptoms such as leaf curling, drooping shoots, higher proline, catalase, malondialdehyde (MDA), and hydrogen peroxide (H_2O_2) levels, along with electrolyte leakage and loss of membrane integrity (Tiwari et al. 2022). Under drought stress finger millet exhibits enhanced activity of antioxidant enzymes including glutathione reductase (GR), superoxide dismutase (SOD), ascorbate peroxidase (APX), glutathione peroxidase (GPX), and catalase (CAT) (Bhatt et al. 2011). When plants face water stress (like drought), their normal cell functions get disturbed. This can weaken cell membranes, make proteins and genetic material unstable, and reduce overall growth and yield. During drought, plants close their stomata (tiny pores on leaves) to save water, but this often causes a build-up of harmful molecules called reactive oxygen species (ROS). These molecules create oxidative stress, which damages fats, proteins, and other important cell parts (Bhatt et al. 2011). Plants rely on special hormones to cope with stress. For example, abscisic acid (ABA) and ethylene (ET) are key messengers that signal drought responses, while jasmonic acid (JA) and salicylic acid (SA) help plants tolerate drought, heat, and salty soils (Tiwari et al. 2017). Certain genes, especially those controlled by transcription factors like *NAC*, *MYB*, and *WRKY*, also play significant roles in helping plants respond to drought (Cao et al. 2022).

Still, even with these defence mechanisms, drought can greatly reduce yields in many millet species—such as wild millet (*Setaria glauca*), little millet, foxtail millet, and proso millet—especially if it happens before flowering (Tadele et al. 2016). With drought and other stresses becoming more frequent in the twenty-first century, keeping millet yields stable is a major challenge (Ashok et al. 2018). Nevertheless, their resilience in low-input and marginal farming systems, combined with their nutritional value, makes them strong candidates for integration into modern cropping systems, particularly with the aid of advanced genomic tools (Rani et al. 2023). Overall, while millets show diverse drought-adaptive strategies at morphological, physiological, biochemical, and molecular levels, their yield stability under severe climate stress is still inconsistent. Critical research is required to link these traits with long-term productivity and to explore how genomics can accelerate the development of truly climate-resilient millet varieties.

11.3.2 Impact of Thermal Stress on Millets

The major cereals projected to suffer significant yield losses under a 1–2 °C rise in temperature (Sultan et al. 2013). Although millets are traditional crops with strong adaptation to tropical and temperate climates, their potential remains underutilized

(Kumar et al. 2021). Heat stress triggers complex physiological and molecular responses in plants, and millets offer a useful model for studying these mechanisms. Plants cope with high temperatures through acclimation, avoidance, and tolerance strategies. These include antioxidant defence, activation of transcription factors, heat-shock protein production, regulation of ion transporters, osmoprotectant accumulation, and signaling molecule activity (Hatfield and Prueger 2015). However, heat stress disrupts electron transport, damages photosystem II, increases reactive oxygen species (ROS), and induces lipid peroxidation, compromising membrane stability and yield (Hasanuzzaman et al. 2013).

Pearl millet demonstrates strong adaptability to heat, making it an important model for heat tolerance research (Yan et al. 2023). For example, pearl millet cv. BK 560 germinates between 10 °C and 12 °C, peaks at 33–34 °C, and fails above 45 °C. Leaf initiation, spikelet development, and tillering show similar temperature dependence, with leaf extension increasing linearly up to 34 °C. Although dry matter production per unit of radiation remains stable across 20–36 °C, yield is highest at cooler temperatures (22 °C) due to longer crop duration (Ong and Monteith 1985). Despite these adaptations, heat stress above 36/26 °C significantly reduces seed number, seed weight, and yield, with the reproductive stage being most sensitive (Gupta et al. 2015; Djanaguiraman et al. 2018). Stress just before anthesis (10–12 and 2–0 days) causes sharp declines in pollen germination and high floret sterility. Pistils are more sensitive than pollen, showing greater ROS accumulation and weaker antioxidant activity, explaining higher susceptibility. Molecular studies support these findings. In pearl millet heat stress–induced differential gene expression (DEGs) was observed under different interval of time. These DEGs are specially found in protein folding pathways. Many genes encoding ROS scavenging enzymes and flavonoid biosynthesis pathway are activated under heat stress (Huang et al. 2021).

Foxtail millet also shows heat sensitivity. Growth traits such as plant height, leaf area, and dry matter decline with increasing temperatures (Panneerselvam et al. 2019). In foxtail millet, *Setaria italica, calcium-dependent protein kinase 7* (*SiCDPK7*) was found to increase heat stress potential mainly responsible for catalase activity and decrease malonaldehyde (Wei et al. 2023). Yet heat-tolerant (38 °C) foxtail varieties display adaptive metabolic shifts, including lower photosynthesis and stomatal conductance, reduced root respiration, accumulation of protective metabolites (amino acids, sugars, organic acids), and more efficient carbon and nitrogen use (Aidoo et al. 2016). The metabolite shift, in the root, under the influence of high temperature is genotype specific. Overall, millets possess multiple physiological, biochemical, and genetic strategies to withstand heat stress.

11.3.3 Impact of Waterlogging Stress on Millets

When soil becomes waterlogged, air spaces fill with water, which restricts gas exchange and leads to the build-up of harmful compounds. As a result, photosynthesis, stomatal conductance, and root function are negatively affected (Linkemer et al.

1998). Grain yield losses of about 18% in wild millet and 16% in proso millet have been reported under waterlogging stress from 2 weeks after planting until maturity (Linkemer et al. 1998). Waterlogging stress occurs due to hypoxia (low oxygen) or anoxia (no oxygen). To survive, plants switch to anaerobic respiration (formation of lysigenous aerenchyma-air channel), a process also observed in finger millet (*Eleusine coracana*) and little millet (*Panicum sumatrense*) (Hossain et al. 2011; Raj et al. 2025). Since this pathway requires more sugar than aerobic metabolism, waterlogging-tolerant crops like rice and finger millet adjust their carbohydrate metabolism accordingly (Kulkarni et al. 2013).

Finger millet can adapt to waterlogging, although its response varies with growth stage and duration of stress (Maai 2024). Short-term waterlogging reduces photosynthesis and plant height, while long-term exposure during summer improves photosynthesis due to favourable physiological adjustments. Despite shorter plants, waterlogged finger millet produces more tillers and panicles, with only a slight reduction (non-significant) in grain yield, showing its good tolerance and potential as a reliable crop (Maai 2024). Studies by Matsuura et al. (2016) showed clear differences in waterlogging tolerance among millet species. Grain yield declined in *Panicum miliaceum*, *Setaria glauca*, and *Setaria italica* under continuous waterlogging, while *Panicum sumatrense* showed a remarkable 210% yield increase. This tolerance was associated with vigorous root growth, greater root biomass, more crown roots, and a higher proportion of lysigenous aerenchyma. Thus, *P. sumatrense* is highly suited for waterlogged conditions. Further research by Matsuura et al. (2022) found that *Echinochloa utilis* and *Eragrostis tef* also display strong hypoxia tolerance compared to *S. italica*. Their resilience is linked to stable root growth, balanced nutrient status, and constitutive development of lysigenous aerenchyma in adventitious roots, which help deliver oxygen to the root tip. A smaller stele area reduces oxygen demand, while root hairs near the tip may enhance nutrient uptake. Together, these traits explain their better survival under low-oxygen environments.

11.3.4 Impact of Elevated CO_2 on Millets

Elevated CO_2 increased net photosynthetic rate, transpiration rate, stomatal conductance, intracellular concentration of CO_2, spontaneous transpiration efficiency, and effective quantum yield of PSII photochemistry. In broomcorn millet, elevation of CO_2 leads to improved yield as well as photosynthesis boosted by increased stomatal conductance (Hao et al. 2017). Under drought conditions of semi-arid regions, high CO_2 level in broomcorn millet induced increases in carbon uptake and water use efficiency (Zhang et al. 2021). Plants exposed to both elevated CO_2 and temperature were 19% taller and had 22% more shoot biomass (Lima et al. 2022). Regardless, grain production remained unaffected by any of the conditions. Elevated CO_2 significantly improved the concentration of Mg (27.3%), Mn (14.6%), with showing no effect on the concentration of P, K, Fe, Ca, Cu, and Zn in broomcorn millet (Shi et al. 2022). In foxtail millet, increased CO_2 had no effect on the concentration of K, Ca, Cu, and Zn but significantly reduced Fe concentration. Also considerably enhanced P concentration and accumulation therefore slightly affect the nutritional value of foxtail millet (Gong et al. 2021). Pearl millet shows fast photosynthesis rate at high level CO_2 (Vanaja et al. 2013). In sorghum, eCO_2 had minimal adverse impact on photosynthesis under water stress and heat stress combination but an impressive impact on biomass production (Al-Salman et al. 2023). A study conducted on pearl millet in Gujarat using the CERES-MILLET model revealed that an elevated CO_2 level (at 530 ppm) increases grain yield up to 18.2%, anthesis up to 11 days, and maturity up to 17 days (Kanjiya et al. 2021).

11.3.5 Impact of Salinity Stress on Millets

In arid and semi-arid regions, salinity and water scarcity are two major stresses that affect crop growth and quality. Salinity stress causes several harmful effects, such as reduced grain yield, toxic ion accumulation, and disturbed osmotic balance. In foxtail millet, soil salinity levels reached 1.87 g/kg under drought and 1.97 g/kg under well-watered conditions, leading to lower productivity and survival rates. Salinity and water stress also reduced leaf water potential and the K^+/Na^+ ratio (Ma et al. 2023). In pearl millet, the number of tillers and shoot dry weight remains stable under saline conditions, making it suitable for growth in such environments. However, reductions in plant height and chlorophyll content have been observed (Qaoud et al. 2023). In proso millet, silicon (Si) helps protect plants under salt stress by improving chlorophyll stability, water content, and photosynthetic efficiency. Salt and silicon treatments also increased phenylalanine ammonia lyase (PAL), tyrosine ammonia lyase (TAL), and polyphenol oxidase (PPO) enzyme activities, which produce phenolic compounds that neutralize reactive oxygen species, thereby reducing oxidative damage (Mushtaq et al. 2025). Proso millet also uses several mechanisms for sodium (Na^+) tolerance, from germination to seedling growth, root to leaf, and even at cellular and molecular levels (Yuan et al. 2022). In finger millet, salinity causes tissue damage, changes in lignin content in coleoptile and coleorhizae, and reduced growth (Satish et al. 2016). Although biochemical compounds such as proline, protein, and sugars increase under salt stress, seed germination, root length, shoot length, and relative water content decrease (Thrupthi et al. 2023).

11.3.6 Impact of Change in Soil Due to Climate Change on Millets

Climate change disrupts soil-plant-atmosphere interactions, reducing ecosystem stability and resilience (Oishy et al. 2025). Adverse weather events alter soil structure, hinder nutrient cycling, and affect plant growth processes. These disturbances weaken natural balance, making ecosystems more vulnerable to degradation and reducing their capacity to sustain agricultural productivity and ecological health in the long term. Across all millet crops, plant height was greatly affected by soil fertility and also showed an enormous effect on production of grains, with higher yields observed in alluvial soils. Compared to foxtail millet and sorghum, pearl millet produced more grain in both laterite and alluvial soils. The organic carbon content of alluvial soil was 4.8%, whereas that of laterite soil was 0.3% (Devi and Punnoose 2025). In the Sahel region, removing more than 2.5 cm of topsoil, and applying fertilizer greatly decreased pearl millet productivity, and the productivity loss caused by the removal of up to 10 cm top soil cannot be restored even with fertilization (Nishigaki et al. 2025). In general, there has been a declining trend in the activities of soil urease, alkaline phosphatase, sucrase, and catalase; this decline became more promised as the number of years of continuous cropping increased (Zhang et al. 2024). The effects of brassinosteroids on millet production, soil enzyme activity, soil microbiota, and residues of tribenuron-methyl were investigated. While sucrase and alkaline phosphatase activities were suppressed, catalase and dehydrogenase activities were enhanced by brassinolide, which raised millet yield up to 2.4% and 13.6% (Song et al. 2023). Due to insufficient nutrient replenishment, poor soil fertility is a significant issue limiting agricultural output in smallholder farms in sub-Saharan Africa (Ebanyat et al. 2021).

11.4 Genetic and Genomic Resources of Millets

Genetic diversity is the key to developing better crop varieties with traits like higher yields, stress resistance, and improved nutrition. To preserve this diversity, scientists conserve germplasm resources—collections of seeds and plant materials that represent the genetic richness of crops (Salgotra et al. 2022). The International Crops Research Institute for the Semi-Arid Tropics (ICRISAT) plays a major role in this effort. It maintains over 10,193 small millet germplasm samples gathered from 50 countries (Goron and Raizada 2015; Choudhary et al. 2023). Most of these come from Asia (64.4%), followed by Africa (13.8%) and Europe (13.5%) (Kumar et al. 2024). Globally, about 29,000 samples of proso millet are conserved, with 849 stored at ICRISAT (Vetriventhan et al. 2020). Likewise, more than 8000 samples of kodo and barnyard millet, and around 3000 samples of little millet, are preserved in different gene banks (Singh and Upadhyaya 2015). Finger millet has particularly strong representation: India's National Bureau of Plant Genetic Resources (NBPGR) holds the largest collection, while ICRISAT conserves over 7500 finger millet samples from 26 countries (Mukherjee et al. 2025). ICRISAT gene bank has overall 82,077 samples including finger millet (7536), pearl millet (25794), sorghum (44469), barnyard millet (749), little millet (473), proso millet (849), foxtail millet (1542), and kodo millet (665) (http://genebank.icrisat.org/, accessed 2 September 2025). These large collections of millet genetic resources form a strong foundation for crop improvement. They are essential for ensuring food and nutrition security in the future, especially as climate change continues to affect agriculture. A number of potential genes have been identified in millets, which helps in providing tolerance under abiotic stress condition (Table 11.3).

QTLs for stay-green traits and biomass output under early drought stress conditions had been identified by a genome-wide association analysis (GWAS) in pearl

Table 11.3 An overview of the genes found and described in various millets that contribute to abiotic stress tolerance

Millet	Gene	Finding	References
Pearl millet	*PgVDAC* (*Pennisetum glaucum voltage-dependent anion channel*)	Heterologous overexpression demonstrate tolerance to salt stress	Desai et al. (2006)
	PgLEA (*Pennisetum glaucum late embryogenesis abundant*)	Shows resistance against drought, cold, heat, and salt stress	Reddy et al. (2012)
	PgGR (*Pennisetum glaucum glutathione reductase*)	Provides tolerance against methyl viologen–induced oxidative stress and salt stress	Achary et al. (2015)
Foxtail millet	*SiPHGPX* (*Setaria italica phospholipid hydroperoxide glutathione peroxidase*)	Salinity tolerance	Sreenivasulu et al. (2004)
	SiOPR1 (*Setaria italica 12-oxophytodienoic acid reductase*)	Plays a role in osmotic stress/drought tolerance	Zhang et al. (2007)
	SiALR (*Setaria italica aldose reductase*)	Involved in sorbitol biosynthesis, salinity stress responsive pathway	Veeranagamallaiah et al. (2009)
	Si69 (homologous of *wheat aluminium-induced protein 7*)	Aluminium tolerance/ detoxification	Zhao et al. (2009)
	SiACCase (*Setaria italica acetyl-CoA carboxylase*)	Provides resistance to sethoxydim herbicide	Dong et al. (2011)
	SiDREB2 (*Setaria italica dehydration response element binding proteins 2*)	Upregulated during drought and salt stress	Lata et al. (2011)
	SiNAC	Showed salinity and dehydration tolerance	Puranik et al. (2011)
	SiLEA14 (*Setaria italica late embryogenesis abundant 14*)	Provides resistance to salinity and drought stress	Wang et al. (2014)
	SiASR1 (*Setaria italica abscisic acid stress ripening*)	Provides tolerance against drought and oxidative stress	Feng et al. (2016)
	SiAGO1b (*Setaria italica argonaute 1*)	Plays a role on ABA response and drought stress	Liu et al. (2016)
Finger millet	*EcDehydrin 7*	Tolerance to drought stress	Singh et al. (2015)
	EcNAC67	Provides drought and salinity tolerance	Rahman et al. (2016)
	EcPT (*Eleusine coracana phosphate transporter*)	Inorganic phosphate (Pi) stress tolerance	Pudake et al. (2017)

millet (Debieu et al. 2018). In foxtail millet, transcriptome analysis was used to identify differentially expressed genes (DEGs) under drought stress conditions. DEGs were highly enriched in pathways related to photosynthesis, amino acid metabolism, carbohydrate metabolism, and chlorophyll metabolism, as identified by the GO and KEGG pathway analyses (Guo et al. 2023). A total of 32 drought tolerance–related genes were identified. RNA sequencing (RNA-seq) was used to identify (aluminium) Al-tolerant and Al-susceptible genes from genotypes of finger millet. RNA-seq data analysis also identified 119,073 single nucleotide polymorphism (SNP) markers of Al-tolerant and-susceptible finger millet genotypes which are very informative. These genomic databases provide new area for future research (Brhane et al. 2022). Transcriptional stability under drought stress by transcriptomics profiling represents a new idea on the underlying biological mechanism of the little millet genotype OLM-203/Tarini (Narayanrao et al. 2023). Molecular cloning in foxtail millet for salt stress identified NAC family gene member (Puranik et al. 2011).

11.5 Increase Millets Resilience to Stress Via Transgenic/CRISPR

Traditional breeding methods—such as classical, mutant, and transgenic breeding—are often slow and difficult to apply. To speed up crop improvement, scientists are now turning to modern techniques like targeted mutagenesis, epigenetic modification, and genome editing (Zhang et al. 2020). Thanks to advanced genome-editing tools and modern "omics" technologies, it is now possible to make very precise changes to plant genes. One of the most powerful tools is the CRISPR/Cas9 system, which has already been effectively applied to crops like rice, wheat, and maize (Li et al. 2020, 2021; Namata et al. 2025; Zhang et al. 2025). So far, CRISPR has been applied in many plant species to improve yield, stress tolerance, and resistance to diseases (Jaganathan et al. 2018; Ricroch et al. 2017; Ahmad 2023; Pehlivan et al. 2025). For example, scientists used CRISPR to target the *SlAGAMOUS-LIKE 6* (*SlAGL6*) gene in tomatoes, which helped the plants produce more fruit under heat stress (Klap et al. 2017). Newer methods, such as single-base editing systems, have made CRISPR even more precise. They allow scientists to modify single DNA bases (alleles) to improve key crop traits (Eid et al. 2018). For millets, which are still underutilized but naturally climate-resilient, these modern genome-editing technologies offer huge potential. They could help enhance yield, stress tolerance, and nutritional quality, making millets even more valuable for the future (Ceasar 2022). To enhance genetic improvement for tolerance to abiotic stresses such as drought and salinity, *Agrobacterium*-mediated transformation method using shoot apex explants was established for finger millet (Antony Ceasar and Ignacimuthu 2011). Similarly, in foxtail millet, phosphate transporter genes (*SiPHT1;2, SiPHT1;3, SiPHT1;4*) have been downregulated through *Agrobacterium*-mediated transformation method to study the levels of inorganic phosphorus in both shoot and root tissues and found that decrease in inorganic phosphorus content leads to a higher number of lateral roots and root hairs (Ceasar et al. 2017). Foxtail millet was also the first millet to be sequenced (Bennetzen et al. 2012). Recently, Liang et al. (2022) developed a CRISPR/Cas9 system in foxtail millet targeting *SiFMBP*, *SiDof4*, *SiBADH2*, *SiGBSS1*, and *SiIPK1* for functional studies, while base editing of *SiALS* and *SiACC* generated herbicide-tolerant mutant (Liang et al. 2022). Similarly, Hao et al. (2025) revealed that *SiEPF2* regulates stomatal density and panicle development, balancing drought resistance and yield (Hao et al. 2025). These advances highlight the potential of CRISPR/Cas9 in functional genomics and breeding climate-resilient, high-yielding millet varieties.

11.6 Conclusion

Millets are resilient crops that thrive in drylands and can withstand many environmental stresses, both biotic and abiotic. They have a low carbon footprint and high climate resilience and play a vital role in sustainable farming. Researchers emphasize that more research and development, especially in genomic technologies, are essential to unlock their full potential for sustainable agriculture, food security, and climate adaptation. Often called “wonder grains”, millets are a perfect combination of sustainability, nutrition, and resilience. Their natural traits—like tolerance to heat and drought, efficient use of water and nutrients, and ability to grow in poor soils—make them especially suited to the challenges of climate change. Rich in essential amino acids, minerals, and dietary fibre, millets are powerful allies in the fight against hunger and malnutrition. Recent advances in CRISPR, marker-assisted selection, and genomics are opening new possibilities to develop improved, nutritious, and climate-resilient millet varieties. With growing pressure on global food systems and increasing climate variability, millets stand out as a smart, climate-friendly, and nutrient-rich choice for future food security and better global health.

Acknowledgements Thanks are due to CSIR-IHBT (MLP-201) and DST-SHRI (GAP-345) for funding. CSIR-IHBT communication number is 5942.

References

Abedin MJ, Abdullah ATM, Satter MA, Farzana T (2022) Physical, functional, nutritional and antioxidant properties of foxtail millet in Bangladesh. Heliyon 8(10):e11068

Achary VM, Reddy CS, Pandey P, Islam T, Kaul T, Reddy MK (2015) Glutathione reductase a unique enzyme- molecular cloning, expression and biochemical characterization from the stress adapted C4 plant, *Pennisetum glaucum* (L.) R. Br. Mol Biol Rep 42(5):947–962

Ahmad M (2023) Plant breeding advancements with “CRISPR-Cas” genome editing technologies will assist future food security. Front Plant Sci 14:1133036

Aidoo MK, Bdolach E, Fait A, Lazarovitch N, Rachmilevitch S (2016) Tolerance to high soil temperature in foxtail millet (*Setaria italica* L.) is related to shoot and root growth and metabolism. Plant Physiol Biochem 106:73–81

Ajithkumar IP, Panneerselvam R (2014) ROS scavenging system, osmotic maintenance, pigment and growth status of *Panicum sumatrense* Roth. under drought stress. Cell Biochem Biophys 68(3):587–595

Al-Salman Y, Ghannoum O, Cano FJ (2023) Elevated [CO_2] negatively impacts C4 photosynthesis under heat and water stress without penalizing biomass. J Exp Bot 74(9):2875–2890

Amadou I, Gounga ME, Le GW (2013) Millets, nutritional composition, some health benefits and processing. Emir J Food Agric 25(7):501–508

Antony Ceasar S, Ignacimuthu S (2011) *Agrobacterium*-mediated transformation of finger millet (*Eleusine coracana* (L.) Gaertn.) using shoot apex explants. Plant Cell Rep 30(9):1759–1770

Antony Ceasar S, Maharajan T, Ajeesh Krishna TP, Ramakrishnan M, Victor Roch G, Satish L, Ignacimuthu S (2018) Finger millet [*Eleusine coracana* (L.) Gaertn.] improvement: current status and future interventions of whole genome sequence. Front Plant Sci 9:1054

Ashok S, Senthil A, Sritharan N, Punitha S, Divya K, Ravikesavan R (2018) Yield potential of small millets under drought condition. Mad Agric J 105:370–372

Aubry S, Brown NJ, Hibberd JM (2011) The role of proteins in C3 plants prior to their recruitment into the C4 pathway. J Exp Bot 62(9):3049–3059

Babypriyanka S, Hepziba JS, Pushpam KA, Pillai AM, Vijayalakshmi R, Theradimani M (2025) Exploring the bioactive components of millets for their nutraceutical potential. Food Sci Biotechnol 34(3):563–575

Banerjee P, Maitra S, Banerjee P (2020) The role of small millets as functional food to combat malnutrition in developing countries. Indian J Nat Sci 10(60):20412–20417

Bennetzen JL, Schmutz J, Wang H, Percifield R, Hawkins J, Pontaroli AC, Devos KM (2012) Reference genome sequence of the model plant *Setaria*. Nat Biotechnol 30(6):555–561

Bhatt D, Negi M, Sharma P, Saxena SC, Dobriyal AK, Arora S (2011) Responses to drought induced oxidative stress in five finger millet varieties differing in their geographical distribution. Physiol Mol Biol Plants 17(4):347–353

Bhatt R, Kukal SS, Busari MA, Arora S, Yadav M (2016) Sustainability issues on rice–wheat cropping system. Int Soil Water Conserv 4(1):64–74

Bidinger FR, Nepolean T, Hash CT, Yadav RS, Howarth CJ (2007) Quantitative trait loci for grain yield in pearl millet under variable postflowering moisture conditions. Crop Sci 47(3):969–980

Brhane H, Haileselassie T, Tesfaye K, Ortiz R, Hammenhag C, Abreha KB, Geleta M (2022) Finger millet RNA-seq reveals differential gene expression associated with tolerance to aluminum toxicity and provides novel genomic resources. Front Plant Sci 13:1068383

Cao X, Hu Y, Song J, Feng H, Wang J, Chen L, Qiao Z (2022) Transcriptome sequencing and metabolome analysis reveals the molecular mechanism of drought stress in millet. Int J Mol Sci 23(18):10792

Ceasar A (2022) Genome-editing in millets: current knowledge and future perspectives. Mol Bio Rep 49(1):773–781

Ceasar SA, Baker A, Ignacimuthu S (2017) Functional characterization of the *PHT1* family transporters of foxtail millet with development of a novel *Agrobacterium*-mediated transformation procedure. Sci Rep 7(1):14064

Choudhary P, Shukla P, Muthamilarasan M (2023) Genetic enhancement of climate-resilient traits in small millets: a review. Heliyon 9(4):e14621

Debieu M, Sine B, Passot S, Grondin A, Akata E, Gangashetty P, Laplaze L (2018) Response to early drought stress and identification of QTLs controlling biomass production under drought in pearl millet. PLoS One 13(10):e0201635

Desai MK, Mishra RN, Verma D, Nair S, Sopory SK, Reddy MK (2006) Structural and functional analysis of a salt stress inducible gene encoding voltage dependent anion channel (VDAC) from pearl millet (*Pennisetum glaucum*). Plant Physiol Biochem 44(7–9):483–493

Deshpande SS, Mohapatra D, Tripathi MK, Sadvatha RH (2015) Kodo millet—nutritional value and utilization in Indian foods. J Grain Process Storage 2(2):16–23

Devi VS, Punnoose A (2025) Effect of soil fertility on growth and yield attributes of different millets in the agro-ecological unit of Kuttanad (AEU-4) in Kottayam district, Kerala. Int J Agric Nutr 7(2):01–04

Devos KM, Qi P, Bahri BA, Gimode DM, Jenike K, Manthi SJ, Lule D, Lux T, Martinez-Bello L, Pendergast TH IV, Plott C (2023) Genome analyses reveal population structure and a purple stigma color gene candidate in finger millet. Nat Commun 14(1):3694

Djanaguiraman M, Perumal R, Ciampitti IA, Gupta SK, Prasad PV (2018) Quantifying pearl millet response to high temperature stress: thresholds, sensitive stages, genetic variability and relative sensitivity of pollen and pistil. Plant Cell Environ 41(5):993–1007

Dong Z, Zhao H, He J, Huai J, Lin H, Zheng J, Liu Y, Wang G (2011) Overexpression of a foxtail millet *Acetyl-CoA carboxylase* gene in maize increases sethoxydim resistance and oil content. Afr J Biotechnol 10(20):3986–3995

Duque AS, de Almeida AM, da Silva AB, da Silva JM, Farinha AP, Santos D, de Sousa Araújo S (2013) Abiotic stress responses in plants: unraveling the complexity of genes and networks to survive. In: Abiotic stress - plant responses and applications agri. IntechOpen, pp 49–74

Ebanyat P, de Ridder N, Bekunda M, Delve RJ, Giller KE (2021) Efficacy of nutrient management options for finger millet production on degraded smallholder farms in eastern Uganda. Front Sustain Food Syst 5:674926

Eid A, Alshareef S, Mahfouz MM (2018) CRISPR base editors: genome editing without double-stranded breaks. Biochem J 475(11):1955–1964

Feng ZJ, Xu ZS, Sun J, Li LC, Chen M, Yang GX, He GY, Ma YZ (2016) Investigation of the ASR family in foxtail millet and the role of ASR1 in drought/oxidative stress tolerance. Plant Cell Rep 35(1):115–128

Gerrano AS, Labuschagne MT, Van Biljon A, Shargie NG (2016) Quantification of mineral composition and total protein content in sorghum [*Sorghum bicolor (L.)* Moench] genotypes. Cereal Res Commun 44(2):272–285

Gong Z, Dong L, Lam S, Zhang D, Zong Y, Hao X, Li P (2021) Nutritional quality in response to elevated CO2 concentration in foxtail millet (*Setaria italica*). J Cereal Sci 102:103318

Goron TL, Raizada MN (2015) Genetic diversity and genomic resources available for the small millet crops to accelerate a new green revolution. Front Plant Sci 6:157

Guo L, Qiu J, Ye C, Jin G, Mao L, Zhang H, Fan L (2017) Echinochloa crus-galli genome analysis provides insight into its adaptation and invasiveness as a weed. Nat Commun 8(1):1031

Guo Y, Hao D, Wang X, Wang H, Wu Z, Yang P, Zhang B (2023) Comparative transcriptomics reveals key genes contributing to the differences in drought tolerance among three cultivars of foxtail millet (*Setaria italica*). Plant Growth Regul 99(1):45–64

Gupta SK, Rai KN, Singh P, Ameta VL, Gupta SK, Jayalekha AK, Mahala RS, Pareek S, Swami ML, Verma YS (2015) Seed set variability under high temperatures during flowering period in pearl millet (*Pennisetum glaucum* L. (R.) Br.). Field Crop Res 171:41–53

Hadimani NA, Malleshi NG (1993) Studies on milling, physico-chemical properties, nutrient composition and dietary fibre content of millets. J Food Sci Technol (India) 30(1):17–20

Hao XY, Li P, Li HY, Zong YZ, Zhang B, Zhao JZ, Han YH (2017) Elevated CO_2 increased photosynthesis and yield without decreasing stomatal conductance in broomcorn millet. Photosynthetica 55(1):176–183

Hao JH, Kang X, Zhang L, Chen J, Wang D, Dong S, Li X, Gao L, Yang G, Yuan X, Chu X (2025) CRISPR/Cas9-mediated *SiEPF2* mutagenesis attenuates drought tolerance and yield in foxtail millet (*Setaria italica*). Plant Cell Environ 48(8):6043–6046

Hariprasanna K (2017) Foxtail millet, *Setaria italica* (L.) P Beauv. In: Millets and sorghum: biology and genetic improvement, pp 112–149

Hasanuzzaman M, Nahar K, Alam MM, Roychowdhury R, Fujita M (2013) Physiological, biochemical, and molecular mechanisms of heat stress tolerance in plants. Int J Mol Sci 14(5):9643–9684

Hassan ZM, Sebola NA, Mabelebele M (2021) The nutritional use of millet grain for food and feed: a review. Agric Food Secur 10(1):16

Hatfield JL, Prueger JH (2015) Temperature extremes: effect on plant growth and development. Weather Clim Extrem 10(Part A):4–10

Hossain MA, Uddin SN (2011) Mechanisms of waterlogging tolerance in wheat: morphological and metabolic adaptations under hypoxia or anoxia. Aust J Crop Sci 5(9):1094–1101

Huang D, Sun M, Zhang A, Chen J, Zhang J, Lin C, Huang L (2021) Transcriptional changes in pearl millet leaves under heat stress. Genes 12(11):1716

Jaganathan D, Ramasamy K, Sellamuthu G, Jayabalan S, Venkataraman G (2018) CRISPR for crop improvement: an update review. Front Plant Sci 9:985

Jain N, Arora P, Tomer R, Mishra SV, Bhatia A, Pathak H, Chakraborty D, Kumar V, Dubey DS, Harit RC, Singh JP (2016) Greenhouse gases emission from soils under major crops in Northwest India. Sci Total Environ 542:551–561

Kanjiya HN, Chopada MC (2021) Calibrate and validate CERES-MILLET model under variant environmental condition for different cultivars of summer pearl millet in south Saurashtra region. J Phyto 10(3):211–214

Karkannavar SJ, Shigihalli S, Nayak G, Bharati P (2021) Physico-chemical and nutritional composition of proso millet varieties. Pharma Innov J 10:136–140

Khatoon H, Singh V (2016) Impact of water stress on physiological and biochemical parameters of finger millet (*Eleusine coracana* L.). Res Environ Life Sci 9(12):1474–1477

Klap C, Yeshayahou E, Bolger AM, Arazi T, Gupta SK, Shabtai S, Barg R (2017) Tomato facultative parthenocarpy results from *SlAGAMOUS-LIKE 6* loss of function. Plant Biotechnol J 15(5):634–647

Kulkarni SS, Chavan PD (2013) Study of some aspects of anaerobic metabolism in roots of finger millet and rice plants subjected to waterlogging stress. Int J Bot 9(2):80–85

Kumar AA, Reddy BV, Ramaiah B, Sahrawat KL, Pfeiffer WH (2013) Gene effects and heterosis for grain iron and zinc concentration in sorghum [*Sorghum bicolor* (L.) Moench]. Field Crop Res 146:86–95

Kumar A, Tomar RSS, Chandra A, Joshi D, Tiwari S, Singh P, Kumar V (2021) Genomics-assisted improvement of grain quality and nutraceutical properties in millets. Springer, pp 333–343

Kumar S, Kumar A, Sen H, Janeja HS, Maity S, Banerjee S, Singh P, Arun MC (2024) Small millets: a multifunctional crop for achieving sustainable food security under climate change. World 718:863

Kusaka M, Lalusin AG, Fujimura T (2005) The maintenance of growth and turgor in pearl millet (*Pennisetum glaucum* [L.] Leeke) cultivars with different root structures and osmo-regulation under drought stress. Plant Sci 168(1):1–14

Laishram B, Dutta R, Devi OR, Ngairangbam H (2023) Importance of millets for food and nutritional security in the context of climate resilient agriculture. Adv Agron 1

Lata C, Bhutty S, Bahadur RP, Majee M, Prasad M (2011) Association of an SNP in a novel DREB2-like gene *SiDREB2* with stress tolerance in foxtail millet [*Setaria italica* (L.)]. J Exp Bot 62(10):3387–3401

Li P, Brutnell TP (2011) *Setaria viridis* and *Setaria italica*, model genetic systems for the Panicoid grasses. J Exp Bot 62(9):3031–3037

Li J, Wang Z, He G, Ma L, Deng XW (2020) CRISPR/Cas9-mediated disruption of *TaNP1* genes results in complete male sterility in bread wheat. J Genet Genomics 47(5):263–272

Li J, Li Y, Ma L (2021) Recent advances in CRISPR/Cas9 and applications for wheat functional genomics and breeding. aBIOTECH 2(4):375–385

Liang Z, Wu Y, Ma L, Guo Y, Ran Y (2022) Efficient genome editing in *Setaria italica* using CRISPR/Cas9 and base editors. Front Plant Sci 12:815946

Lima GVO, Oki Y, Bordignon L, Siqueira WK, França MGC, Boanares D, Fernandes GW (2022) Interaction between increased CO_2 and temperature enhances plant growth but does not affect millet grain production. Acta Sci Agron 44:e53515

Linkemer G, Board JE, Musgrave ME (1998) Waterlogging effects on growth and yield components in late-planted soybean. Crop Sci 38(6):1576–1584

Liu X, Tang S, Jia G, Schnable JC, Su H, Tang C, Zhi H, Diao X (2016) The C-terminal motif of *SiAGO1b* is required for the regulation of growth, development and stress responses in foxtail millet (*Setaria italica* (L.) P. Beauv). J Exp Bot 67(11):3237–3249

Ma Y, Liu W, Qiao Y, Qiao W, Yang H, Zhong Y, Yang H, Wang H, Li Y, Dong B, Liu M (2023) Effects of soil salinity on foxtail millet osmoregulation, grain yield, and soil water utilization under varying water conditions. Agric Water Manag 284:108354

Maai E (2024) Effect of waterlogging on photosynthesis and growth of finger millet (*Eleusine coracana*). Agrosyst Geosci Environ 7(1):e20467

Maharajan T, Antony Ceasar S, Ajeesh Krishna TP, Ignacimuthu S (2021) Finger millet [*Eleusine coracana* (L.) Gaertn]: an orphan crop with a potential to alleviate the calcium deficiency in the semi-arid tropics of Asia and Africa. Front Sustain Food Syst 5:684447

Maitra S, Shankar T, Hossain A, Sairam M, Sagar L, Sahoo U, Gaikwad DJ, Pramanick B, Mandal TK, Sarkar S, Gitari HI (2024) Climate-smart millets production in future for food and nutritional security. In: Adapting to climate change in agriculture—theories and practices: approaches for adapting to climate change in agriculture in India. Springer, pp 11–41

Matsuura A, An P, Murata K, Inanaga S (2016) Effect of pre- and post-heading waterlogging on growth and grain yield of four millets. Plant Prod Sci 19(3):348–359

Matsuura A, Kato Y, Suzuki T, Murata K, An P (2022) Hypoxia tolerance of four millet species is attributable to constitutive aerenchyma formation and root hair development of adventitious roots. Plant Prod Sci 25(2):157–171

Meena RP, Joshi D, Bisht JK, Kant L (2021) Global scenario of millets cultivation. In: Millets and millet technology. Springer Singapore, pp 33–50

Mickelbart MV, Hasegawa PM, Bailey-Serres J (2015) Genetic mechanisms of abiotic stress tolerance that translate to crop yield stability. Nat Rev Genet 16(4):237–251

Mukherjee B, Jha RK, Sattar A, Dutta S, Bhattacharya U, Samanta S, Bal SK (2025) Harnessing the potential of millets for climate-resilient and sustainable agriculture. Front Plant Sci 16:1574699

Mushtaq NU, Saleem S, Manzoor A, Tahir I, Seth CS, Rehman RU (2025) Physiological and biochemical perspective on silicon induced salt stress tolerance in proso millet. SILICON 17(6):1413–1426

Namata MJ, Xu J, Habyarimana E, Palakolanu SR, Wang L, Li J (2025) Genome editing in maize and sorghum: a comprehensive review of CRISPR/Cas9 and emerging technologies. Plant Genome 18(2):e70038

Narayanrao DR, Tomar RS, Padhiyar SM, Jasminkumar K, Ashish G, Chauhan NM, Singh SC, Upadhye V, Kuddus M, Kamble L, Hajare ST (2023) De novo transcriptome sequencing of drought tolerance-associated genes in little millet (*Panicum sumatrense* L.). Funct Integr Genomics 23(4):303

Nishigaki T, Ikazaki K, Shinjo H, Tanaka U, Fatondji D, Funakawa S (2025) Pearl millet yield reduction by soil erosion and its recovery potential through fertilizer application on an Arenosol in the Sahel. Soil Tillage Res 246:106324

Numan M, Serba DD, Ligaba-Osena A (2021) Alternative strategies for multi-stress tolerance and yield improvement in millets. Genes 12(5):739

Oishy MN, Shemonty NA, Fatema SI, Mahbub S, Mim EL, Raisa MB, Anik AH (2025) Unravelling the effects of climate change on the soil-plant-atmosphere interactions: a critical review. Soil Environ Health 3(1):100130

Ong CK, Monteith JL (1985) Response of pearl millet to light and temperature. Field Crop Res 11:141–160

Panneerselvam S, Dheebhakaran G, Kokilavani S (2019) Effect of elevated temperature on growth parameters of foxtail millet (*Setaria italica*). Madras Agric J 106:349–352

Paterson AH, Bowers JE, Bruggmann R, Dubchak I, Grimwood J, Gundlach H, Rokhsar DS (2009) The *Sorghum bicolor* genome and the diversification of grasses. Underw Nat 457(7229):551–556

Pehlivan N, Altaf MT, Emamverdian A, Ghorbani A (2025) Beyond the lab: future-proofing agriculture for climate resilience and stress management. Front Plant Sci 16:1565850

Prasad PV, Staggenborg SA (2009) Growth and production of sorghum and millets. In: Soils, plant growth and crop production, vol 2. EOLSS Publications, pp 1–27

Pudake RN, Mehta CM, Mohanta TK, Sharma S, Varma A, Sharma AK (2017) Expression of four phosphate transporter genes from finger millet (*Eleusine coracana* L) in response to mycorrhizal colonization and Pi stress. 3 Biotech 7(1):17

Puranik S, Bahadur RP, Srivastava PS, Prasad M (2011) Molecular cloning and characterization of a membrane-associated NAC family gene, *SiNAC* from foxtail millet [*Setaria italica* (L.) P. Beauv.]. Mol Biotechnol 49(2):138–150

Qaoud HA, Ali IA, Al-Fares H, Qubbaj T, Shtaya MJ (2023) Effect of salinity on the growth and some morphological traits of pearl millet. Pak J Bot 55(3):807–811

Rahman H, Ramanathan V, Nallathambi J, Duraialagaraja S, Muthurajan R (2016) Over-expression of a NAC 67 transcription factor from finger millet (*Eleusine coracana* L.) confers tolerance against salinity and drought stress in rice. BMC Biotechnol 16(Suppl 1):35

Raj R, Kumar S, Mishra S, Sherpa D (2025) Development of Lysigenous Aerenchyma in root alleviates the waterlogging stress in little millet. Trop Plant Biol 18(1):1–12

Ramachandra G, Virupaksha TK, Shadaksharaswamy M (1977) Relation between tannin levels and *in vitro* protein digestibility in finger millet (*Eleusine coracana* Gaertn.). J Agric Food Chem 25(5):1101–1104

Rani V, Joshi DC, Joshi P, Singh R, Yadav D (2023) "Millet models" for harnessing nuclear factor-Y transcription factors to engineer stress tolerance in plants: current knowledge and emerging paradigms. Planta 258(2):29

Ray DK, West PC, Clark M, Gerber JS, Prishchepov AV, Chatterjee S (2019) Climate change has likely already affected global food production. PLoS One 14(5):e0217148

Reddy PS, Reddy GM, Pandey P, Chandrasekhar K, Reddy MK (2012) Cloning and molecular characterization of a gene encoding late embryogenesis abundant protein from *Pennisetum glaucum*: protection against abiotic stresses. Mol Biol Rep 39(6):7163–7174

Reddy PS, Satyavathi CT, Khandelwal V, Patil HT, Gupta PC, Sharma LD et al (2021) Performance and stability of pearl millet varieties for grain yield and micronutrients in arid and semi-arid regions of India. Front Plant Sci 12:670201

Ricroch A, Clairand P, Harwood W (2017) Use of CRISPR systems in plant genome editing: toward new opportunities in agriculture. Emerg Top Life Sci 1(2):169–182

Sachan DS, Kumar R, Kumar P, Pal V, Yadav KK (2023) Millets production and consumption in India. Eur Chem Bull 8:20

Sage RF, Zhu XG (2011) Exploiting the engine of C_4 photosynthesis. J Exp Bot 62(9):2989–3000

Salgotra RK, Thompson M, Chauhan BS (2022) Unravelling the genetic potential of untapped crop wild genetic resources for crop improvement. Conserv Genet Resour 14(1):109–124

Satish L, Rathinapriya P, Rency AS, Ceasar SA, Prathibha M, Pandian S, Rameshkumar R, Ramesh M (2016) Effect of salinity stress on finger millet (*Eleusine coracana* (L.) Gaertn): histochemical and morphological analysis of coleoptile and coleorhizae. Flora-Morphol Distrib Funct Ecol Plants 222:111–120

Satyavathi CT, Ambawat S, Khandelwal V, Srivastava RK (2021) Pearl millet: a climate-resilient nutricereal for mitigating hidden hunger and provide nutritional security. Front Plant Sci 12:659938

Selvi VM, Nirmalakumari A (2023) Flowering mechanism in kodo millet (Paspalum scrobiculatum). Int J Environ Clim Change 13(9):3009–3019

Sharma S, Saxena DC, Riar CS (2016) Analysing the effect of germination on phenolics, dietary fibres, minerals and γ-amino butyric acid contents of barnyard millet (*Echinochloa frumentaceae*). Food Biosci 13:60–68

Sharma R, Sharma S, Dar BN, Singh B (2021) Millets as potential nutri-cereals: a review of nutrient composition, phytochemical profile and techno-functionality. Int J Food Sci Technol 56(8):3703–3718

Sheethal HV, Baruah C, Subhash K, Ananthan R, Longvah T (2022) Insights of nutritional and anti-nutritional retention in traditionally processed millets. Front Sustain Food Syst 5:735356

Shi J, Ma X, Zhang J, Zhou Y, Liu M, Huang L, Lai J (2019) Chromosome conformation capture resolved near complete genome assembly of broomcorn millet. Nat Commun 10(1):464

Shi X, Shen J, Niu B, Lam SK, Zong Y, Zhang D, Hao X, Li P (2022) An optimistic future of C4 crop broomcorn millet (*Panicum miliaceum* L.) for food security under increasing atmospheric CO2 concentrations. PeerJ 10:e14024

Shweta Saloni SS, Sindhu Sujata SS, Sushma Kumari SK, Sandhya Suman SS (2018) Little millets: properties, functions and future prospects. J Pharmacogn Phytochem 7(2):337–341181

Singh M, Upadhyaya HD (2015) Genetic and genomic resources for grain cereals improvement. Academic Press

Singh F, Rai KN, Reddy BV, Diwakar B (1997) Development of cultivars and seed production techniques in sorghum and pearl millet

Singh RK, Singh VK, Raghavendrarao S, Phanindra MLV, Raman KV, Solanke AU, Kumar PA, Sharma TR (2015) Expression of finger millet *EcDehydrin7* in transgenic tobacco confers tolerance to drought stress. Appl Biochem Biotechnol 177:207–216

Song XE, Cao J, Guo S, Wang H, Dong Q, Guo P, Yuan X (2023) Effect of brassinolide on soil microorganisms in millet field polluted by tribenuron-methyl. Microorganisms 11:1829

Sood S, Joshi DC, Pattanayak A (2020) Breeding advancements in barnyard millet, vol 1. Springer Cham, pp 391–409

Sreenivasulu N, Miranda M, Prakash HS, Wobus U, Weschke W (2004) Transcriptome changes in foxtail millet genotypes at high salinity: identification and characterization of a *PHGPX* gene specifically up-regulated by NaCl in a salt-tolerant line. J Plant Physiol 161:467–477

Srivastava S, Arya C (2021) Millets: malnutrition and nutrition security. In: Millets and millet technology. Springer, Singapore, pp 81–100

Sultan B, Roudier P, Quirion P, Alhassane A, Muller B, Dingkuhn M, Baron C (2013) Assessing climate change impacts on sorghum and millet yields in the Sudanian and Sahelian savannas of West Africa. Environ Res Lett 8(1):014040

Tadele Z (2016) Drought adaptation in millets. In: Shanker AK, Shanker C (eds) Abiotic and biotic stress in plants: recent advances and future perspectives. InTech, Rijeka, pp 639–662

Thrupthi GN, Anusha KK, Shilpa MD, Krishna MA (2023) Study of salinity tolerance in 3 varieties of finger millets (*Eleusine coracana* (L.) Gaertn.). J Med Plants Stud 11:99–105

Tiwari S, Lata C, Chauhan PS, Prasad V, Prasad M (2017) A functional genomic perspective on drought signalling and its crosstalk with phytohormone-mediated signalling pathways in plants. Curr Genomics 18:469–482

Tiwari A, Kesarwani K, Sharma A, Ghosh T, Bisht N, Punetha S (2022) Drought stress in millets and its response mechanism. Adv Plant Def Mech:1–35

Tripathi T, Vyas S (2023) From ancient grains to modern solutions: a history of millets and their significance in agriculture and food security. Int J Home Sci 9(2):72–78

Uppal RK, Wani SP, Garg KK, Alagarswamy G (2015) Balanced nutrition increases yield of pearl millet under drought. Field Crops Res 177:86–97

Vanaja M, Srinivas K, Lal M, Satish T, Reddy P (2013) Effect of elevated CO2 concentration on groundnut (*Arachis hypogaea*) (C3) and pearl millet (*Pennisetum glaucum*) (C4) crop plants and some implications on growth and photosynthetic activity. Indian J Agric Sci 83:7

Varshney RK, Shi C, Thudi M, Mariac C, Wallace J, Qi P, Xu X (2017) Pearl millet genome sequence provides a resource to improve agronomic traits in arid environments. Nat biotechnol 35(10):969–976

Veeranagamallaiah G, Ranganayakulu GS, Thippeswamy M, Sivakumar M, Eswaranarayana Reddy K, Pandurangaiah M, Sridevi V, Sudhakar C (2009) Aldose reductase expression contributes in sorbitol accumulation and 4-hydroxynon-2-enal detoxification in two foxtail millet (*Setaria italica* L.) cultivars with different salt stress tolerance. Plant Growth Regul 59(2):137–143

Vetriventhan M, Azevedo VC, Upadhyaya HD, Nirmalakumari A, Kane-Potaka J, Anitha S, Tonapi VA (2020) Genetic and genomic resources, and breeding for accelerating improvement of small millets: current status and future interventions. Nucleus 63:217–239

Wang M, Li P, Li C, Pan Y, Jiang X, Zhu D, Zhao Q, Yu J (2014) *SiLEA14*, a novel atypical LEA protein, confers abiotic stress resistance in foxtail millet. BMC Plant Biol 14:290

Wei JT, Hou ZH, Wang Y, Hao JM, Wang J, Wang W, Wang W, Wang DM, Xu ZS, Song X, Wang F (2023) Foxtail millet *SiCDPK7* gene enhances tolerance to extreme temperature stress in transgenic plants. Environ Exp Bot 207:105197

Wheeler T, Von Braun J (2013) Climate change impacts on global food security. Science 341(6145):508–513

Yan H, Sun M, Zhang Z, Jin Y, Zhang A, Lin C, Huang L (2023) Pangenomic analysis identifies structural variation associated with heat tolerance in pearl millet. Nat Genet 55:507–518

Yenagi NB, Handigol JA, Ravi SB, Mal B, Padulosi S (2010) Nutritional and technological advancements in the promotion of ethnic and novel foods using the genetic diversity of minor millets in India. Indian J Plant Genet Resour 23:82–86

Yuan Y, Wu C, Liu L, Ma Q, Yang Q, Feng B (2022) Unravelling the distinctive growth mechanism of proso millet (*Panicum miliaceum L.*) under salt stress: from root-to-leaf adaptations to molecular response. Glob Change Biol Bioenergy 14(2):192–214

Zemp et al (2025) Community estimate of global glacier mass changes from 2000 to 2023. Nature 639:382–388

Zhang JP, Liu TS, Zheng J, Wang GY (2007) Cloning and characterization of a putative 12-oxophytodienoic acid reductase cDNA induced by osmotic stress in roots of foxtail millet. DNA Seq 18(2):138–144

Zhang G, Liu X, Quan Z, Cheng S, Xu X, Pan S et al (2012) Genome sequence of foxtail millet (*Setaria italica*) provides insights into grass evolution and biofuel potential. Nat Biotechnol 30:549–554

Zhang DZ, Panhwar RB, Liu JJ, Gong XW, Liang JB, Liu MX, Feng BL (2019) Morphological diversity and correlation analysis of phenotypes and quality traits of proso millet (*Panicum miliaceum L.*) core collections. J Integr Agric 18(5):958–969

Zhang J, Zhang X, Chen R, Yang L, Fan K, Liu Y, Liu Y (2020) Generation of transgene-free semidwarf maize plants by gene editing of gibberellin-oxidase20–3 using CRISPR/Cas9. Front Plant Sci 11:1048

Zhang D, Li A, Lam SK, Li P, Zong Y, Gao Z, Hao X (2021) Increased carbon uptake under elevated CO2 concentration enhances water-use efficiency of C4 broomcorn millet under drought. Agric Water Manage 245:106631

Zhang P, Xia L, Sun Y, Gao S (2024) Soil nutrients and enzyme activities based on millet continuous cropping obstacles. Sci Rep 14:17329

Zhang T, Wang Z, Liu Q, Zhao D (2025) Genetic improvement of rice grain size using the CRISPR/Cas9 system. Rice 18(1):3

Zhao L, Zhao Q, Ao G, Yu J (2009) The foxtail millet *Si69* gene is a *Wali7* (*wheat aluminum-induced protein 7*) homologue and may function in aluminium tolerance. Chin Sci Bull 54:1697–1706

Zou C, Li L, Miki D, Li D, Tang Q, Xiao L, Zhang H (2019) The genome of broomcorn millet. Nat Commun 10(1):436

MIX
Papier aus verantwortungsvollen Quellen
Paper from responsible sources
FSC® C105338

If you have any concerns about our products,
you can contact us on
ProductSafety@springernature.com

In case Publisher is established outside the EU,
the EU authorized representative is:
Springer Nature Customer Service Center GmbH
Europaplatz 3, 69115 Heidelberg, Germany

Printed by Libri Plureos GmbH
in Hamburg, Germany